D0849130

DEVELOPMENTS IN BLOCK COPOLYMERS—1

THE DEVELOPMENTS SERIES

Developments in many fields of science and technology occur at such a pace that frequently there is a long delay before information about them becomes available and usually it is inconveniently scattered among several journals.

Developments Series books overcome these disadvantages by bringing together within one cover papers dealing with the latest trends and developments in a specific field of study and publishing them within *six months* of their being written.

Many subjects are covered by the series including food science and technology, polymer science, civil and public health engineering, pressure vessels, composite materials, concrete, building science, petroleum technology, geology, etc.

Information on other titles in the series will gladly be sent on application to the publisher.

DEVELOPMENTS IN BLOCK COPOLYMERS—1

Edited by

I. GOODMAN

Professor of Polymeric Materials, University of Bradford, UK

APPLIED SCIENCE PUBLISHERS
LONDON and NEW YORK

APPLIED SCIENCE PUBLISHERS LTD
Ripple Road, Barking, Essex, England

Sole Distributor in the USA and Canada
ELSEVIER SCIENCE PUBLISHING CO., INC.,
52 Vanderbilt Avenue, New York, NY 10017, USA

British Library Cataloguing in Publication Data

Developments in block copolymers.—1.—(The
Developments series)
1. Block copolymers—Periodicals
I. Series
547.8′4 QD382.B5

ISBN 0-85334-145-1

WITH 47 TABLES and 99 ILLUSTRATIONS

Printed in Great Britain by Galliard (Printers) Ltd, Great Yarmouth

PREFACE

Block copolymers, which are substances composed of two or more types of polymeric segments combined linearly in the macromolecular chain, have been known for a number of years and, together with the related families of graft copolymers, they form a well-established discipline within polymer science and technology.

The individual segment species of block copolymers frequently exercise independent roles in the overall behaviour of the materials, of which the most familiar are the thermoplastic elastomers of the block copolyurethane, copolyether-ester and styrene–diene copolymer classes, and the spandex elastic fibres. These groups of materials are produced industrially on a substantial scale; their useful elastomeric properties are due to a different structural principle to that operating in covalently crosslinked rubbers. Other block copolymers, of a variety of structures, display compatibility of a constituent segment with appropriate substrates and thereby find important, though smaller-scale, applications as specialised surfactants, blending agents, antistatic agents, impact-resistance modifiers and in other ways.

This span of technical interests, as well as the body of fundamental scientific investigation in the field, has generated a considerable specialist literature together with two excellent monographs[1,2] which covered progress up to the mid-1970s. The publishers' suggestion that a volume in their *Developments Series* might be devoted to recent work on block copolymers was therefore timely and, as editor, I have been fortunate in securing the collaboration of leading specialists in the subject as authors of the various chapters. The international character of the authorship, with contributions from France, the Netherlands, the UK and the USA and representation from industrial and academic laboratories, is a measure of the widespread activity in the field. It is hoped that the contents will prove of interest to practitioners and students in a diversity of areas of polymer and materials science and technology, as well as to those concerned with chemical technology in general.

Within a single volume it was not thought practicable to cover the whole range of the subject. Instead, we have dealt in some depth with a limited number of themes that have evolved actively during recent years. The treatment thus falls into two major parts, one (Chapters 1–4) mainly concerned with the physico-chemical and structural aspects of amorphous and relatively non-polar copolymers having a small number of blocks per molecule, and the second (Chapters 5–8) with the synthesis, characteristics and technology of the more polar and frequently crystalline classes of multiblock copolymers. A major insight emerging from the book is indeed the primary difference in qualities that distinguishes the two groups.

An important advance in block copolymer science was the recognition in the 1960s that, in certain ranges of composition, the polymeric sub-species may separate into more or less distinct microphases to give a texture or morphology that is often highly organised and whose detailed nature, in relation to the molecular structure and prior history of the materials, can exercise a powerful effect upon their macroscopic properties and utility. The elucidation of block copolymer morphology, its rationalisation in fundamental terms and its correlation with bulk properties, has been a central topic of investigation during the past decade and forms a recurrent theme throughout this volume. A notable aspect, discussed in several chapters, is the wide range of advanced physical techniques now employed for the exploration of morphology and interphase character in block copolymers.

Turning to the specific subject coverage, Chapters 1 and 2 deal primarily with block copolymer–solvent systems, showing, for example, that the treatment of a given copolymer with selective solvents can yield varied morphologies and widely different mechanical and thermal properties, and that the association of block copolymers into the micellar-aggregated state can be characterised in detail and employed for such purposes as viscosity index improvers for lubricating oils and as colloidal stabilisers for dispersion polymerisation processes. Chapters 3 and 4 reflect advances in the investigation and fundamental understanding of the phase-separation phenomenon. The first considers the self-organising characteristics of flexible-chain block copolymers into mesomorphic (liquid-crystalline) states and shows how the various phase topologies can be identified and assessed quantitatively by the parallel use of small angle X-ray scattering and electron-microscopical techniques; the second presents a statistical thermodynamic theory of the size and shape of the microphases and the interface between them, and explores the relationship between phase state and composition in amorphous block copolymers.

The next two chapters form a bridge leading to the final technological contributions. Chapter 5 provides a review of the synthesis of block copolymers containing heteroatoms in their chains and examines in some detail the often bewildering evidence concerning the influence of molecular structure on the phase state of various types of constituent blocks combined in such copolymers. Chapter 6 commences with a survey of the experimental methods available for the investigation of texture (and its variation with deformation) in bulk block copolymer materials, especially those which, because of the presence of a crystalline phase, have more complex structures than the polystyrene–polydiene type, and proceeds to a thorough account of recently gained knowledge of the morphology of linear-chain polyurethane block copolymer elastomers.

The final chapters, both by groups of authors from industrial organisations, are concerned with some newer block copolymers and processes of growing commercial significance. Chapter 7 describes the preparative technology, properties, processing and applications of elastoplasts composed of aromatic polyester units combined with aliphatic polyether or aliphatic polyester segments; a detailed comparison is made of the characteristics of the two groups and of those of analogous block copolyurethanes. Chapter 8 deals with the development of reaction injection moulding technology for the direct conversion of liquid precursors into shaped solid branched-chain block copolyurethane microcellular articles which are already of importance in automotive construction, and which, because of the facility and energy-economy of production, seem likely to have a promising future in numerous other fields. The contents of both chapters lead to an appreciation of the many, varied, and often elaborately interrelated factors that have to be resolved in transposing laboratory concepts to technically acceptable fruition.

My thanks are due to my coauthors for their authoritative, critical and forward-looking contributions. Inevitably in a work of this kind, certain topics are dealt with in more than one chapter; the presentation of different and often complementary viewpoints upon particular matters does, however, confer its own advantages.

Lastly, attention must be drawn to a problem besetting all writing on polymers (and especially that on block copolymers), namely that of choosing explicit chemical names for the numerous complex structures involved. The difficulties are discussed more fully in Chapter 5. For the purposes of this volume it has been thought best to aim at ready comprehensibility and a reasonable measure of internal consistency rather than at rigorous but often unfamiliar formal accuracy. For this reason, only limited recourse has been made to IUPAC nomenclature and considerable use has been made of the abbreviations and conventions listed before the main text.

I. GOODMAN

REFERENCES

1. ALLPORT, D. C. and JANES, W. H. (Eds), *Block Copolymers*, Applied Science Publishers, London, 1973.
2. NOSHAY, A. and MCGRATH, J. E., *Block Copolymers: Overview and Critical Survey*, Academic Press, New York, 1977.

CONTENTS

LIST OF CONTRIBUTORS

D. C. ALLPORT

Technical Manager, ICI Polyurethanes, Organics Division, Blackley, Manchester M9 3DA, UK

C. BARKER

Formerly ICI Polyurethanes, Organics Division, Blackley, Manchester M9 3DA, UK

J. F. CHAPMAN

Project Leader, Microcellular Elastomers, ICI Polyurethanes, Organics Division, Blackley, Manchester M9 3DA, UK

S. L. COOPER

Professor, Department of Chemical Engineering, University of Wisconsin, Madison, Wisconsin 53706, USA

J. M. G. COWIE

Professor of Chemistry, Department of Chemistry, University of Stirling, Stirling FK9 4LA, Scotland, UK

S. A. G. DE GRAAF

Research Coordinator, Industrial Colloids, Enka Research Institute, PO Box 60, Arnhem, The Netherlands

P. E. GIBSON

Research Assistant, Department of Chemical Engineering, University of Wisconsin, Madison, Wisconsin 53706, USA

I. GOODMAN

Professor of Polymeric Materials, University of Bradford, Bradford BD7 1DP, UK

E. HELFAND

Technical Staff Member, Bell Laboratories, Murray Hill, New Jersey 07974, USA

F. J. HUNTJENS

Research Chemist, Corporate Research Department, Akzo Research, PO Box 60, Arnhem, The Netherlands

C. PRICE

Reader in Chemistry, Department of Chemistry, University of Manchester, Manchester M13 9PL, UK

A. E. SKOULIOS

Directeur de Recherches, Centre de Recherches sur les Macromolécules, Centre National de la Recherche Scientifique, 6, rue Boussingault, 67083 Strasbourg Cedex, France

M. A. VALLANCE

Research Assistant, Department of Chemical Engineering, University of Wisconsin, Madison, Wisconsin 53706, USA

R. W. M. VAN BERKEL

Research Chemist, Department of Plastics, Enka Research Institute, PO Box 60, Arnhem, The Netherlands

C. M. F. VROUENRAETS

Research Chemist, Polymerization Department, Enka Research Institute, PO Box 60, Arnhem, The Netherlands

Z. R. WASSERMAN

Technical Staff Member, Bell Laboratories, Murray Hill, New Jersey 07974, USA (Present address: E. I. du Pont de Nemours & Co., Experimental Station, Wilmington, Delaware 19898, USA)

ABBREVIATIONS

GENERAL

IR	infrared (spectroscopy)
UV	ultraviolet (spectroscopy)
NMR	nuclear magnetic resonance (spectroscopy)
GPC	gel permeation chromatography
DSC	differential scanning calorimetry
WAXD	wide-angle X-ray diffraction
SAXS	small-angle X-ray scattering
SALS	small-angle light scattering
M.W.	molecular weight

MONOMERS, SIMPLE CHEMICAL SUBSTANCES AND DERIVED STRUCTURE-FRAGMENTS

Bd	1,3-butadiene or in-chain C_4H_6 unit
BPA	bisphenol A (synonyms: 4,4′-isopropylidenediphenol; 2,2-di(4-hydroxyphenyl)propane; 4,4′-(1-methylethylidene)bisphenol)
DMAc	*N,N*-dimethylacetamide
DMF	*N,N*-dimethylformamide
*n*G	a diol of structure $HO(CH_2)_nOH$, or the derived $O(CH_2)_nO$ unit; thus, 4G represents 1,4-butanediol (synonyms: 1,4-butylene glycol; tetramethylene glycol) or the $O(CH_2)_4O$ unit
HMPA	hexamethylphosphorotriamide

Ip	isoprene (2-methyl-1,3-butadiene); in-chain C_5H_8 unit
MDI	diphenylmethane-4,4′-diisocyanate (synonyms: methylenediphenylene diisocyanate, di(4-isocyanatophenyl)methane; 1,1′-methylenebis[4-isocyanatobenzene])
MEK	ethyl methyl ketone (synonyms: methyl ethyl ketone; 2-butanone)
Ph	phenyl
St	styrene (phenylethene) or in-chain —$CH_2CH(C_6H_5)$— unit
TDI	toluene diisocyanate (synonym: diisocyanatomethylbenzene) (isomeric substitution positions shown where appropriate)
THF	tetrahydrofuran (synonyms: oxacyclopentane; oxolane)

POLYMERS OR POLYMER-STRUCTURE FRAGMENTS*

BPAC	bisphenol-A polycarbonate (synonym: poly[oxycarbonyloxy-1,4-phenylene(1-methylethylidene)-1,4-phenylene])
PBd	polybutadiene or polybutadiene unit (mono- or divalent)
PCL	poly(ε-caprolactone) (synonyms: poly(6-hexanolactone); poly[oxy(1-oxo-1,6-hexanediyl])
PDMS	poly(dimethylsiloxane) (synonym: poly(oxydimethylsilylene)); $[OSi(CH_3)_2]_n$ unit
PIp	polyisoprene or polyisoprene unit (mono- or divalent)
PαMeSt	poly(α-methylstyrene)
PMMA	poly(methyl methacrylate) (synonym: poly[1-(methoxycarbonyl)1-methylethylene])
POE	poly(oxyethylene) (synonyms: poly(ethylene oxide); polyoxirane); $[(CH_2)_2O]_n$ unit
POP	poly(oxypropylene) (synonyms: poly(methyloxirane); poly(propylene oxide)); $[CH_2CH(CH_3)O]_n$ unit
POTM	poly(oxytetramethylene) (synonyms: polytetrahydrofuran; poly(butylene oxide)); $[(CH_2)_4O]_n$ unit
PSt	polystyrene or polystyrene unit (mono- or divalent)
PTMA	poly(tetramethylene adipate) (synonyms: poly(butylene adipate), poly[oxy-1,4-butanediyloxy(1,6-dioxo-1,6-hexanediyl)])

(Other abbreviations are defined at their first occurrence.)

* Certain of the structure fragments may require an additional linking atom or group for correct stoichiometry.

Chapter 1

CARBON-CHAIN BLOCK COPOLYMERS AND THEIR RELATIONSHIP WITH SOLVENTS

J. M. G. COWIE

Department of Chemistry, University of Stirling, UK

SUMMARY

One of the characteristic features of block copolymers is the tendency for microphase separation to occur in the solid state. It is this separation which imparts many of the interesting and desirable properties displayed by this group of macromolecules. The morphology of such multiphase systems influences their behaviour and can be controlled by altering the ratio of block lengths, the molecular weight distribution and the nature of the component blocks. The presence of a liquid during sample preparation cannot be neglected, however, as the differential solvation of the component blocks by a solvent can have a profound influence on the distribution of the phases in the solid. This influence is manifest initially in the anomalous dilute solution behaviour of some block copolymer solutions caused by the formation of definite structures which are maintained into the solid state. The effect of these 'preset' structures is ultimately reflected in changes in the dynamic mechanical response, the tensile strength, and the stress–strain behaviour of the materials and demonstrates that the presence of a liquid must also be considered as a structure-controlling parameter.

1. GENERAL INTRODUCTION

The most commonly encountered block copolymer structures consist of extended sequences of one monomer covalently bonded to extended

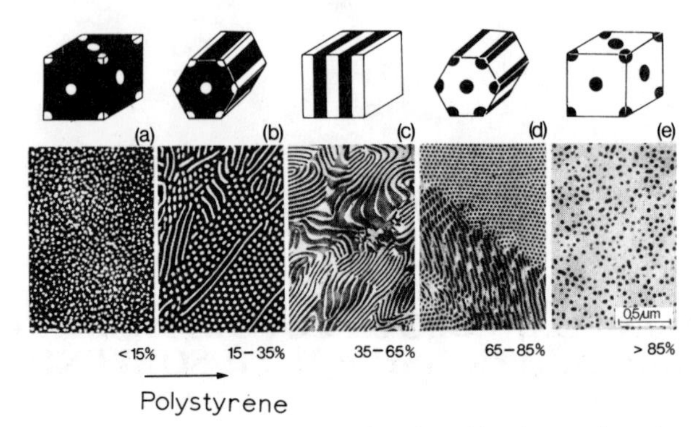

FIG. 1. Morphological changes brought about by increasing the polystyrene content of an AB poly(styrene-*b*-butadiene) copolymer. Schematic diagrams are shown above the electron micrographs and indicate (a) PSt spheres in a PBd matrix, (b) PSt cylinders in a PBd matrix, (c) lamellae, (d) PBd cylinders in a PSt matrix and (e) PBd spheres in a PSt matrix. (From Schmitt, B. J., *Angew. Chem. Int. Ed. Engl.*, 1979, **18**, 287; with permission Verlag Chemie GMBH, Weinheim.)

sequences of another in AB diblock, ABA or BAB triblock, or $[AB]_n$ multiblock arrangements. The component blocks are usually thermodynamically incompatible to varying degrees and this has an important influence on the morphology of the bulk phase and on the dilute solution behaviour. Many of the unique and fascinating properties exhibited by these structures can be attributed to the tendency for microphase separation to take place, thereby imparting a heterogeneous structure to the sample.

The thermodynamic criteria necessary for phase separation in the bulk have been defined by a number of workers[1–10] and include parameters such as the interaction energy between the unlike segments, χ_{AB}, and the M.W. of each block. In general, it has been shown that for phase separation to occur in a block copolymer, component block M.W.'s must be higher than those required to produce phase separation in a blend of the parent homopolymers.

Evidence from electron micrographs indicates that a number of distinguishable and characteristic morphological units are created during microphase separation. These have been formalised[11] in terms of (i) alternating lamellae of A and B entities, (ii) cylinders or rods and (iii) spheres of one component embedded in a continuous matrix of the second. The various structures, and the morphological changes which are brought about by altering the relative compositions of A and B, are illustrated for a sample of AB poly(styrene-*b*-butadiene) in Fig. 1. When this copolymer

TABLE 1
INTERDEPENDENCE OF MORPHOLOGY AND M.W.D.

MWD of block	*% Composition of PSt in copolymer*	*Morphology*
Broad in PSt	72	Lamellae
	50	PSt cylinders
	31	PSt spheres
Broad in PBd	75	PBd spheres
	44	Lamellae
	28	PSt cylinders
Both broad	88	PBd spheres
	59	Lamellae
	27	Irregular PSt domains in PBd matrix

contains less than 15% of PSt, spherical domains of PSt in a PBd matrix characterise the structure. An increase in the PSt content to 15–35% changes the PSt domains to a cylindrical form, and then to a lamellar structure when the PSt content is further increased to between 35% and 65%. At higher PSt contents a phase inversion takes place with the formation of cylinders followed by spheres of PBd in a PSt matrix. Thus the relative composition of the blocks has a fundamental bearing on the bulk morphology.

Many of the samples studied, which are prepared by anionic techniques, have approximately monodisperse structures, but the influence of a broader molecular weight distribution (M.W.D.) has recently been examined[12] and the summary of results in Table 1 again shows that the domain structure can be altered in a significant way.

Finally, the observations that block copolymer films cast from various solvents exhibit different physical properties, and that the results of dilute solution studies can be more significantly dependent on solvent than for homopolymers, both highlight the fact that the interaction of a liquid with a chemically heterogeneous block copolymer introduces novel features to these systems.

Thus systematic changes in the domain morphology can apparently be effected by altering

(a) the relative composition of the blocks;
(b) their molecular weight distribution;
(c) the solvent used to dissolve and cast a specimen.

It is with the last aspect that this chapter will mainly be concerned.

2. DISSOLUTION OF BLOCK COPOLYMERS

2.1. Selective and Non-selective Solvents

When a homopolymer is dissolved in a liquid the process is aided by a favourable entropy of mixing, but the extent to which the polymer is soluble is determined predominantly by the energy of interaction between the polymer segments and the molecules of the liquid. The situation becomes somewhat more complex when dealing with copolymers. For random copolymers it has been suggested that the dissolution behaviour will reflect the mean solubility of the parent homopolymers, but this situation may not necessarily pertain to block copolymers. As the individual blocks are probably mutually incompatible, there is, in essence, a ternary system in which the component blocks may interact with the liquid to different degrees. Consequently, one must now consider that there are three interaction parameters which will influence the solubility. Solution of a block copolymer then depends on the relative magnitude and interplay of these interaction parameters, which was expressed by Stockmayer *et al.*,[13] in the form

$$\chi_{app} = \phi_A \chi_{1A} + \phi_B \chi_{1B} - \phi_A \phi_B \chi_{AB} \qquad (1)$$

where ϕ_A and ϕ_B are the volume fractions of blocks A and B, χ_{1A} and χ_{1B} are the interaction parameters for each block with solvent [1], and χ_{AB} is the energy difference when AB contacts are formed from AA and BB contacts.

Prospective solvents can be categorised as either 'selective' or 'non-selective' depending on the extent of liquid interaction with each block. A non-selective solvent is then one which is equally 'good' or 'poor' for both blocks whereas a selective solvent will be 'good' for one block but a 'poor' or non-solvent for the other block. In between these two extremes there will of course be a spectrum of relative block–liquid interactions. The solubility behaviour of a block copolymer will then range from that of a random or homopolymer to a state where it appears to have two separate affinities for the liquid. One might expect the χ_{AB} parameter to play a dominant role in the solubility behaviour, i.e. when χ_{AB} is small, solvents should tend to be non-selective, and when χ_{AB} is large (very incompatible blocks) the liquids are more likely to be selective, with each block exhibiting the solubility characteristics of the individual homopolymers in a simultaneous manner. This generalisation does not always hold true and even for large χ_{AB}, the χ_{1A} and χ_{1B} parameters may be of compatible magnitude, imparting a non-selective character to the liquid.

Values of χ_{1A} and χ_{1B} can be estimated from dilute solution studies or inverse gas chromatography,[14,15] while a useful approximation for χ_{AB} can be obtained from solubility parameters, using[16]

$$\chi_{AB} = \frac{V_1}{RT}(\delta_A - \delta_B)^2 \qquad (2)$$

Here V_1 is the molar volume of the solvent and δ_A, δ_B the solubility parameters of the respective homopolymers A and B.

2.2. Dilute Solution Studies

Hydrodynamic studies on block copolymers can be carried out using the techniques which are applicable to homopolymers, if it is recognised that the potentially incompatible nature of the blocks will influence the results, and will be accentuated in selective solvents. Number-average molecular weights ($\bar{M}_n$) are readily measured by osmometry (or other colligative properties) but the application of light scattering to determine the weight-average molecular weight ($\bar{M}_w$) normally requires that measurements be made in three different solvents, in order to overcome the problem of variations in copolymer composition. Stockmayer *et al.*[13] derived a relation between the intensity of scattered light and the composition of the block copolymer which was verified experimentally by Bushuk and Benoit[17] and Krause.[18] The experimental methods have been fully described in other publications[19] and will not be treated here.

One of the most intensively studied groups of diblock and triblock copolymers[20-31] has been the AB poly(styrene-*b*-methyl methacrylate) and BAB poly(methyl methacrylate-*b*-styrene-*b*-methyl methacrylate). Thermodynamic parameters obtained from dilute solution studies were generally found to be higher for the block structures than for the parent homopolymers, and anomalies in intrinsic viscosity–M.W. relations were observed, particularly in selective solvents favouring the PSt block. This was found to be due to the tendency of the blocks to aggregate and form micellar structures in these solvents. The concept of a θ-state for block copolymers has also had to be modified as it was found it did not always correspond to that pertaining to a homopolymer. In the latter case, all types of excluded volume interactions between segment pairs, both inter and intra, vanish at a unique temperature. In block copolymer solutions, the θ-state, as determined by the normal methods such as location of the temperature at which the second virial coefficient is zero, may only represent the temperature at which intermolecular interactions disappear. In many instances, depending on the type of solvent, the temperature must be lowered before the intramolecular interactions vanish. This dependence on solvent is clearly illustrated when viscosity data are plotted according to equations typified by the Stockmayer-Fixman relation[32]

$$[\eta]/M^{1/2} = K_\theta + C\Phi M^{1/2} \tag{3}$$

where M is the molecular weight, $[\eta]$ is the intrinsic viscosity, K_θ is the Flory-Fox viscosity constant under θ conditions related to the unperturbed dimensions, Φ is the Flory universal constant and C contains the solvent–polymer interaction parameter $(\frac{1}{2} - \chi)$.

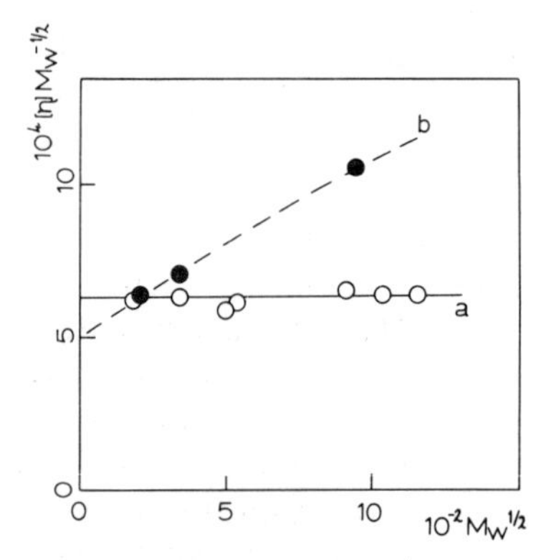

FIG. 2. Stockmayer-Fixman plots for poly(methyl methacrylate-*b*-styrene-*b*-methyl methacrylate) triblock copolymers dissolved in apparent θ solvents (a) cyclohexanol and (b) 2-ethoxyethanol at $\theta = 354$ K.

For true θ-solvents, the ratio $([\eta]/M^{1/2})$ in eqn (3) should be independent of $M^{1/2}$, because $(\frac{1}{2} - \chi) = 0$ and so C is zero. Viscosities of the BAB (PMMA-*b*-PSt-*b*-PMMA) triblocks have been measured in the apparent θ-solvents cyclohexanol and 2-ethoxyethanol, both of which form solutions in which the second virial coefficient vanishes at a temperature of 354 K. Copolymer solutions in cyclohexanol exhibit the expected θ-behaviour (see Fig. 2), but large deviations are observed for 2-ethoxyethanol solutions. As 2-ethoxyethanol is a solvent for PMMA but a non-solvent for PSt, it seems likely that aggregation occurs and that the random flight conformation of the copolymer chain is not attained at the same temperature at which the second virial coefficient is zero.[33,34] Anomalous viscosity behaviour has also been detected in good solvents.

Shima *et al.*[35,36] have prepared triblock copolymers of PSt and poly(*p*-chlorostyrene) with block arrangements ABA and BAB (A = PSt) and investigated their dilute solution properties in four solvents. For non-selective solvents, toluene and MEK, no difference in conformational behaviour could be detected between the ABA and BAB structures, but in selective solvents, cumene and CCl_4, the chains of ABA triblocks were more extended than their BAB counterparts. No apparent difference could be found in the behaviour of the samples in selective solvents when plotted according to eqn (3). In this case the Stockmayer-Fixman plot was equally valid in all solvents.

Normal behaviour was also observed for toluene solutions of BAB (PMMA-*b*-PSt-*b*-PMMA) when plotted according to eqn (3) but when $[\eta]$ was measured in 1-chlorobutane or *p*-xylene unreasonably high values of K_θ were obtained.[37] Toluene is a good solvent for both blocks, but the others are both poor solvents for PMMA and moderately good for PSt; so again selective solvents are responsible for the abnormalities.

A conformational transition in this type of BAB block copolymer has been detected by Dondos *et al.*[38] in *p*-xylene and several other solvents. In the selective solvent *p*-xylene, the transition occurred at 311 K, but was observed at lower temperatures (near 303 K) in less selective solvents such as tetrahydrofuran, benzene, dioxan and chloroform. The authors have suggested that the transition represents the change from a segregated conformation at lower temperatures to one with a significant number of heterocontacts which increase as the temperature is raised. A similar conclusion was arrived at[39] when the temperature dependence of the conformation of the triblock (PSt-*b*-P*p*ClSt-*b*-PSt) was studied. This implies that heterocontacts form less readily in selective solvents. The question of describing the precise conformation of di and triblock copolymers in solution has proved to be one of the controversial issues to arise from these studies and also from a substantial body of work on AB poly(styrene-*b*-isoprene).[40-43]

Two plausible conformations have been proposed:

(a) The segregated model, in which there are few if any heterocontacts and the molecular size of the blocks in solution is close to that expected for comparable homopolymer chains.

(b) The random structure, where the different blocks overlap to some extent resulting in heterocontacts which will expand the component coils because of the additional repulsive forces between unlike segments.

Hydrodynamic studies alone do not appear capable of resolving the problem in a satisfying manner, although some ingenious techniques have been tried, such as the suggestion by Leng and Benoit[44] that a block copolymer should be dissolved in a solvent which is isorefractive with one block. Light-scattering measurements to determine the radius of gyration of the coil will then only 'see' the other block, thereby allowing individual block dimensions to be measured directly in the presence of the other 'invisible' block. This has been applied to diblocks of AB poly(styrene-*b*-isoprene)[45] and AB poly(styrene-*b*-methyl methacrylate).[30] In both cases expansion of the PSt chain was observed which was taken to be indicative of the presence of heterocontacts.

While opinions differ, the majority of workers favour the idea that there is some degree of interpenetration of the component blocks, but that the number of heterocontacts formed will depend on both solvent and temperature. Many of the dilute solution studies are open to some flexibility of interpretation and a new approach to this problem is required. In this context, an interesting study on the conformation of AB poly(styrene-*b*-methyl methacrylate) in which the PSt block was deuterated has been performed by Han and Mozer[46] using neutron scattering. The experiments, performed in toluene solutions, produced the rather unexpected observation that while the *d*-PSt block was slightly expanded, the PMMA block had actually contracted and had a more tightly coiled conformation than one would expect for a comparable free homopolymer. This led the authors to speculate that the conformation could best be described by a partially segregated core and shell model, with the tightly coiled PMMA as the core surrounded by a PSt shell, with repulsive forces pushing the PSt coil outwards. The model is similar in many ways to the monomolecular micelle structure proposed by Sadron[47] and suggests that even in potentially non-selective solvents there may be a tendency for limited micellisation to take place. It also suggests that even so-called non-selective solvents may tend to behave selectively even if the difference between χ_{1A} and χ_{1B} is small. Use of selective solvents for PMMA must then cause a conformational inversion to occur, which might be one way of interpreting the results reported by Utiyama *et al.*[31] although these authors were not looking specifically for such an effect.

Triblocks may behave differently, as it was observed[28] that the central PSt block in the BAB copolymers was expanded much more than in a comparable diblock, whose overall dimensions were also smaller than the triblock sample. Values of $[\eta]$ in different solvents have also been used to determine the presence of heterocontacts in poly(dimethylsiloxane-*b*-styrene-*b*-dimethylsiloxane) triblocks.[48] As viscosity measurements figure largely in such studies, the several extrapolation procedures used to evaluate $[\eta]$ have been assessed for block copolymer solutions.[49]

It is evident from these data that the non-uniformity of interaction of a liquid with each block in the copolymer can give rise to some interesting effects. Aggregation and micellisation are important features of such systems and will be treated fully in Chapter 2, but in as much as this behaviour has an important bearing on the physical properties of solvent-cast block copolymer films, some relevant aspects will be discussed in the next section.

2.3. Selective Solvents

The rather special behaviour of block (and graft) copolymers in solutions prepared from selective solvents, was first demonstrated by Merrett[50,51] who studied poly(*cis*-isoprene-*g*-methyl methacrylate). In benzene solution both segments of the copolymer were solvated non-selectively, but when methanol was added in quantities which would have precipitated the free PIp, no phase separation was detected and the solution was apparently stabilised by the presence of the PMMA side chains. While the resulting solution was distinctly turbid, it exhibited colloidal characteristics and remained quite stable. Merrett proposed that as PMMA chains tend to expand slightly on addition of up to 25% methanol, they are sufficiently well solvated to prevent the PIp portion of the copolymer from undergoing phase separation, and although further addition of methanol precipitates the entire copolymer, a stable solubility region can be created. The reverse situation was achieved by adding petroleum ether to the copolymer dissolved in benzene. This time the PMMA moieties collapse and are stabilised by the solvated PIp portion of the molecule.

The ability to form colloidal systems in selective solvents seems to be a general property of block copolymers, within the limitations imposed by the incompatibility and molecular weights of the individual blocks. The resemblance between the behaviour of block copolymers in selective solvents and surfactant molecules has prompted the use of the term micelle to describe the state of the macromolecule in solution. Two forms have been proposed: (i) a simple monomolecular micelle, formed from one block copolymer molecule with the tightly coiled core of one block surrounded by the expanded coil of the second, and (ii) polymolecular micelles formed by the aggregation of many monomolecular micelles.

Evidence has been found for the formation of polymolecular micelles of AB poly(styrene-*b*-methyl methacrylate) in acetone[52] and toluene/furfuryl alcohol mixtures,[23,31] both selective for PMMA, and in triethylbenzene,[52] *p*-xylene and toluene/*p*-cymene mixtures[23,31] which are selective for PSt. The data are less conclusive for the BAB triblocks. Kotaka *et al.*[23] found no aggregation in selective solvents for PSt, only in those selective for PMMA, whereas Krause[52] observed aggregation in both. Periard and Riess[53] reported that association may depend mainly on copolymer composition which could explain the observed discrepancies, as Krause's samples were heterogeneous in both composition and M.W. More direct evidence for the shape of the micellar structures comes from electron microscopy[54,56] and confirms the idea of a spherical entity with the collapsed core surrounded by the solvated block. These studies assume that the structure formed in solution is carried through to the solid state on evaporation and so influences the morphology of the solid. This is an important and critical extrapolation whose validity is established by the work of Price and Woods.[56] They used an elegant 'freeze-etching-replication' technique which avoids the possible structural changes which might be induced during solvent evaporation. Their results confirm those of other workers and thereby lend credence to the assumption that the structure of the solution is retained on removal of the solvent. This effect was first recognised by Merrett[51] who observed different physical properties for copolymer films cast from the two types of selective solvent. Films cast from the solvent which expanded the PMMA and collapsed the PIp were non-tacky and had a higher modulus and hardness than the tacky films cast from the selective solvent for PIp. This was interpreted as indicating that the PMMA was more dispersed and formed an extended load-bearing phase with PIp inclusions when a PMMA selective solvent was used, while in the latter case the PIp would form the continuous phase. Electron microscopy[57] of poly(styrene-*b*-butadiene) samples confirmed this structural analysis.

These observations suggest that while the size, composition, and χ_{AB} parameter for the blocks in a copolymer can determine the solid state morphology of samples prepared in the absence of a liquid, use of a solvent to prepare a film by casting may alter this equilibrium morphology to a significant degree and so change the physical properties of the sample. The implication is that a structure is preset by the solvent which is maintained and 'locked' in during the evaporation process to produce a metastable state with its own characteristic physical properties. The remarkable stability observed for micelles formed in solutions of selective solvents will certainly help this process, but, as has been demonstrated by Kawai and his coworkers,[58] even solvents which are apparently non-selective will influence the morphology of copolymer films to some extent.

3. SOLVENT EFFECTS IN DOMAIN FORMATION

3.1. Experimental Observations

The influence of the solvent on the distribution in the solid state of the components of a block copolymer after microphase separation has been established beyond doubt by the work of Kawai *et al.*[58–60] They observed changes in the domain structure for a 40/60 wt % poly(styrene-*b*-isoprene) diblock copolymer with casting solvent, which followed the pattern shown in Fig. 3. The lamellar structure, obtained from toluene-cast films, was altered to PSt rods or islands in a continuous PIp matrix by changing the solvent to either cyclohexane or heptane. The several solvents used gave varying sizes and shapes of PSt domain depending on the relative magnitude of the solvent–polymer interactions. Good solvents for PSt, such as toluene and MEK, produced extended domains, while poor solvents such as hexane restricted the PSt to small, discrete, collapsed domains.

Toluene, which is an equally good solvent for both blocks gives regular, well-defined, lamellar structures for the 40/60 wt % and 50/50 wt % poly(styrene-*b*-isoprene) samples. This is altered as shown in Fig. 4 by the addition of a selective solvent, MEK, which is much poorer for the PIp blocks. The PSt component becomes more dispersed as the MEK content increases until the PIp is eventually reduced to isolated domains in a PSt matrix. Triblock copolymers behave similarly as can be seen from the electron micrographs in Fig. 5 of a 35/30/35 wt % poly(styrene-*b*-isoprene) triblock cast from four different solvents. Dispersed rods or spheres of PIp in a continuous PSt matrix are formed when the film is cast from toluene. The PIp domains are contracted further when MEK is used as solvent, but cyclohexane expands the PIp to form an inter-connected domain structure. This is the behaviour one would expect from a consideration of the relative solvating power of the solvents for each block.

These general observations have been confirmed by Leary and Williams.[61,62] Their statistical thermodynamic treatment of microphase separation allowed them to predict the type of domain formation favoured by the thermodynamic interactions in the absence of solvent. The predicted structure for a 25/75/25 wt % poly(styrene-*b*-butadiene-*b*-styrene) triblock was lamellar at the temperature of melt preparation, and this was confirmed from electron micrographs. Use of a non-selective solvent, toluene, to cast a sample led to a structure comprising well-defined cylinders of PSt in a matrix of PBd. One can then conclude that any solvent must be considered as a potential structure-controlling parameter.

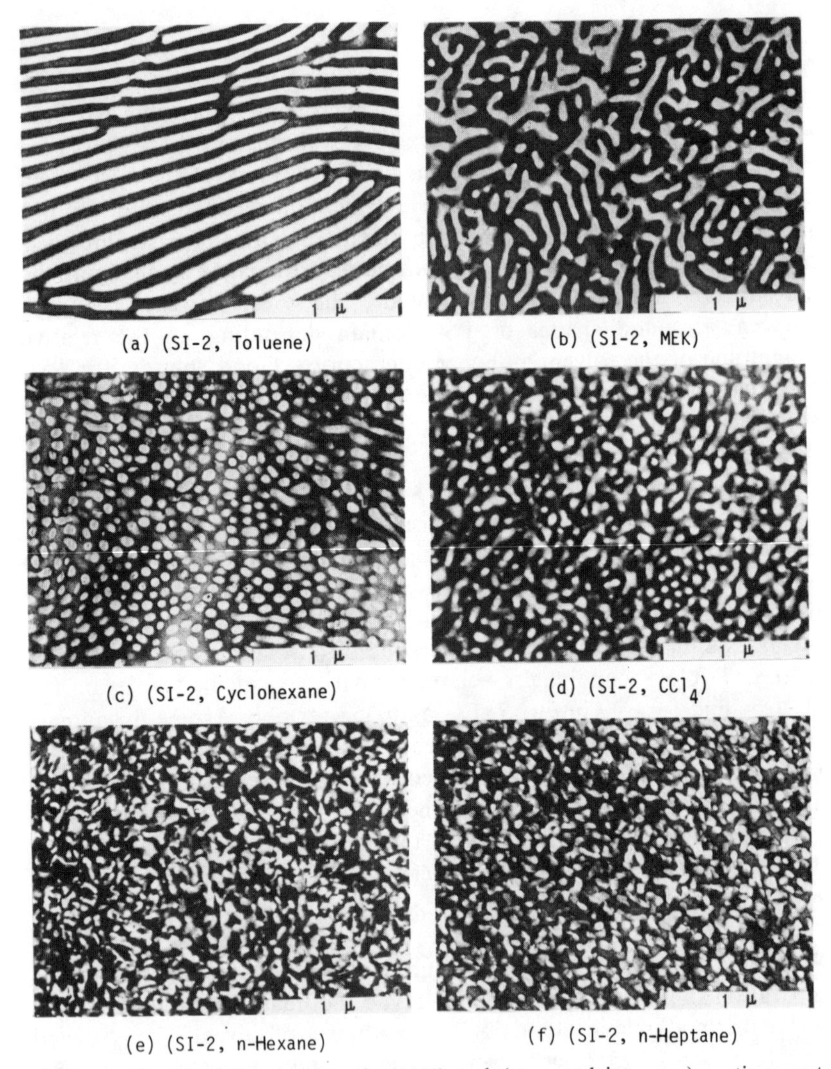

FIG. 3. Electron micrographs of (40/60) poly(styrene-*b*-isoprene) sections cut from films cast from the solvents indicated. (With permission, H. Kawai; from Kawai, H., *Memoirs of the Faculty of Engineering, Kyoto University*, Kyoto University Press, 1971, p. 392.)

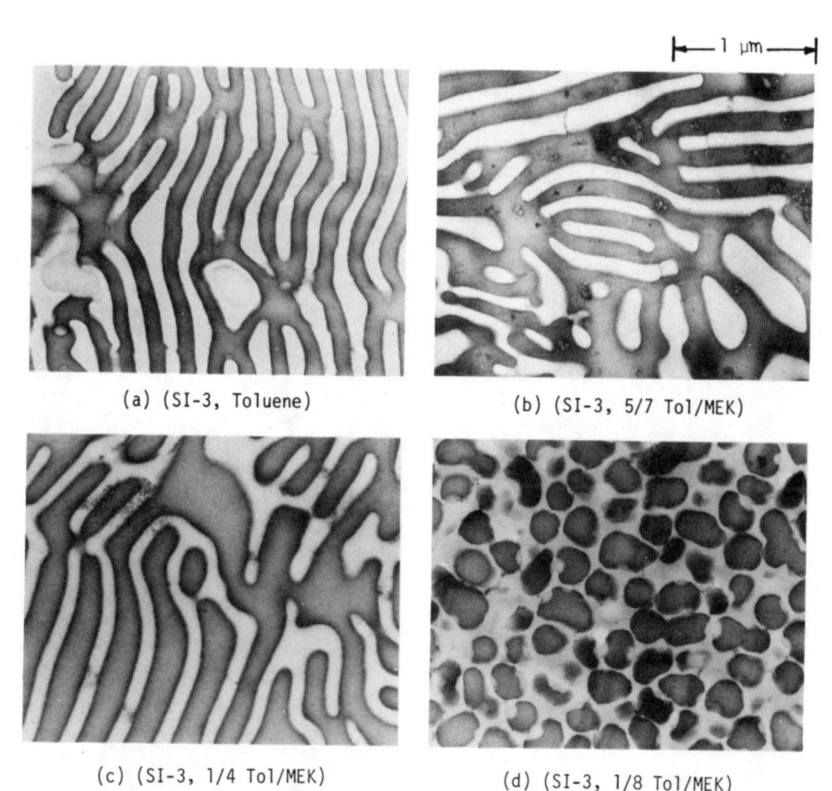

FIG. 4. Electron micrographs of structure modification produced by the addition of a selective solvent MEK to toluene, in films of (50/50) poly(styrene-*b*-isoprene). (With permission, H. Kawai; from Kawai, H., *Memoirs of the Faculty of Engineering, Kyoto University*, Kyoto University Press, 1971, p. 393.)

Films for such studies are normally prepared by slow evaporation from solutions of approximately 5% initial concentration; rapid evaporation tends to produce less well-defined morphologies. It has been argued that there must be a certain critical concentration, at which the final structure of the copolymer is determined during evaporation. This idea is similar to that of a critical micelle concentration. Some workers[59] believe that this is much higher than 10% as films of poly(styrene-*b*-isoprene) cast from toluene solutions ranging from 1% to 10% concentration gave essentially the same final structure in the bulk state. Sadron and Gallot[63] have postulated that there is a progressive aggregation of monomolecular micellar species as the concentration of the solution increases, until at a certain critical concentration between 20% and 60%, depending on the particular system, a

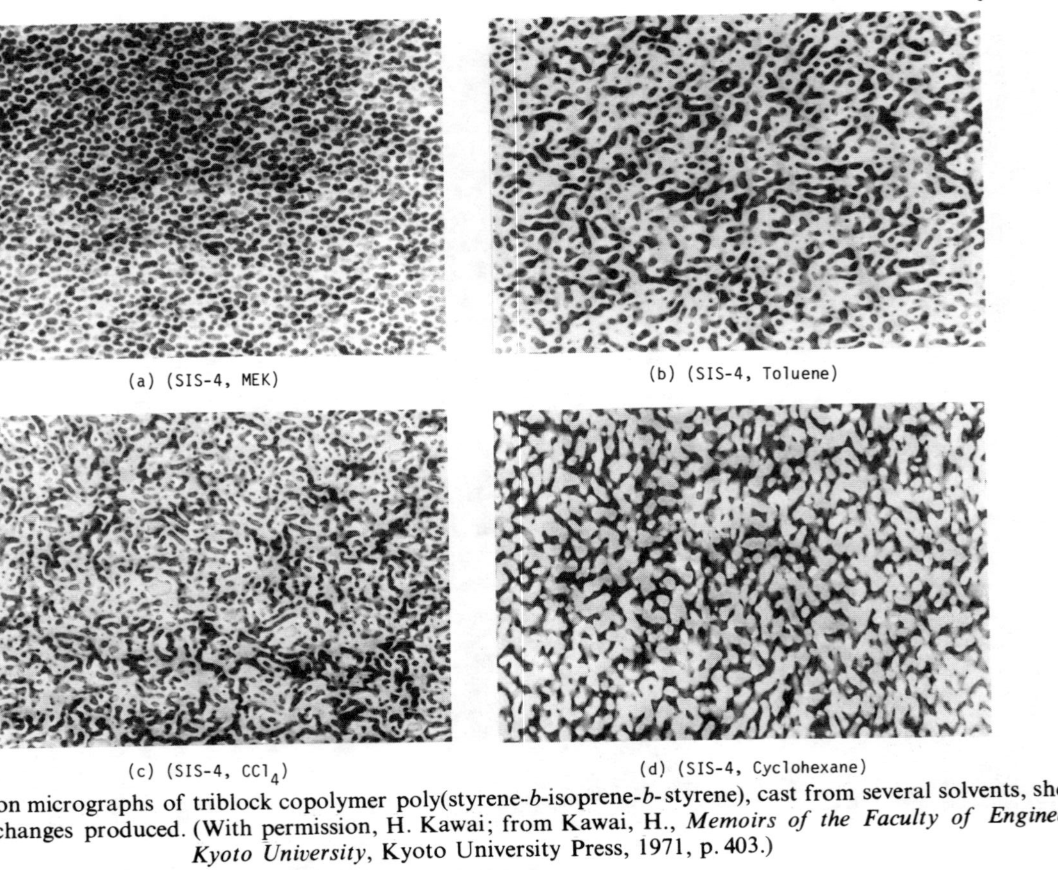

FIG. 5. Electron micrographs of triblock copolymer poly(styrene-*b*-isoprene-*b*-styrene), cast from several solvents, showing the structural changes produced. (With permission, H. Kawai; from Kawai, H., *Memoirs of the Faculty of Engineering, Kyoto University*, Kyoto University Press, 1971, p. 403.)

regular ordered structure is obtained which persists into the solid state. In one group of polymers they studied, poly(styrene-*b*-2-vinylpyridine), and poly(styrene-*b*-4-vinylpyridine) they observed that the systems passed through more than one structure, i.e. spheres to cylinders or cylinders to lamellae, depending on the composition and preferential solvent used. This was not observed in poly(styrene-*b*-isoprene) or poly(styrene-*b*-butadiene) which appeared uninfluenced by solvent. The range of casting solvents was rather restricted and this renders the observations inconclusive in the latter group.

The explanation offered for the structural changes seen in the first group is based on the idea that the presence of a preferential solvent, for say an A block in an AB diblock copolymer will in effect increase the 'size' of that sequence. Consequently the morphology would change in the manner one would expect for a 'dry state' sample if the block length was increased; i.e. for a lamellar structure in an AB diblock, addition of a preferential solvent for A produces successively cylinders then spheres of B in an A matrix. This is an attractive but fairly simplistic picture and although it is a good first approximation, attempts to derive a more rigorous thermodynamic description of solvent effects on domain formation have been made.[59,64] The question of how to determine the critical concentration remains at present largely unresolved.

3.2. Theoretical Approaches

While several authors[1-10,61,62] have tackled the problem of developing a theory to explain domain formation in block copolymers when no solvent is present, only Meier[64] and Kawai[59] have seriously attempted to consider the influence a liquid may have on microstructure.

The Meier[64] model assumes that (i) the chains obey random flight statistics, (ii) the Flory-Huggins equation can be used to evaluate the free energies of mixing, (iii) only three domain structures are possible, viz. spheres, cylinders and lamellae, and (iv) the segment densities in the domains are uniform. The model also assumes that there is a volume restriction imposed by the requirement that A segments remain in the domain space of A, and B segments cannot enter that space but must remain in B domains; that there is a positional restriction of A–B bonds which must be located at the domain interfaces; and that chain perturbation is modified in the domains. Consideration of these requirements leads to a difference in entropy calculated between a block copolymer and that of a corresponding mixture of homopolymers. The free energy of domain formation is then derived by summing these contributions along with the enthalpy change and the interfacial energy of the system. The equations developed cover both non-selective and selective solvents and the latter are worth examining in greater detail.

If one considers for simplicity that a selective solvent for an A block resides solely in A domains, then the relative free energies of domain formation from solution can be expressed for the different morphologies as

Lamellae:

$$\frac{\Delta G_{(\mathrm{rel})}}{N_{\mathrm{AB}}\mathrm{k}T} = -\ln\left[\frac{6(l+q+r)^2}{r(l+q)}\exp-\frac{\pi^2}{6}\right] = \ln\left(\frac{1+\mu_1^2}{2\mu_1}\right)$$

Cylinders:

$$\frac{\Delta G_{(\mathrm{rel})}}{N_{\mathrm{AB}}\mathrm{k}T} = -\ln\left[\frac{3\pi}{4}M\exp\left\{-\frac{1}{6}(\beta_1^2+r\mu_c^2\gamma^2)\right\}+\ln\left(\frac{1+\mu_c^2}{2\mu_c}\right)\right.$$

and spheres:

$$\frac{\Delta G_{(\mathrm{rel})}}{N_{\mathrm{AB}}\mathrm{k}T} = -\ln\frac{2}{3}\left(\frac{R'+R}{R'-R}\right)\exp\left\{-\frac{9\pi^2}{96}\left(1+\frac{rR^2}{(R'-R)^2}\right)\right\}+\ln\left(\frac{1+\mu_s^2}{2\mu_s}\right)$$

where N_{AB} is the number of block copolymer molecules, $q = V_s/V_{\mathrm{A}}$ is the ratio of volume of added solvent to that of the A component, $r = M_{\mathrm{B}}/M_{\mathrm{A}}$ is the ratio of block M.W.'s, R is the domain radius, R' the outer constraining surface of a domain, γ is the interfacial tension, M and β_1 are complex functions defined in the original paper[64] and the μ values are equal to the ratio of the expansion factors $(\alpha_\beta/\alpha_{\mathrm{A}})$ for each structure, the exact value being dependent on the morphological model.

These equations have recently been refined[65] but the general form is the same. The free energy–solvent relationships have been calculated for several block ratios and the predicted morphologies are illustrated in Figs. 6 and 7 for the block ratios $(M_{\mathrm{B}}/M_{\mathrm{A}}) = 2$ and 4. These predict that for $M_{\mathrm{B}}/M_{\mathrm{A}} = 2$ a lamellar structure will be found in the absence of any solvent, while for $M_{\mathrm{B}}/M_{\mathrm{A}} = 4$ the corresponding equilibrium structure will be spheres of A in a B matrix. The influence of a selective solvent for A can be seen by approaching from the right of each diagram. A morphology of cylinders of A in a B matrix will be formed if the critical concentration at which the structure is locked in during solvent evaporation is less than $\simeq 50\%$ for $M_{\mathrm{B}}/M_{\mathrm{A}} = 2$ or $\simeq 60\%$ for $M_{\mathrm{B}}/M_{\mathrm{A}} = 4$. If the morphologies are not determined until solution concentrations exceed these values, then a lamellar structure will form. The use of a preferential solvent for B changes

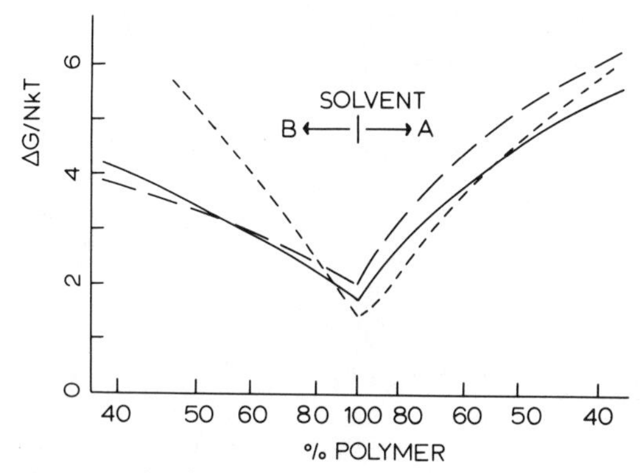

FIG. 6. Free energy and solvent content relationships for the various domain morphologies in block copolymers with a block length ratio of $M_B/M_A = 2$. (———), spheres; (— — —), cylinders; (- - - -), lamellae.

this picture as can be seen by now moving across the diagrams (Figs. 6 and 7) from the left.

The diagrams demonstrate that a logical transformation through the three main morphological forms is not necessary as suggested by Sadron and Gallot,[63] and that the final form can depend on the type of selective solvent used. It should also be noted that if a structure has to be created by diffusion of polymer chains through a high energy state, then that transformation is unlikely to take place. Consequently the critical solution concentration is an important factor, but a difficult one to determine precisely.

Kawai and coworkers[59] have also calculated the free energy of domain

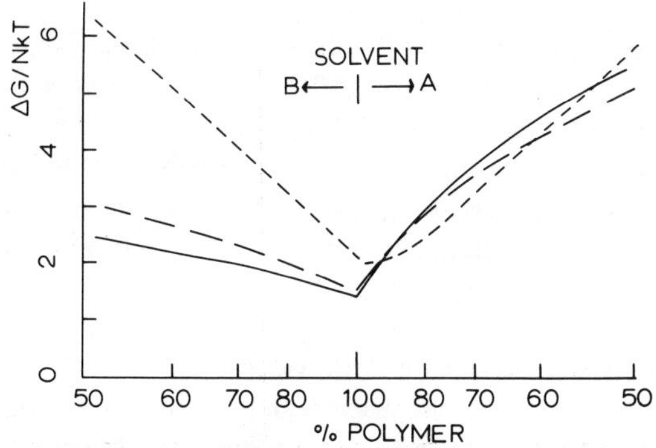

FIG. 7. Free energy and solvent content relationships for the various domain morphologies in block copolymers with a block length ratio of $M_B/M_A = 4$. (———), spheres; (— — —), cylinders; (- - - -), lamellae.

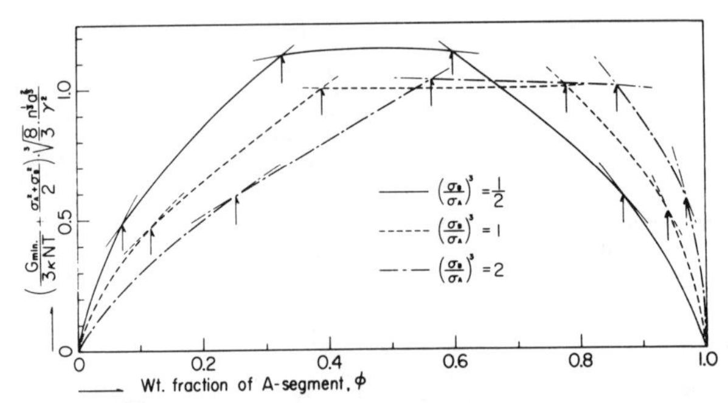

FIG. 8. Influence of solvent on the relation between the minimum free energy of micelle formation and the weight fraction of A blocks in an AB block copolymer. The relative solvating power of the solvent is denoted by (σ_B/σ_A). (With permission, H. Kawai; from Kawai, H., *Memoirs of the Faculty of Engineering, Kyoto University*, Kyoto University Press, 1971, p. 422.)

formation in the presence of solvent but have used a different approach when estimating chain perturbation which tends to produce unrealistic segment densities. The general trends are somewhat similar to those of Meier as shown in Fig. 8 for AB block copolymers in selective solvents.

It is now obvious that one should be able to predetermine the morphology in block copolymers and hence their properties by exercising control of block length ratios and the nature of the solvent. The extent to which one can achieve this goal is illustrated in the sections which follow.

4. THE EFFECT OF SOLVENT ON THE PHYSICAL PROPERTIES OF BLOCK COPOLYMERS

The striking variations in physical behaviour which can be obtained when block copolymers are cast from different solvents was demonstrated qualitatively by Merrett[50,51] for the PMMA/natural rubber graft copolymers. As we have seen, he obtained a hard form when the casting solvent was selective for PMMA and a soft form when the solvent was selective for the rubber portion. More detailed and quantitative studies have since been carried out by a number of workers.

4.1. Dynamic Thermomechanical Behaviour

Beecher *et al.*[66] studied the temperature dependence of the damping in an SBS (14–72–14) sample (Kraton 101) cast from three different solvents. The resulting spectra showed significant differences in the intensities of the

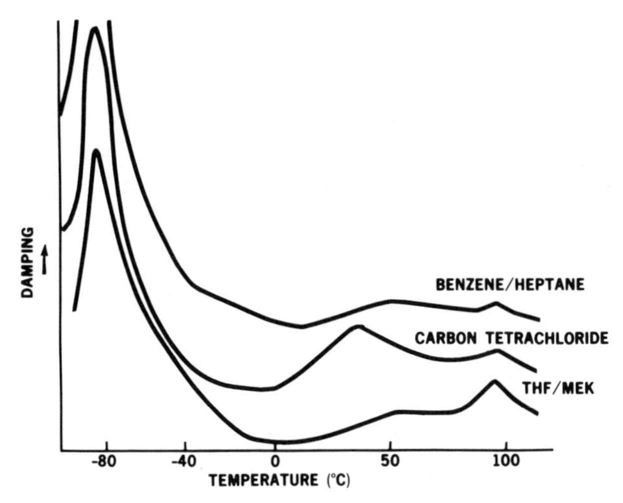

FIG. 9. Damping thermograms for poly(styrene-*b*-butadiene-*b*-styrene) triblock copolymer cast from three solvents. (From *Polymer*, 1976, **17**, 938; with permission IPC Business Press, Ltd. ©)

damping maxima. The loss peak observed at 193 K, attributed to the glass transition of the PBd phase, was much larger than the corresponding loss peak located at 373 K for the PSt phase. This one would expect from the copolymer composition, but the PBd peak also decreased in area and intensity as the solvent changed from a benzene/heptane mixture, through carbon tetrachloride (CCl_4) to a THF/MEK mixture (Fig. 9). The effect is more clearly seen in Fig. 10 for an SBS triblock of similar composition (Thermolastic 125) cast from four solvents.[67] The area of the PBd damping peak is seen to decrease while the area of the PSt peak increases in the solvent order CCl_4 > toluene > ethyl acetate > MEK. The modulus of the copolymer, in the plateau region lying between the two glass transition temperatures (T_g), is also solvent dependent and increases in the same solvent order as above. The modulus is highest when the PSt peak area is large. The same general behaviour was found by Kalfoglou[68] using the same solvents but a different range of block lengths for the SBS samples.

The range of casting solvents has been extended for SBS by Beamish *et al.*[69] and by Cowie *et al.*[70] The temperature dependence of the dynamic storage modulus for various films, displayed in Fig. 11, shows the obvious differences resulting from a change of casting solvent. The principle feature in each is the decrease in the elastic storage modulus E' associated with the glass transitions of both blocks (at 192 K and 373 K). The drop in E' is

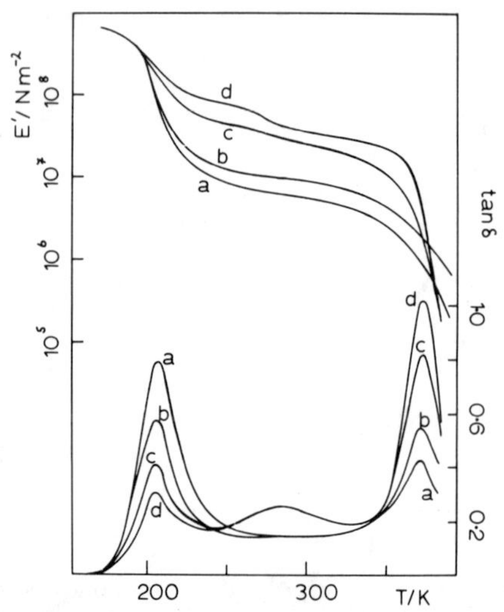

FIG. 10. Temperature dependence of the storage modulus (E') and tan δ for polystyrene-*b*-butadiene-*b*-styrene) samples (Thermolastic 125) cast from four solvents: (a) carbon tetrachloride, (b) toluene, (c) ethyl acetate and (d) methyl ethyl ketone. (From reference 67, with permission John Wiley & Sons, Inc.)

dependent on the casting liquid and mirrors the change in area of the corresponding damping peaks. This effect is most easily monitored by inspecting the modulus value in the plateau region lying between the two T_g values, which is the normal 'working temperature range' for the copolymer. The data (Table 2) show a ten-fold difference between the two extreme

TABLE 2
COMPLEX MODULUS (MEASURED AT 293 K) FOR SBS BLOCK COPOLYMER FILMS CAST FROM VARIOUS SOLVENTS

Solvent	δ ((*cal cm*$^{-3}$)$^{1/2}$)	$10^{-8}E^*_{(293)}$ (Nm^{-2})
Cyclohexane	8·2	0·115
1-Chlorobutane	8·4	0·090
Carbon tetrachloride	8·6	0·128
Cyclopentane	8·7	0·103
Toluene	8·9	0·273
Ethyl acetate	9·1	1·180
Propylene oxide	9·2	0·820
Cyclohexanone	9·0	0·760
Dioxane	10·0	0·280

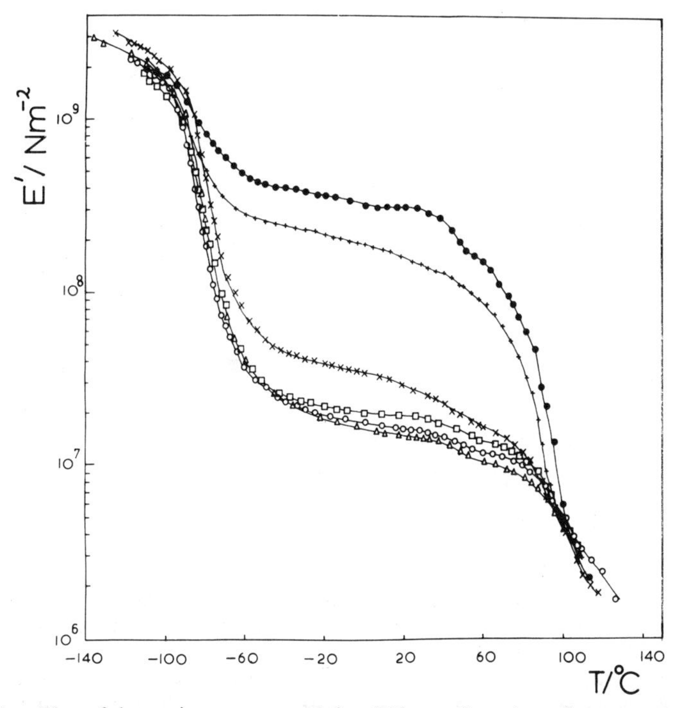

FIG. 11. Plot of dynamic storage modulus (E') as a function of temperature for poly(styrene-*b*-butadiene-*b*-styrene) films cast from (●) ethyl acetate, (+) cyclohexanone, (×) tetralin (□), toluene, (○) carbon tetrachloride and (△) cyclohexane. (From *Polymer*, 1977, **18**, 49; with permission IPC Business Press, Ltd. ©)

values found in films cast from 1-chlorobutane and ethyl propyl ketone.[70] An even wider variation in the modulus at 293 K was found by Cowie *et al.*[71] in films formed from Kraton G-1650, a poly(styrene-*b*-ethene-*co*-butene-*b*-styrene) triblock copolymer with a PSt weight fraction of 0·29. The complex modulus (E^*)–temperature curves in Fig. 12(a) illustrate the solvent dependence quite clearly and E^* values[20] range from $4{\cdot}4 \times 10^6$ Nm^{-2} for the heptane-cast film to $2{\cdot}05 \times 10^8$ Nm^{-2} for a THF film. A sample pressed in the absence of solvent had an intermediate value of $1{\cdot}4 \times 10^7$ Nm^{-2}.

An interesting feature of the damping spectra for films prepared from MEK and ethyl acetate[67] is a damping peak centred at 283 K which is not associated with the main glass transitions. This has an apparent activation energy of 126 kJ mol^{-1} which is lower than those measured for both glass

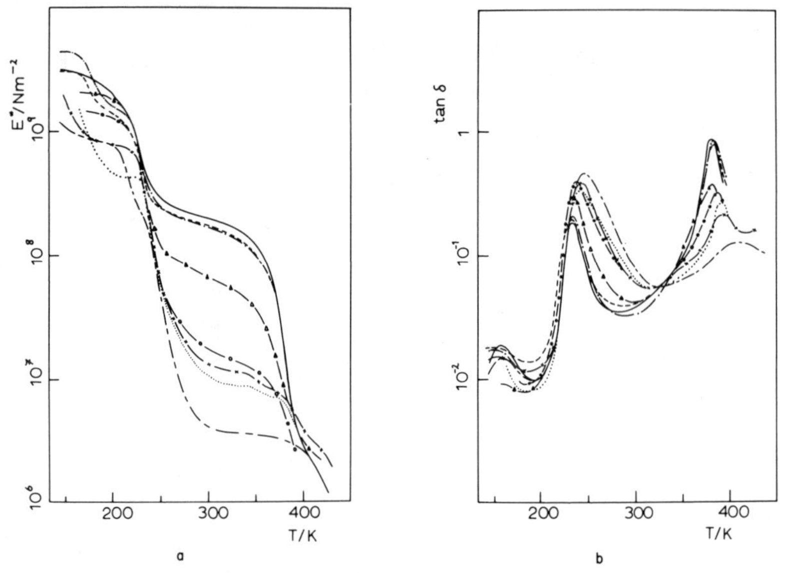

FIG. 12. (a) Temperature dependence of the complex modulus E^* for films of Kraton G-1650 cast from (——) tetrahydrofuran, (—·—·) chloroform, (----) toluene, (○—○—) cyclohexene, (△—△—) bromobutane, (×—×—) 'pressed', (······) cyclohexane, and (— — —) *n*-heptane. (b) Temperature dependence of tan δ for the same samples as in (a). (From reference 71, with permission American Chemical Society.)

transitions. Similar damping features have been observed in films cast from CCl_4,[66] cyclohexane and tetralin,[69] ethyl propyl ketone, propylene oxide and cyclopentane,[70] but at much higher temperatures. DSC thermograms show slope changes at 259 K (Fig. 13) which may be associated with these relaxations but not necessarily as we shall see later. It has been suggested that these damping maxima indicate a degree of phase mixing in the samples and arise from the relaxations occurring in the interfacial regions separating the different domains in the copolymer. Another suggestion is that local chain motion in the PSt, such as phenyl ring rotation or the more general β-relaxation, causes the damping to occur. The apparent activation energy for the PSt β-relaxation, which is observed at 300 K, is about 147 kJ mol^{-1}, slightly higher than that estimated for the intermediate damping process in SBS.[67] Also the fact that the temperature at which these subsidiary damping peaks are observed is not constant makes the first explanation more attractive. It should be emphasised that these observations are not consistent; for example not all workers detect a peak in

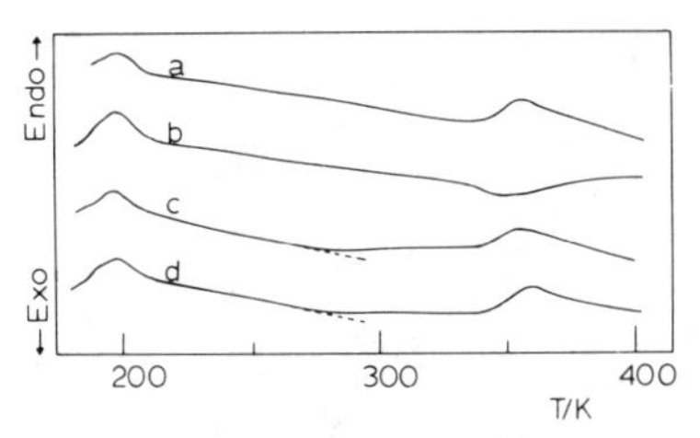

FIG. 13. DSC curves for poly(styrene-*b*-butadiene-*b*-styrene) samples (Thermolastic 125) cast from (a) carbon tetrachloride, (b) toluene, (c) ethyl acetate and (d) methyl ethyl ketone. (From reference 67, with permission John Wiley & Sons, Inc.)

CCl_4-cast films and Wilkes *et al.*[72] could find no intermediate relaxations in SBS in the temperature range 220–320 K at all.

The evidence for the existence of these intermediate relaxations is fairly convincing but the inconsistencies make precise interpretation difficult and it may be necessary to ensure that samples studied have been subjected to the same treatment and have similar thermal histories. Shen *et al.*[73] found it necessary to remove plasticiser from one of their samples before obtaining reproducible results and this could account for some of the temperature differences observed.

Structural changes in the SBS films were clearly seen in electron micrographs;[66] the fundamental morphology of the Kraton 101 was a matrix of PBd containing spherical PSt domains which changed size with casting solvent (Table 3). Use of MEK or ethyl acetate as solvent caused a phase inversion to take place, with the PSt forming the continuous matrix containing PBd domains. An electron micrograph of an ethyl acetate-cast film (Fig. 14) appears to exhibit a fine structure which has been interpreted[64] as a multimolecular micellar formation with a PBd core surrounded by a PSt shell encased in a layer of PBd. The Kraton G-1650 morphology cannot be studied so easily by electron microscopy because no selective staining technique is available and small angle X-ray scattering has

TABLE 3

EFFECT OF SOLVENT ON THE SIZE OF PSt DOMAINS IN Kraton 101 FILMS

Solvent	*Domain size (nm)*
(90/10) Tetrahydrofuran/methyl ethyl ketone	12·1 ± 2·0
Carbon tetrachloride	11·0 ± 2·5
(90/10) Benzene/heptane	9·4 ± 2·4

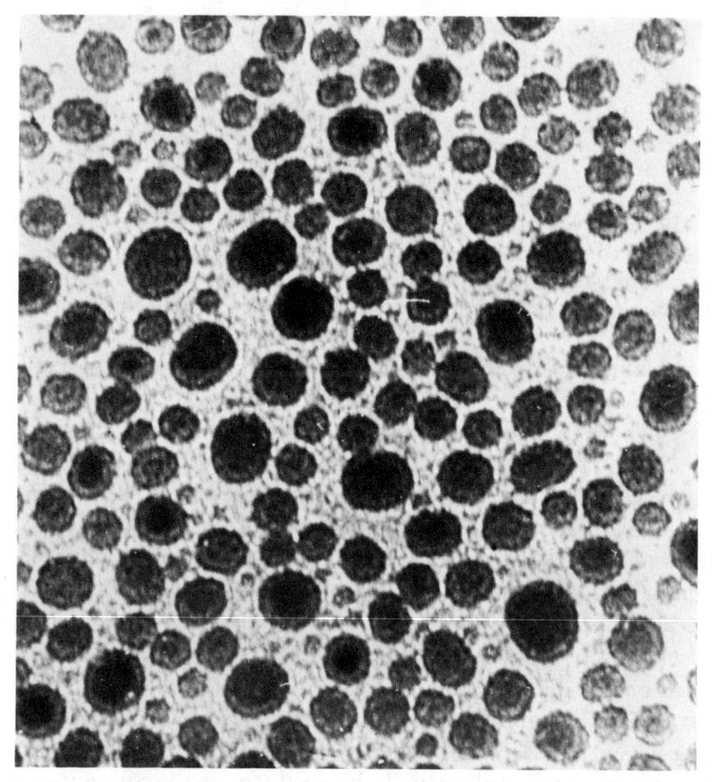

FIG. 14. Electron micrograph of a poly(styrene-*b*-butadiene-*b*-styrene) sample (26 wt % polystyrene), cast from ethyl acetate. Magnification ×160 000. (From *Polymer*, 1977, **18**, 49; with permission IPC Business Press, Ltd. ©)

proved more useful in identifying the solvent related structures.[74] For toluene-cast films the diffraction patterns suggest a lamellar structure, while the heptane-cast film is best described by PSt spheres in a rubbery matrix. For cyclohexane, hexagonally packed cylinders of PSt in a continuous rubbery phase seems the most likely structure.

Having established some of the morphological changes brought about by the presence of a solvent, the viscoelastic behaviour can now be explained in terms of the distribution of phases within the system. It can be assumed that in a multiphase system the modulus will be controlled, predominantly, by the ability of the continuous phase to transmit the applied stress. Most of the samples of SBS and Kraton G-1650 studied have had compositions which produce a continuous rubber phase with discontinuous PSt glassy domains, when prepared in the absence of a solvent. The modulus should then have a value similar to that of an elastomer as this will be the main load-bearing phase at ambient temperatures. A film of Kraton G-1650, prepared in the absence of a solvent, had a modulus $\simeq 1{\cdot}4 \times 10^7\,\mathrm{Nm}^{-2}$, which is comparable to that of a filled crosslinked elastomer, and tends to substantiate the assumption. Use of truly non-selective solvents to prepare the copolymer films should leave the morphology and physical behaviour essentially unchanged as the blocks will interact with the liquid to the same extent. This situation will change when selective solvents are used.

Ethyl acetate is a fairly good solvent for PSt and a non-solvent for PBd and, as we have seen, this produces films with a continuous phase of PSt incorporating discontinuous PBd micellar domains. The strain can now be transmitted through the glassy PSt phase and the modulus of this film approaches a value similar to that of pure polystyrene. A similar situation prevails in Kraton G films cast from THF and $CHCl_3$.[71] If a selective solvent for the PBd phase is used a reversal takes place. Heptane is a precipitant for PSt but a good solvent for PBd or poly(ethene-*co*-butene), and in films cast from this liquid the PSt domains tend to contract into even smaller units than those formed in the equilibrium state. The elastomeric phase is well dispersed and the modulus at 298 K drops to a value approaching that of an unfilled vulcanised elastomer.

Thus, by altering the equilibrium morphology with a selective solvent and changing the continuity of a particular phase in the matrix, one can exercise considerable control over the modulus of the resulting material. The quasi-equilibrium state is relatively stable and is not easily changed by annealing at higher temperatures, a feature which may not be too surprising when one remembers that the average rate of diffusion for a polymer in the melt is of the order 10^{-10}–$10^{-13}\,\mathrm{cm}^2\,\mathrm{s}^{-1}$. Thus a re-equilibration of the block distribution could be a slow process.

4.1.1. *Solvent–Structure Correlation*

As the morphological units formed in block copolymer films can depend on the casting solvent, it would be useful to correlate solvent–polymer interaction with the potential structure. For a block copolymer in which the blocks are very incompatible (χ_{AB} is large), the solubility parameter approach may assist in predicting the copolymer behaviour. For Kraton G-1650, the $E^*_{(298)}$ values were observed to vary with the solubility parameter (δ) of the solvent in the manner shown in Fig. 15. If the δ value for the central block is taken to be 8·1 $((\mathrm{cal\,cm}^{-3})^{1/2})$ and that for PSt is 9·15

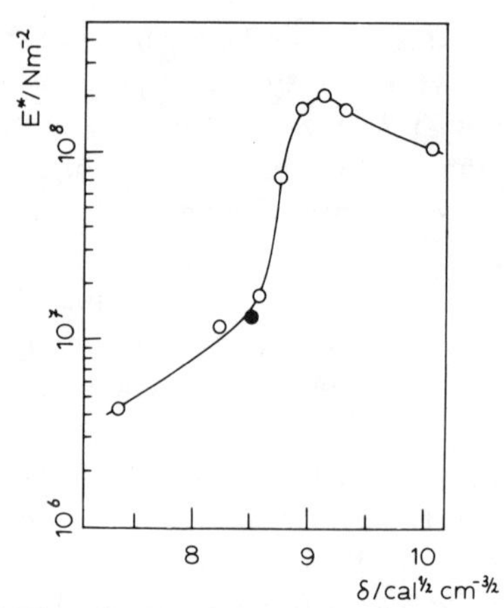

FIG. 15. Variation of $E^*_{(298)}$ for Kraton G-1650 with the solubility parameter of the casting solvent. The point marked (●) is for a sample prepared in the absence of solvent. (From reference 17, with permission American Chemical Society.)

((cal cm^{-3})$^{1/2}$) then solvents with δ values close to that for the central block will disperse it preferentially and produce low modulus films, whereas solvents with δ values similar to PSt will solvate this block preferentially and result in films with a much higher modulus. The extent to which this phase dispersion occurs is reflected in the relative areas of the damping peaks (Fig. 12(b)). The variation in the area with δ for the solvent is shown in Fig. 16 and follows the expected pattern. The solvents with high δ disperse the PSt phase more efficiently. The area under the damping curve maximum corresponding to the T_g of the PSt block is proportionally large as is the $E^*_{(298)}$ values and these features indicate that PSt is the continuous load-bearing phase. The situation is reversed when solvents with low δ values are used. The film pressed in the absence of a solvent has an estimated δ of 8·5 ((cal cm^{-3})$^{1/2}$), and from the curves in Fig. 15 one can predict that $E^*_{(298)} = 1{\cdot}6 \times 10^7$ Nm^{-2}, which agrees closely with the experimental value of $1{\cdot}4 \times 10^7$ Nm^{-2}. This simple approach suggests that the modulus of a Kraton G film at 298 K can be predicted and then produced by casting from the appropriate solvent.

For SBS copolymers, the solubility parameter method does not work quite so well. There is a general trend of modulus with solvent δ similar to that observed for Kraton G, but occasionally solvents such as tetrahydronaphthalene ($\delta = 9{\cdot}5$ ((cal cm^{-3})$^{1/2}$)), which one would assume to be a selective solvent for PSt, appear to act as non-selective solvents.[69] As suggested earlier, a more accurate representation of the polymer–solvent

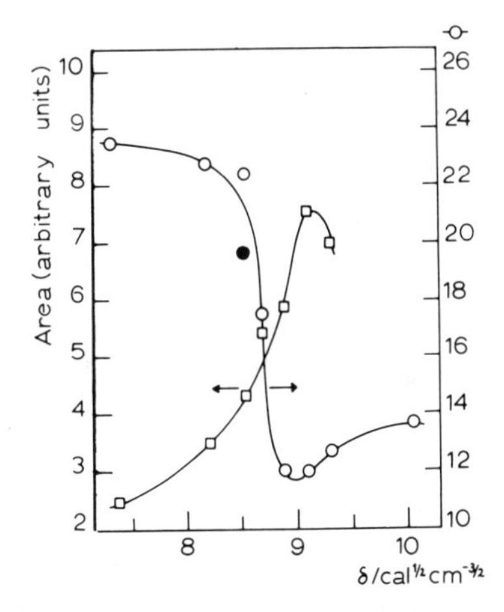

FIG. 16. Area under tan δ peak at the T_g of the central block in Kraton G-1650 (cf. Fig. 12(b)) plotted as a function of the solubility parameter of the casting solvent (○). The curve delineating points (□), is for the areas under the tan δ curve at T_g in the polystyrene block of Kraton G-1650. (From reference 17, with permission American Chemical Society.)

interaction is given by the χ values and the difference, $\Delta\chi = (\chi_{1A} - \chi_{1B})$, is more likely to determine whether a solvent is selective or non-selective. Small values of $\Delta\chi$ would indicate non-selective solvents and larger values would suggest selective solvents with the $\pm$ sign showing which block was preferred. Unfortunately, values of χ_{1A} and χ_{1B} are not always available, but with the advent of inverse gas chromatographic techniques[14,15] relative values could be obtained quite rapidly.

4.2. Tensile Properties

The morphological changes which take place when copolymer composition and casting solvent are varied have been correlated with the tensile and yield strengths of several SBS samples.[75] The main features observed are summarised in Table 4 where one can see that when MEK (the preferential solvent for PSt) was used, the PBd phase became restricted and discontinuous resulting in films which were too brittle for specimen preparation. Comparison of compression-moulded samples (large PBd particles in a PSt matrix) with those cast from cyclohexane revealed that the cast films with the cylindrical and lamellar morphologies had much higher tensile strengths than the compression-moulded sample (except for sample 1 which was lower). Thus, when some continuity of the PBd phase was

TABLE 4
MORPHOLOGICAL FEATURES IN THE POLYBUTADIENE PHASE OF SEVERAL SBS BLOCK COPOLYMERS AS A FUNCTION OF BOTH COMPOSITION AND SOLVENT

Mol ratio of SBS copolymers	*Structure formed in solvents*		
	Toluene	*Cyclohexane*	*Methyl ethyl ketone*
1. 40–20–40	Spheres	Rough network	Spheres
2. 30–40–30	Cylinders	Cylinders	Spheres
3. 20–60–20	Lamellae	Lamellae	Short rods

encouraged the samples became less brittle and the tensile properties were improved.

Tensile strengths at break were also improved in SB diblocks (62 % St) if films were cast from benzene rather than from benzene–isooctane mixtures. In the mixtures, the addition of isooctane, which is a selective solvent for PBd, causes the disruption of the continuous PSt phase in places, resulting in an improved dispersion of the PBd.[76]

So far, the data discussed have been acquired predominantly from PSt–polydiene systems, but other block copolymers exhibit similar properties. Bajaj and Varshney[77] studied poly(dimethylsiloxane-*b*-styrene-*b*-dimethylsiloxane) and the pertinent results are given in Table 5. The highest tensile strengths are again achieved with lamellar structures, while the lowest initial modulus is found in films composed of rods of PSt in a continuous PDMS matrix. Similar solvent-dependent changes in structure were obtained for PSt–PDMS diblock copolymers.[78]

Poly(sulphone-*b*-dimethylsiloxane) $[AB]_n$ block copolymers whose properties are susceptible to casting solvent have been synthesised by Robeson *et al.*[79] The variation of tensile strength and % elongation with casting

TABLE 5
PHYSICAL PROPERTIES OF POLY(DIMETHYLSILOXANE-*b*-STYRENE-*b*-DIMETHYLSILOXANE) FILMS CAST FROM SEVERAL SOLVENTS

Solvent	*Solvent selective for block*	*Tensile strength* (kNm^{-2})	*Initial modulus* (kNm^{-2})	*Morphology of DMS phase*
Cyclohexane	DMS	5 700	22 474	Continuous matrix
Methyl ethyl ketone	PSt	6 100	44 831	Discrete domains
Tetrahydrofuran	PSt	6 500	39 833	Discrete domains
Toluene	Both	6 740	27 838	Lamellar

TABLE 6
TENSILE STRENGTH OF POLY(SULPHONE-*b*-DIMETHYLSILOXANE) $[AB]_n$ BLOCK COPOLYMERS (5 K–5 K) CAST FROM SEVERAL SOLVENTS

Casting solvent	*Tensile strength* (kNm^{-2})	*% elongation*
Tetrahydrofuran	22 819	410
Chlorobenzene	20 820	355
Chloroform	20 475	355
Benzene	20 781	517
Toluene	15 580	310
Xylene	12 892	350
Ethyl acetate	13 030	250
Moulded	11 858	70

solvent is shown in Table 6. High tensile strengths were obtained when selective solvents for the polysulphone block were used, and in all cases those for the solvent-cast films were greater than for the compression-moulded sample. A significant change in tensile strength could be effected if a film cast from THF was subsequently swollen in CCl_4 or *n*-butyl acetate, then dried and retested. The tensile strength dropped from the initial 22 819 kNm^{-2} (THF cast) to 14 408 kNm^{-2} (CCl_4 treated) or 9307 kNm^{-2} (*n*-butyl acetate treated). The most effective swelling agent was *n*-butyl acetate and this seems to disrupt the quasi-equilibrium structure produced by the THF, resulting in a film with properties similar to those of the compression-moulded sample.

The tensile stress of block copolypeptides has also been found to be solvent dependent.[80]

4.3. Stress Hardening and Softening

It was recognised by Beecher *et al.*[66] that if the dynamic thermomechanical behaviour of SBS films varied when cast from different solvents then the stress–strain response should also be affected. Such changes had been reported briefly for poly(styrene-*b*-isoprene) by Bresler *et al.*[81] who observed striking differences when the casting solvents were heptane and MEK.

Films of SBS (Kraton 101) cast from a THF/MEK mixture[66] had a high modulus and displayed a yield point during the first stress–strain cycle. On recycling the sample there was evidence of substantial stress softening, with the specimen exhibiting a much lower modulus and behaving like a vulcanised rubber (see Fig. 17). A study of the electron micrographs of the

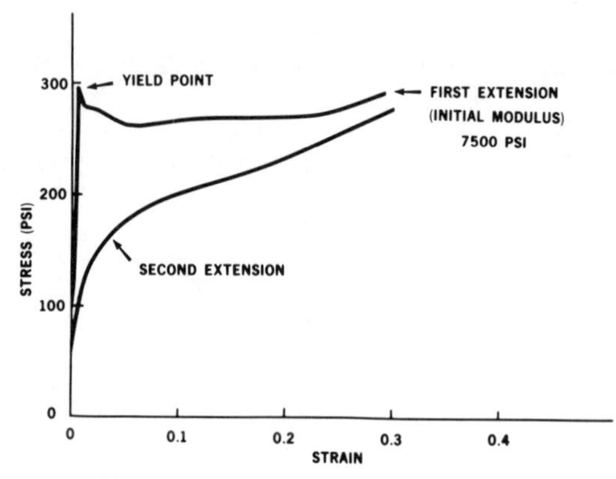

FIG. 17. Stress–strain curves for poly(styrene-*b*-butadiene-*b*-styrene) triblock (Kraton 101) films cast from tetrahydrofuran/methyl ethyl ketone (90/10) mixture. (From *Polymer*, 1976, **17**, 938; with permission IPC Business Press, Ltd. ©)

sample shown in Figs. 18 and 19 revealed that the interconnected PSt domains had acted as though they were ductile and had deformed after 300% stretching to become ovoid rather than spherical in shape. The copolymer also demonstrated an elastic memory as the original stress–strain behaviour was recovered after annealing. Samples of Kraton 101 cast from CCl_4 showed no yield point but again stress softening was detected on recycling.

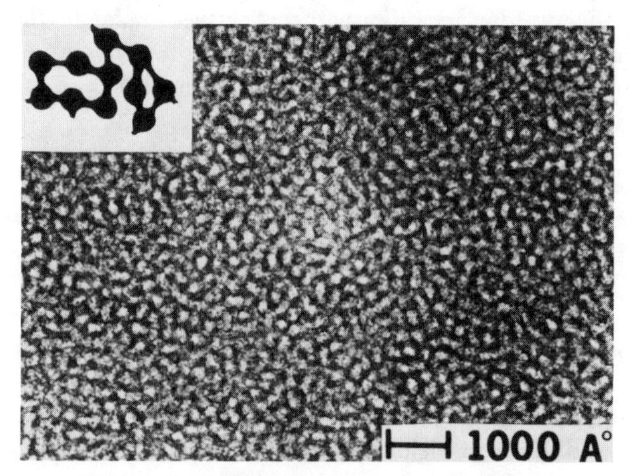

FIG. 18. Electron micrograph of Kraton 101 cast from tetrahydrofuran/methyl ethyl ketone ((90/10) mixture), undeformed. (From *Polymer*, 1976, **17**, 938; with permission IPC Business Press, Ltd. ©)

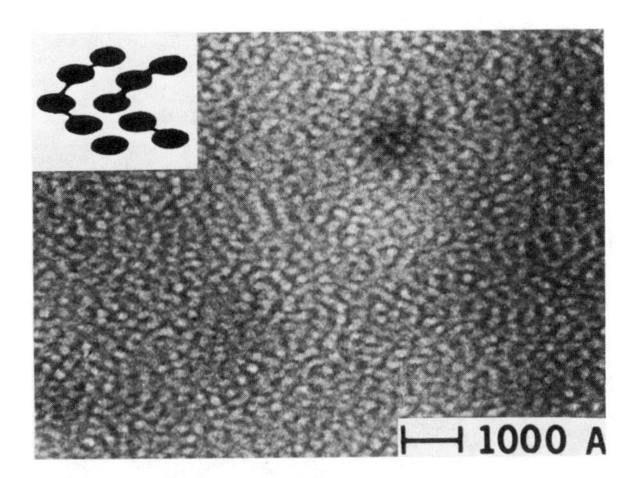

FIG. 19. Electron micrograph of Kraton 101 cast from tetrahydrofuran/methyl ethyl ketone ((90/10) mixture), after 300% stretching. (From *Polymer*, 1976, **17**, 938; with permission IPC Business Press, Ltd. ©)

Pedemonte and Alfonso[82] extended the stress range studied for SBS cast from the same MEK/THF mixture. A yield point was observed on the first stress–strain cycle, and on recycling stress softening was found as before. A further increase of the stress led to stress hardening at high deformations which was also reversible[83] for both SBS and SIS block copolymers.

A comprehensive investigation of the solvent-dependent stress hardening and stress softening qualities of Kraton G-1650 films has been reported by Cowie and McEwen.[84] Stress–strain diagrams for films, cast from eight solvents, in Fig. 20, show the results of three stress–strain cycles on each film and can be compared with the response of a melt-pressed sample ($\delta = 8{\cdot}5/(\text{cal cm}^{-3})^{1/2}$). Plastic yield was observed in films cast from solvents with $\delta \geq 8{\cdot}7$ $((\text{cal cm}^{-3})^{1/2})$, which on recycling showed stress softening up to $\simeq 100\%$ extension followed by a rapid hardening process. For samples cast from less polar solvents ($\delta < 8{\cdot}7$ $((\text{cal cm}^{-3})^{1/2})$), stress hardening was observed over the complete range of extensions. A third cycle led to even greater degrees of stress hardening. All films regained their original characteristics if allowed to relax unloaded for 48 h.

These results indicate a strong solvent dependence of stress–strain behaviour in Kraton G-1650 which is not observed in the SIS system.[83] This difference in behaviour can be explained by examining the structures of the copolymers. Stress hardening can be accounted for by postulating that the hard glassy domains deform under stress and impinge on one

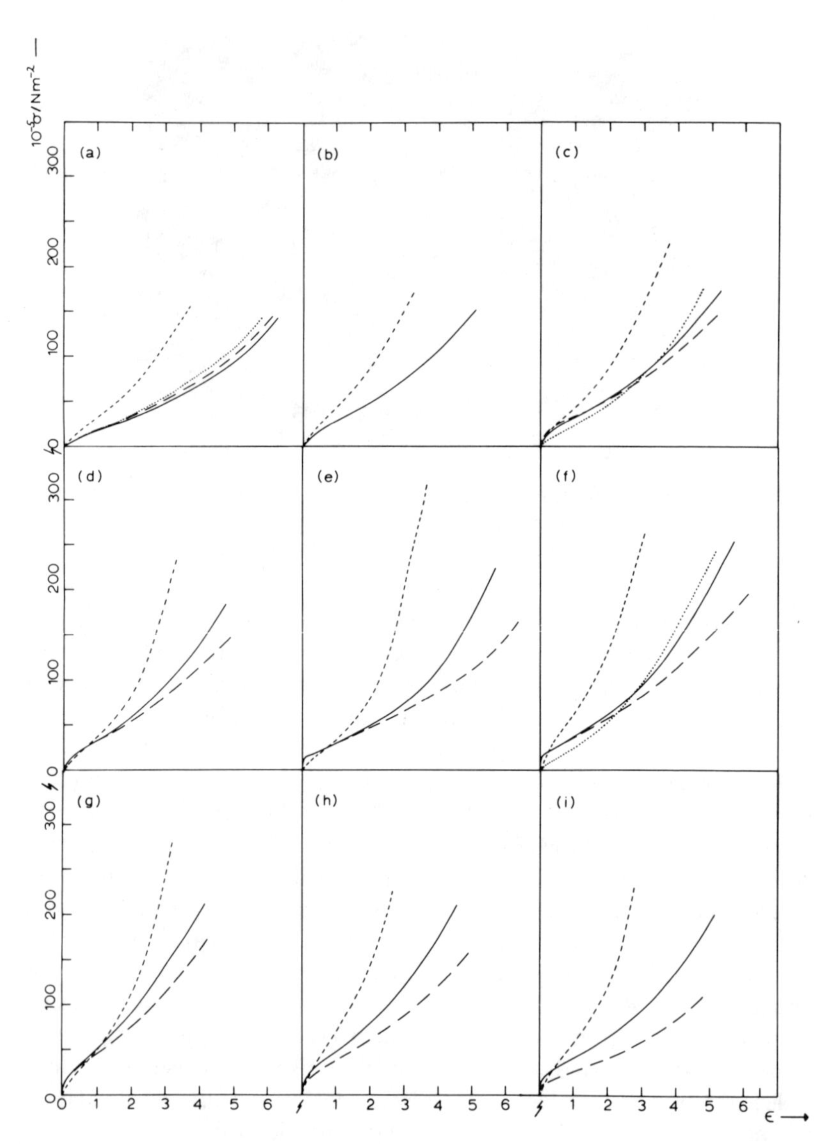

FIG. 20. Stress–strain isotherms (298 K) for Kraton G-1650 cast from (a) *n*-heptane, (b) cyclohexane, (c) melt pressed film, (d) cyclohexene, (e) 1-bromobutane, (f) toluene, (g) tetrahydrofuran, (h) chloroform and (i) dioxane. (——) First stress cycle, (– – – –) second stress cycle, (— —) third stress cycle after annealing at 416 K; (· · · ·) stress cycle for samples held at 400 % elongation for 48 h. (From reference 84, with permission Marcel Dekker, Inc.)

another, thereby improving the continuity of the load-bearing pathway through the copolymer matrix. Stress softening occurs when ties between the hard glassy domains are ruptured by deformation and the continuity of the glassy phase is impaired. This has the effect of lowering the stress necessary to produce the same strain in a sample when it is recycled. The structure of the SIS samples was determined to be discrete PSt spheres imbedded in a continuous PIp phase. If the solvents used were unable to disperse the PSt domains sufficiently to allow them to impinge on one another when a stress was applied, then little change in behaviour with casting solvent would be detected. The appropriate structures[71,74] are present in Kraton G-1650 which allows such changes to take place.

Leblanc[85] has suggested that work hardening and work softening are not entirely due to the ductility of the glassy phase but that the diffuse interfácial regions play a predominant role. This is quite possible but from the existing evidence one cannot determine whether the deformation process is restricted wholly to the interfacial region or includes plastic deformation of the glassy domains as an additional contribution. Electron micrographs[66] tend to suggest that both are involved.

5. OTHER METHODS

Several other techniques have been used to demonstrate that a casting solvent can alter the properties of block copolymers. Djermouni and Ache[86] have examined positron annihilation in SBS films cast from toluene, CCl_4, MEK and ethyl acetate. Pronounced discontinuities were detected at 203 K and 358 K corresponding to the two major glass transitions (Fig. 21). In addition distinct changes were also observed at 259 K for films cast from all solvents, and at 283 K for films cast from MEK and ethyl acetate. These observations agree with the dynamic mechanical and DSC data referred to in Section 4.1.[66,67] These intermediate discontinuities have been interpreted as evidence for the onset of molecular motion in the PBd and PSt rich interfaces, respectively, the latter being seen only when a selective solvent for PSt is used. Thus a distinction is now made between the two processes which was not clearly defined in earlier work.

X-ray photoelectron spectroscopy provides information about polymer surfaces and has been applied[87] to poly(styrene-*b*-oxyethylene). These block copolymers show a distinct structural dependence on the casting solvent[88] which is reflected in the surface distribution of the blocks. By

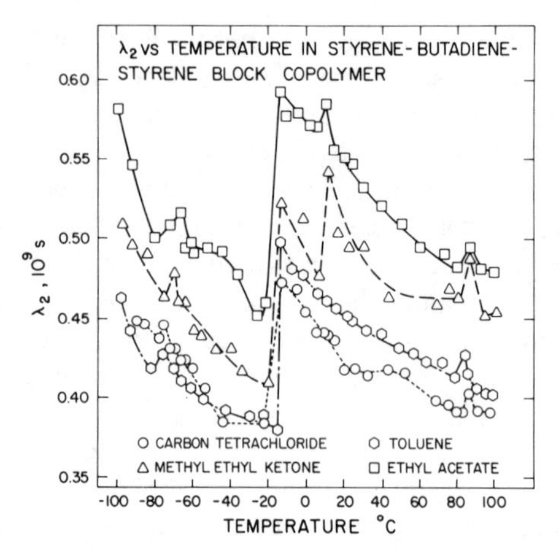

FIG. 21. Plot of the annihilation rate of thermalised *o*-Ps atoms (λ_2) against temperature for poly(styrene-*b*-butadiene-*b*-styrene) samples cast from four solvents. (From reference 86, with permission American Chemical Society.)

using films cast from a non-selective solvent, $CHCl_3$, as a reference, it was found that the PSt content of the surface increased from 21·4 mol % to 70 mol % when ethylbenzene (a preferential solvent for PSt) was used to cast the film. If a preferential solvent for POE was used, e.g. nitromethane, then the PSt surface content was less than in the $CHCl_3$-cast film. This demonstrates that the distribution of a particular block on the surface of a copolymer can be altered by the use of an appropriate solvent and that the surface will reflect the distribution in the bulk.

Small angle X-ray studies[89] of the structure of block copolymers swollen in different solvents has revealed that in a lamellar structure swelling occurs parallel to the interface when the swelling solvent is non-selective. If a selective solvent for a polymer block forming one of the lamellae is used, then swelling takes place in the direction of the thickness of that lamella.

In normal crosslinked elastomers the birefringence developed on the application of a stress is non-linear with strain. The behaviour of block copolymers does not always follow this pattern[90,91] and it has also been found that the stress–strain birefringence response depends on the casting solvent.[90,92,93] Measurement of the strain optical coefficient showed that this was lower for films of Kraton 101 when cast from MEK compared with toluene-cast films and again this reflects the differences in the internal structure. When the casting solvent was toluene the continuity of the PBd phase was improved when compared with the MEK-cast film and this led to higher values of the stress optical coefficient.[92] This behaviour was confirmed in other block copolymer systems.[90,93]

These are only a few examples of the properties of block copolymers which are influenced by the presence of a liquid during sample preparation. Others will undoubtedly be reported in the future, as it is obvious from the foregoing that the presence of a liquid during preparation of a block copolymer sample must be regarded as a structure- and property-controlling parameter.

REFERENCES

1. MEIER, D. J., *J. Polym. Sci., Part C*, 1969, **26**, 81.
2. MEIER, D. J., *Polym. Prepr. Amer. Chem. Soc. Div. Polym. Chem.*, 1970, **11**, 400.
3. KRAUSE, S., *J. Polym. Sci., A2*, 1968, **7**, 249.
4. KRAUSE, S., *Macromolecules*, 1970, **3**, 84.
5. POUCHLY, J., ZIVNY, A. and SIKORA, A., *J. Polym. Sci., Part C*, 1972, **39**, 133.
6. POUCHLY, J., ZIVNY, A. and SIKORA, A., *J. Polym. Sci., A2*, 1972, **10**, 151.
7. BIANCHI, U., PEDEMONTE, E. and TURTURRO, A., *J. Polym. Sci., B*, 1969, **7**, 785.
8. BIANCHI, U., PEDEMONTE, E. and TURTURRO, A., *Polymer*, 1970, **11**, 268.
9. HELFAND, E., *J. Chem. Phys.*, 1975, **62**, 999.
10. HELFAND, E., *Macromolecules*, 1975, **8**, 552.
11. MOLAU, G. E. and KESKKULA, H., *J. Polym. Sci., A1*, 1966, **4**, 1595.
12. GERBERDING, K., HEINZ, G. and HECKMANN, W., *Macromol. Chem. Rapid Commun.*, 1980, **1**, 221.
13. STOCKMAYER, W. H., MOORE, L. D., FIXMAN, M. and EPSTEIN, B. N., *J. Polym. Sci.*, 1955, **16**, 517.
14. TEWARI, Y. B. and SCHREIBER, H. P., *Macromolecules*, 1972, **5**, 329.
15. ITO, K. and GUILLET, J. E., *Macromolecules*, 1979, **12**, 1163.
16. SCOTT, R. L., *J. Chem. Phys.*, 1949, **17**, 279.
17. BUSHUK, W. and BENOIT, H., *Can. J. Chem.*, 1958, **36**, 1616.
18. KRAUSE, S., *J. Phys. Chem.*, 1961, **65**, 1618.
19. BENOIT, H. and FROELICH, D., in *Light Scattering from Polymer Solutions*, Ed. M. B. Huglin, Academic Press, New York, 1972, p. 467.
20. BURNETT, G., MEARES, P. and PATON, C., *Trans. Farad. Soc.*, 1962, **58**, 737.
21. URWIN, J. R. and STEARNE, J. M., *Makromol. Chem.*, 1964, **78**, 204.
22. KOTAKA, T., TANAKA, T., OHNUMA, H., MURAKAMI, Y. and INAGAKI, H., *Polymer J.*, 1970, **1**, 245.
23. KOTAKA, T., TANAKA, T. and INAGAKI, H., *Polymer J.*, 1972, **3**, 327.
24. KOTAKA, T., TANAKA, T. and INAGAKI, H., *Polymer J.*, 1972, **3**, 338.
25. KAMIYAMA, F., INAGAKI, H. and KOTAKA, T., *Polymer J.*, 1972, **3**, 470.
26. TANAKA, T., KOTAKA, T. and INAGAKI, H., *Macromolecules*, 1974, **7**, 311.
27. TANAKA, T., KOTAKA, T. and INAGAKI, H., *Macromolecules*, 1976, **9**, 561.
28. TANAKA, T., KOTAKA, T., BAN, K., HATTORI, M. and INAGAKI, H., *Macromolecules*, 1977, **10**, 960.

29. Kotaka, T., Tanaka, T., Hattori, M. and Inagaki, H., *Macromolecules*, 1978, **11**, 138.
30. Utiyama, H., Takenaka, K., Mizumori, M. and Fujkuda, M., *Macromolecules*, 1974, **7**, 28.
31. Utiyama, H., Takenaka, K., Mizumori, M., Tsunashima, Y. and Kurata, M., *Macromolecules*, 1974, **7**, 515.
32. Stockmayer, W. H. and Fixman, M., *J. Polym. Sci., Part C*, 1963, **1**, 137.
33. Kotaka, T., Ohnuma, H. and Murakami, Y., *J. Phys. Chem.*, 1977, **70**, 4099.
34. Kotaka, T., Tanaka, T. and Inagaki, H., *Polymer*, 1969, **10**, 517.
35. Shima, M., Ogawa, E. and Konoshi, K., *Makromol. Chem.*, 1976, **177**, 241.
36. Shima, M., Ogawa, E., Ban, S. and Sato, M., *J. Polym. Sci., Polym. Phys. Ed.*, 1977, **15**, 1999.
37. Ohnuma, H., Kotaka, T. and Inagaki, H., *Polymer J.*, 1970, **1**, 716.
38. Dondos, A., Rempp, P. and Benoit, H., *Polymer*, 1972, **13**, 97.
39. Shima, M., Ogawa, E. and Sato, M., *Polymer*, 1979, **20**, 311.
40. Cramond, D. N. and Urwin, J. R., *Europ. Polym. J.*, 1969, **5**, 35.
41. Cramond, D. N. and Urwin, J. R., *Europ. Polym. J.*, 1969, **5**, 45.
42. Girolamo, M. and Urwin, J. R., *Europ. Polym. J.*, 1972, **8**, 299.
43. Prud'homme, J., Roovers, J. E. L. and Bywater, S., *Europ. Polym. J.*, 1972, **8**, 901.
44. Leng, M. and Benoit, H., *J. Polym. Sci.*, 1962, **57**, 263.
45. Prud'homme, J. and Bywater, S., *Macromolecules*, 1971, **4**, 543.
46. Han, C. C. and Mozer, B., *Macromolecules*, 1977, **10**, 44.
47. Sadron, C., *Pure Appl. Chem.*, 1962, **4**, 347.
48. Bajaj, P. and Varshney, S. K., *Polymer*, 1980, **21**, 201.
49. Enyiegbulam, M. and Hourston, D. J., *Polymer*, 1979, **20**, 818.
50. Merrett, F. M., *Trans. Farad. Soc.*, 1954, **50**, 760.
51. Merrett, F. M., *J. Polym. Sci.*, 1957, **24**, 467.
52. Krause, S., *J. Phys. Chem.*, 1964, **68**, 1948.
53. Periard, J. and Riess, G., *Europ. Polym. J.*, 1973, **9**, 687.
54. Newman, S., *J. Appl. Polym. Sci.*, 1962, **6**, 515.
55. Horii, F., Ikada, Y. and Sakurada, I., *J. Polym. Sci., Polym. Chem. Ed.*, 1974, **12**, 323.
56. Price, C. and Woods, D., *Europ. Polym. J.*, 1973, **9**, 827.
57. Molau, G. E. and Wittbrodt, W. M., *Macromolecules*, 1968, **1**, 260.
58. Inoue, T., Soen, T., Hoshimoto, T. and Kawai, H., *J. Polym. Sci., A2*, 1969, **7**, 1283.
59. Kawai, H., Soen, T., Inoue, T., Ono, T. and Uchida, T., *Mem. Fac. Engin. Kyoto Univ.*, 1971, **33**, 383.
60. Inoue, T., Soen, T., Kawai, H., Fukatsu, M. and Kurata, M., *J. Polym. Sci., B.*, 1968, **6**, 75.
61. Leary, D. F. and Williams, M. C., *J. Polym. Sci., Polym. Phys. Ed.*, 1973, **11**, 345.
62. Leary, D. F. and Williams, M. C., *J. Polym. Sci., Polym. Phys. Ed.*, 1974, **12**, 265.
63. Sadron, C. and Gallot, B., *Makromol. Chem.*, 1973, **164**, 301.
64. Meier, D. J., in: *Sagamore Conference on Block and Graft Copolymers*, Eds. J. J. Burke and V. Weiss, Syracuse University Press, Syracuse, New York, 1973, p. 105.

65. MEIER, D. J., private communication.
66. BEECHER, J. F., MARKER, L., BRADFORD, R. D. and AGGARWAL, S., *J. Polym. Sci., Part C*, 1969, **26**, 117.
67. MIYAMOTO, T., KODAMA, K. and SHIBAYAMA, K., *J. Polym. Sci., A-2*, 1970, **8**, 2095.
68. KALFOGLOU, N. K., *J. Appl. Polym. Sci.*, 1979, **23**, 2385.
69. BEAMISH, A., GOLDBERG, R. A. and HOURSTON, D. J., *Polymer*, 1977, **18**, 49.
70. COWIE, J. M. G. and RASHID, H., unpublished results.
71. COWIE, J. M. G., LATH, D. and MCEWEN, I. J., *Macromolecules*, 1979, **12**, 52.
72. WILKES, G. L., BAGRODIA, S., OPHIR, Z. and EMERSON, J. A., *J. Appl. Phys.*, 1978, **49**, 5060.
73. SHEN, M., CIRLIN, E. H. and KAELBLE, D. H., in: *Colloidal and Morphological Behaviour of Block and Graft Copolymers*, Ed. G. E. Molau, Plenum, New York, 1971, p. 307.
74. SEGUELA, R. and PRUD'HOMME, J., *Macromolecules*, 1978, **11**, 1007.
75. MATSUO, M. and SAGAYE, S., in: *Colloidal and Morphological Behaviour of Block and Graft Copolymers*, Ed. G. E. Molau, Plenum, New York, 1971, p. 1.
76. ANDRYUSHCHENKO, T. A., ASKADSKII, A. A. and ZUBOV, P. I., *Vysokomol. Soedin. A*, 1979, **21**, 2360.
77. BAJAJ, P. and VARSHNEY, S. K., *Polymer*, 1980, **21**, 201.
78. SAAM, J. C., GORDON, D. J. and LINDSEY, S., *Macromolecules*, 1970, **3**, 1.
79. ROBESON, L. M., NOSHAY, A., MATZNER, M. and MERRIAM, C. N., *Angew. Makromol. Chem.*, 1973, **29**/**30**, 47.
80. ANDERSON, J. M., HAYASHI, T., WONG, M., BARENBERG, S., SIGLER, G., GEIL, P. H., WALTON, A. G. and HILTNER, A., *J. Polym. Sci., Part C*, 1977, **60**, 77.
81. BRESLER, S. YE., PYRKOV, L. M., FRENKEL, S. YA., LAIUS, L. A. and KLENIN, S. I., *Vysokomol. Soedin. A.*, 1962, **2**, 250.
82. PEDEMONTE, E. and ALFONSO, G. C., *Macromolecules*, 1975, **8**, 85.
83. PEDEMONTE, E., ALFONSO, G. C., DONDERO, G., DE CANDIA, F. and ARAIMO, L., *Polymer*, 1977, **18**, 191.
84. COWIE, J. M. G. and MCEWEN, I. J., *J. Macromol. Sci.-Phys.*, 1979, **B16**, 611.
85. LEBLANC, J. L., *J. Appl. Polym. Sci.*, 1977, **21**, 2419.
86. DJERMOUNI, B. and ACHE, H. J., *Macromolecules*, 1980, **13**, 168.
87. THOMAS, R. H. and O'MALLEY, J. J., *Macromolecules*, 1979, **12**, 323.
88. SHORT, J. M. and CRYSTAL, R. G., *Appl. Polym. Symp.*, 1971, **16**, 137.
89. IONESCU, L. M., *Stud. Cercet. Fiz.*, 1979, **31**, 459.
90. HENDERSON, J. F., GRUNDY, K. H. and FISCHER, R., *J. Polym. Sci., Part C*, 1968, **16**, 3121.
91. FISCHER, E. and HENDERSON, J. F., *J. Polym. Sci., Part C*, 1969, **26**, 149.
92. WILKES, G. L. and STEIN, R. S., *J. Polym. Sci., A-2*, 1969, **7**, 1525.
93. NISHIOKA, A., FURUKAWA, J., YAMASHITA, S. and KOTANI, J., *J. Appl. Polym. Sci.*, 1970, **14**, 799.

Chapter 2

COLLOIDAL PROPERTIES OF BLOCK COPOLYMERS

COLIN PRICE

Department of Chemistry, University of Manchester, UK

SUMMARY

When a block or graft copolymer is dissolved in a liquid that is a solvent for only one of the components, the copolymer may aggregate reversibly to form colloidal particles (micelles) consisting generally of a swollen core of the insoluble blocks surrounded by a flexible fringe of soluble blocks. This contribution deals with the properties of dilute micellar solutions of block copolymers in organic solvents and compares them with surfactant micelles in aqueous systems.

In some situations, copolymer micelles dissociate only slowly into free chains. Techniques are described by which their size distributions, molecular weights and sizes, internal structures and hydrodynamic properties can be determined. The results indicate that block copolymer micelles frequently have a very narrow size distribution, implying their formation by a closed-association process, and that under certain conditions simple spherical micelles undergo association forming larger particles. The use of block copolymers as stabilisers in dispersion polymerisations is discussed.

Finally, factors which influence the upper and lower flocculation temperatures of spherically stabilised polymer particles are considered; similar factors may affect the colloidal stability of spherical micelles formed from block copolymers.

1. INTRODUCTION

The reversible aggregation of amphiphiles dissolved in water has been the subject of great interest for many years, the colloidal products of aggregation being known as micelles.[1–3] Amphiphiles are molecules which contain both hydrophobic and hydrophilic parts and amongst the wide variety of such systems which have been investigated in detail are the synthetic surfactants (non-ionic, cationic and anionic), the phospholipids and bile salts.[4–7] Micelle formation requires the presence of two opposing forces:[8] an attractive force between the amphiphiles leading to aggregation, and a repulsive force that prevents unlimited growth of the micelles into a distinct macroscopic phase. When the hydrophobic portion of the amphiphiles is a hydrocarbon, the micelles consist of a hydrocarbon core with the polar or ionic groups at the surface. For many of the systems the cores are liquid-like and contain only a relatively small proportion of water.

In amphiphiles containing ionic head-groups the repulsive force arises primarily from electrostatic interactions, whilst in amphiphiles containing non-ionic head-groups, such as poly(oxyethylene), it is probably the preference for hydration as opposed to self-association which is the main factor.[8] The attractive force is believed to arise predominantly from the 'hydrophobic effect', which is the outcome of a reorganisation of the structure of water which takes place when hydrocarbon units are removed from it.[2,8,9] The effect leads to an overriding increase in entropy and hence a reduction in free energy. An important factor in the association process is the minimum number of amphiphiles which must come together before the hydrocarbon–water interface is sufficiently reduced to give the colloidal particle some degree of thermodynamic stability. With some amphiphiles the process is believed to be sufficiently cooperative to yield colloidal particles with a fairly narrow size distribution. In these cases it is often assumed as a first approximation that there is a single-stage equilibrium[10,11] between micelles, A_N, with a fixed aggregation number, N, and free amphiphiles, A_1:

$$NA_1 \rightleftharpoons A_N \tag{1}$$

If it is also assumed that the solution is ideal except for intramicellar interactions the chemical potentials of the micelles and free amphiphiles can be written, respectively,

$$\mu_N = \mu_N^{\ominus} + \mathrm{R}T\ln[A_N] \tag{2}$$

and

$$\mu_1 = \mu_1^{\ominus} + \mathrm{R}T\ln[A_1] \tag{3}$$

where the standard chemical potentials, $\mu_N^\ominus$ and $\mu_1^\ominus$, may each be considered to contain an internal contribution together with a free energy arising from coupling the species to a pure water environment.[12] From the criterion for thermodynamic stability,

$$N\mu_1 = \mu_N \tag{4}$$

the equilibrium constant

$$^{n}K_{1,N} = [A_N]/[A_1]^N = \exp\,[-(\mu_N^\ominus - N\mu_1^\ominus)/RT] \tag{5}$$

where $\mu_N^\ominus - N\mu_1^\ominus$ is the standard free energy of formation of the micelles. In the polymer literature an association which fits the single-stage equilibrium model is frequently referred to as a closed association (a name coined by Elias).[13] The properties of the model are well known. As the total concentration of amphiphile, $[A]$, is increased there is a continuous increase in both the concentration of free amphiphile, $[A_1]$, and the concentration of micelles, $[A_N]$. A plot of $[A_N]$ against $[A]$ shows that $[A_N]$ increases rapidly above a value of $[A]$ which is termed the critical micelle concentration (c.m.c.) in experimental studies. According to the model, however, micelles exist both above and below the c.m.c. and their formation does not assume the characteristics of a phase transition, except in the limit of infinitely large N.[14] The c.m.c. is effectively the concentration at which micelles are just detectable by the experimental method in use.[15] Each property studied (colligative, physical or spectroscopic) will give a slightly different value for this concentration. Above the c.m.c. $[A_1]$ does not reach a constant value, but continues to increase slowly.

Great difficulty has been encountered in testing the equilibrium model since, in general, experimental techniques are not available for measuring $[A_1]$ and $[A_N]$ directly. However, in principle the model may be tested quantitatively[11] by determining different molecular weight averages over a wide range of $[A]$. The reciprocal weight-average and number-average molecular weights, for instance, would be expected to vary as shown in Fig. 1 for a one stage model. By fitting the data to eqns (6) and (7) separate estimates of N and $^{n}K_{1,N}$ can be obtained

$$^{n}K_{1,N}c^{N-1} = \frac{10^{3(1-N)}(N-1)^{N-1}N[1-(M_1)_{\mathrm{n}}M_{\mathrm{n}}^{-1}]}{N(M_1)_{\mathrm{n}}^{1-N}[N(M_1)_{\mathrm{n}}M_{\mathrm{n}}^{-1}-1]^N} \tag{6}$$

$$^{n}K_{1,N}c^{N-1} = \frac{10^{3(1-N)}(N-1)^{N-1}[M_{\mathrm{w}}(M_1)_{\mathrm{w}}^{-1}-1]}{N(M_1)_{\mathrm{n}}^{1-N}[N-M_{\mathrm{w}}(M_1)_{\mathrm{w}}^{-1}]^N} \tag{7}$$

where $(M_1)_{\mathrm{w}}$ and $(M_1)_{\mathrm{n}}$ are the weight-average and number-average

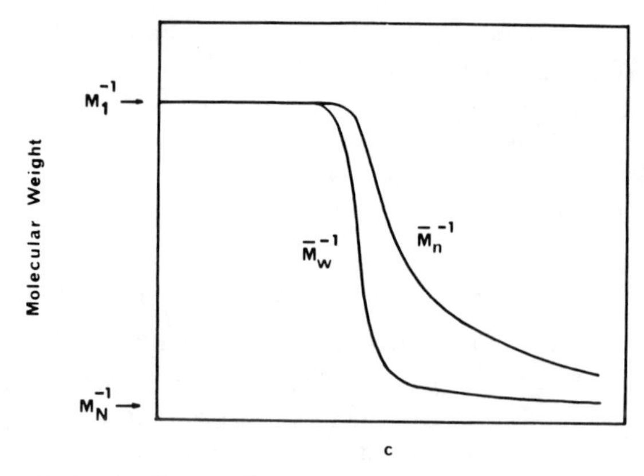

FIG. 1. Variation in $\bar{M}_w$ and $\bar{M}_n$ with total concentration of surfactant for closed association.

molecular weights, respectively, of the solute in the absence of association and the units of c and ${}^nK_{1,N}$ are $g\,cm^{-3}$ and $(dm\ mol^{-1})^{N-1}$, respectively.[13,16,17]

Agreement between the separate estimates for N and ${}^nK_{1,N}$ may be taken as support for the one-stage model. Serious problems can be posed however by the influence of finite virial terms on measured properties. Nevertheless, by using this approach Lässer and Elias[11,18] have been able to gather strong evidence in favour of the use of the one-stage model to describe the micellisation in water of n-octyl-β-D-glucoside and two nonylphenol ethoxylates having, respectively, 42 and 62 oxyethylene units per molecule.

Many surfactants have been found to aggregate loosely in polar solvents such as dimethylformamide.[19] They also form so-called reverse micelles having a polar core in non-polar solvents such as aliphatic hydrocarbons.[20-22] Recent studies have shown that in non-polar solvents the process is probably a consecutive one involving all possible sizes of aggregates starting with dimers.

$$\left.\begin{aligned} A_1 + A_1 &\underset{}{\overset{{}^nK_{1,2}}{\rightleftharpoons}} A_2 \\ A_1 + A_2 &\overset{{}^nK_{2,3}}{\rightleftharpoons} A_3 \\ A_1 + A_3 &\overset{{}^nK_{3,4}}{\rightleftharpoons} A_4 \\ A_1 + A_N &\overset{{}^nK_{N,N+1}}{\rightleftharpoons} A_{N+1} \end{aligned}\right\} \tag{8}$$

This type of association has been given a variety of names amongst the most common of which are open association,[13] indefinite self-association and step-wise aggregation.[22]

The assumptions which are usually made in treating open association are first that the equilibrium constants for each stage are equal, and secondly that the solution may be treated as ideal except for intramicellar interactions.[13,23,24] In open association the c.m.c. phenomenon is absent and the distribution of particle sizes is similar to that obtained in condensation polymerisation (i.e. a most probable distribution). From simple principles of equilibrium it follows that there is a smooth increase in both the extent of association and the average aggregation number (i.e. in the average degree of association) with increase in $[A]$. In contrast to the behaviour shown in Fig. 1, $\bar{M}_n^{-1}$ in open association decreases continuously with increase in $[A]$. A similar but more pronounced decrease is shown by $\bar{M}_w^{-1}$.

It has been recognised for a good number of years that stable colloids can also be formed by block and graft copolymers in selective solvents.[25-38] Studies have shown that when a block or graft copolymer is dissolved in a liquid that is a non-solvent for one of the polymer components but a good solvent for the other, the copolymer may aggregate by a reversible process to form colloidal particles. (In the sense we use the term non-solvent here, a homopolymer/non-solvent mixture would be a mixture which displays a liquid–liquid phase separation.) In spite of the findings concerning the behaviour of surfactants in non-polar solvents, there is now clear evidence[34-43] that block copolymers can undergo essentially closed association in such media (i.e. closed association is able to take place in the absence of water and the hydrophobic effect). The colloidal particles which are formed from block (and graft) copolymers in selective solvents are in many cases spherical in shape and consist of a compact core of insoluble blocks surrounded by a flexible fringe of soluble blocks (see Fig. 2). The use

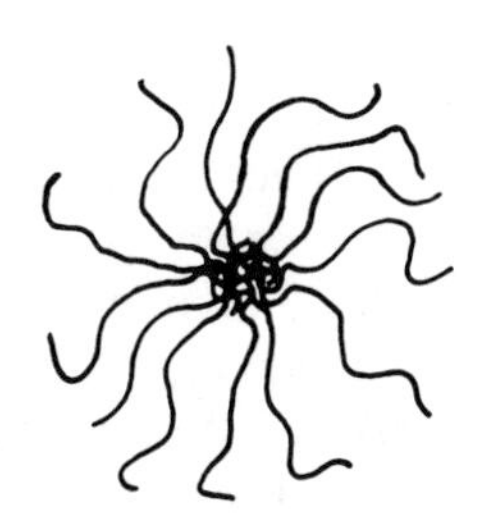

FIG. 2. A sketch of a spherical micelle formed by an AB block copolymer.

of the term micelle has therefore been meaningfully extended to describe such particles.

In this chapter we shall review the structure and properties of micelles formed by block and graft copolymers in selective organic solvents. Only results for dilute solutions will be considered. Attention will be focused mainly on results reported for relatively simple well-characterised copolymers prepared by anionic polymerisation techniques. Similarities and differences between block copolymer micelles in organic solvents and surfactant micelles in aqueous solution will be pointed out. No attempt will be made to consider evidence for so-called unimolecular micelles[44] which are compact configurational states believed to be exhibited by unassociated block and graft copolymer molecules in selective solvents.

2. INVESTIGATIONS OF THE SIZE DISTRIBUTION OF COPOLYMER MICELLES

Micelles formed from simple surfactants in aqueous solution are known to dissociate and reform very rapidly. Experimental investigations have shown that with such systems there are two main relaxation processes involved in the approach to thermodynamic equilibrium.[45,46] From an analysis of the results it has been predicted that the mean residence time of a surfactant molecule in a micelle is of the order of 10^{-5} s.[6]

Kinetic studies have not been carried out on copolymer micelles, but there is considerable evidence to show that in some cases the rate of dissociation of copolymer micelles into free chains is very slow.[43,47] This behaviour has permitted physical methods normally used for determining the molecular weight distribution of unassociated polymers to be applied directly to copolymer micelles.

2.1. Gel Permeation Chromatography (GPC)

In this technique[48] a porous rigid gel is used in the separation process. Fractionation occurs because the amount of solvent required to elute a particle through the gel is related to the hydrodynamic size of the particle;[49,50] since large particles can only enter large pores they are eluted first. When dealing with samples containing large particles care has to be taken that there are a sufficient number of permeable pores to provide adequate resolution. The main disadvantage of GPC is that it requires the use of a calibration procedure since it is not an absolute method. In addition chromatograms require correction for instrumental broadening;

this correction is particularly important for particles with a narrow size distribution.

Studies have been made of an AB polystyrene-*b*-polyisoprene block copolymer, designated here PStPIp(I), for which in the unassociated state $\bar{M}_w = 43\,000$, $\bar{M}_w/\bar{M}_n = 1{\cdot}05$ and $\bar{M}_w$ (polystyrene block) = 34 000. Chromatograms are shown in Fig. 3 for solutions of the block copolymer in *N,N*-dimethylacetamide at 26 °C, 60 °C and 110 °C.[43,51] The column packing material used in the studies was a poly(styrene–divinyl benzene) gel. Since DMAc is a selectively bad solvent for polyisoprene the spherical micelles formed by the block copolymer had a swollen polyisoprene core and an outer flexible fringe of polystyrene chains. At 26 °C the thermodynamic equilibrium was predominantly in favour of micelle formation. The single sharp peak at elution volume 126·5 cm³ (which was well within the range of resolution of the columns) indicated that micelles were able to enter the pores of the gel and elute through the columns as stable species. The reverse flow method of correction for diffusional broadening[52] was used to show that within experimental error the micelles were monodisperse in size ($\bar{M}_w/\bar{M}_n = 1{\cdot}00$, experimental uncertainty = 0·02). Twenty per cent of the copolymer was still left in the columns after the passage of the impurity peaks. It was argued that this loss was due to the continuous adsorption by the columns of the very small concentration of free chains that was always present; adsorption would be expected to occur through the free polyisoprene blocks. Alternatively, adsorption could have occurred whenever a very strong collision brought a micelle core into contact with the gel network. However, an important point is that since the majority of the micelles remained intact as they were eluted through the gel over a period of hours, the rate of dissociation of micelles to form free-chains at 26 °C must have been extremely slow.

In the temperature range 50–65 °C the shape of the chromatograms indicated the presence of both micelles and free chains. As the temperature was raised the equilibrium shifted progressively towards the free-chain form. The chromatogram for 60 °C is shown in Fig. 3(b). The spreading which occurred between the micelle peak ($V_e \simeq 127\,\text{cm}^3$) and the free-chain peak ($V_e \simeq 160\,\text{cm}^3$) could be attributed mainly to the dependence of the micelle/free-chain equilibrium on copolymer concentration, the relaxation time being presumably much shorter than at 26 °C. The extensive spreading which occurred at high elution volumes (up to $V_e = 180\,\text{cm}^3$ and beyond) was consistent with strong absorption of the free chains on to the gel matrix. In the range 50–65 °C the relative amounts of free-chain and micelle forms were dependent on the initial concentration of solution injected into

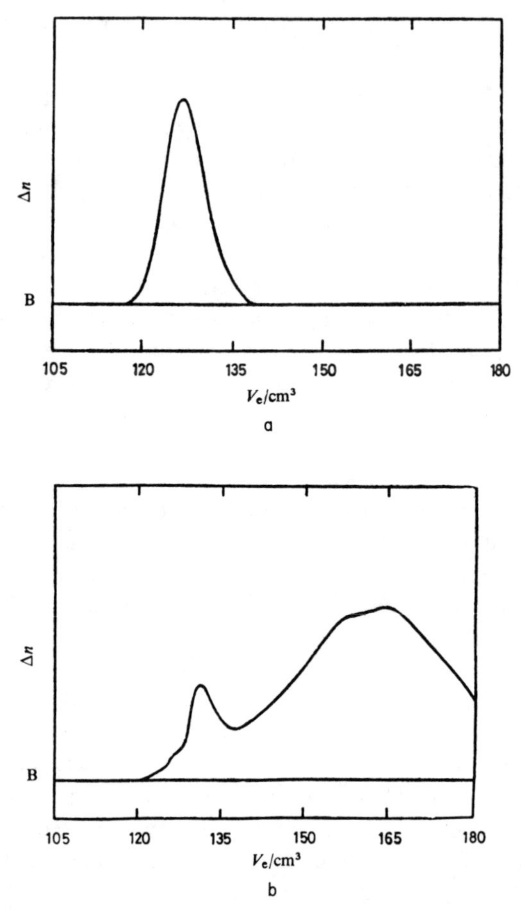

FIG. 3. GPC plots of refractive index difference between solution and pure solvent (Δn) against elution volume (V_e) for block copolymer PStPIp(I) in DMAc at (a) 26 °C, (b) 60 °C and (c) 110 °C. The horizontal line B indicates the baseline. (From reference 43, with permission Royal Society of Chemistry.)

the columns. No attempt however was made to determine equilibrium constants by GPC because of the difficulties posed by dilution and adsorption effects.

At 70 °C and above only a free-chain peak was observed. As the temperature was raised this single peak became progressively sharper and occurred at lower elution volumes. The latter changes were explained in terms of the decreasing importance of adsorption as the solubility of the

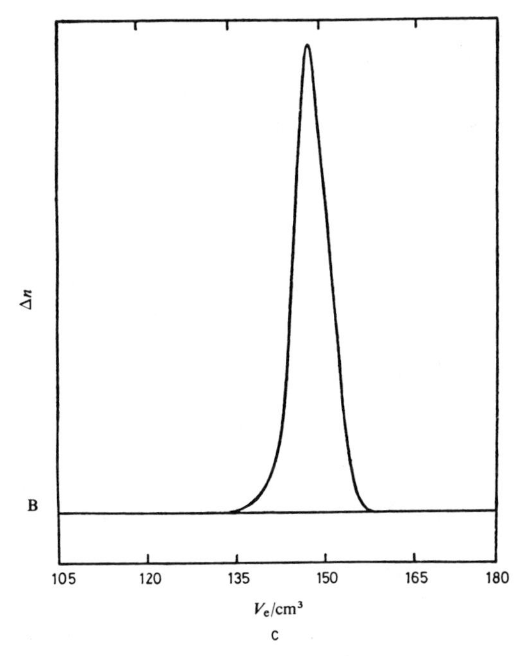

FIG. 3.—*contd.*

free-chains improved with increasing temperature. The chromatogram for 110 °C is shown in Fig. 3(c).

More recently[47] GPC studies were made of an AB polystyrene-*b*-poly(ethylene/propylene) block copolymer, designated here PStPEp(II), for which in the unassociated state $\bar{M}_w = 1{\cdot}06 \times 10^5\,\text{g}\,\text{mol}^{-1}$, $\bar{M}_w/\bar{M}_n = 1{\cdot}14$, and the polystyrene content by weight = 40 %. The block copolymer was dissolved in a base lubricating oil (BLO). As in the previous work the GPC columns were packed with a poly(styrene–divinyl benzene) gel despite the fact that the base lubricating oil was a selectively bad solvent for polystyrene. The spherical micelles which were formed had a polystyrene core and a flexible fringe of poly(ethylene/propylene) blocks. At $T \leq 50\,°\text{C}$ reproducible chromatograms were obtained showing that the equilibrium strongly favoured micelle formation and that the micelles which were formed had a very narrow size distribution ($\bar{M}_w/\bar{M}_n < 1{\cdot}05$). The successful elution of the micelles through the columns demonstrated that the micelle cores were well-shielded from the gel network. Over the temperature range 70–150 °C, however, none of the copolymer could be eluted through the columns because of adsorption.

For both the PStPIp(I)/DMAc and PStPEp(II)/BLO systems the micelles had narrower size distributions than the unassociated copolymers. Elias and Solc[11,53] have shown from statistical arguments that sharpening of the distribution should occur for associations of both the end-to-end and segment-to-segment types. End-to-end association is when one group, or one collection of groups per molecule, is capable of association. Segment-to-segment association is when association may occur at more than one site, the number of sites being roughly proportional to the length of the chain. Micelle formation by copolymers having one block per molecule capable of association can be treated as end-to-end association for which

$$(M_N)_n = N(M_1)_n \tag{9}$$

$$(M_N)_w = (M_1)_w + (N-1)(M_1)_n \tag{10}$$

and (from (9) and (10))

$$(M_N)_w/(M_N)_n = 1 + N^{-1}[((M_1)_w/(M_1)_n) - 1] \tag{11}$$

It follows that provided the association is closed (the association number, N, being the same for all micelles) micelles having a narrow size distribution can be expected even from copolymers which at the molecular level show considerable polydispersity.

2.2. Electron Microscopy

This technique has been used frequently[54–7] as a method of determining the molecular weight characteristics of high molecular weight homopolymers. A typical example is provided by the work of Siegel *et al.*[54] who estimated the molecular weight characteristics of polystyrene samples from electron micrographs of specimens prepared by spraying and evaporation. To calculate the molecular weight from the observed particle size it was assumed that on removing the solvent the polymer coils collapsed until their density was the same as that of the bulk polymer. Fair agreement was obtained between results obtained by electron microscopy and by membrane osmometry.

Newman[58] used electron microscopy to observe spherical particles of an AB polystyrene-*b*-poly(methyl methacrylate) copolymer isolated by evaporation from the selective solvent mixture, benzene/acetonitrile. The spherical particles, which were believed to be collapsed micelles, were found to have a surprisingly wide distribution of sizes.

Price and coworkers have isolated specimens from micelle solutions by two methods. In the first method,[39] freeze-etching, a drop of solution was rapidly frozen by shock-cooling with liquid nitrogen. Solvent was then allowed to evaporate off from a freshly microtomed surface. Finally a replica was made of any collapsed micelles which were left proud of the frozen surface. Four copolymers were studied by the freeze-etching method: an AB polystyrene-*b*-polyisoprene block copolymer, an AB_4 [polystyrene-*b*-polyisoprene]$_4$Si star block copolymer and two polystyrene-*g*-polyisoprene graft copolymers; $\bar{M}_n$ (g mol^{-1}), $\bar{M}_w/\bar{M}_n$ and the % styrene contents for the unassociated copolymers were, respectively, (51 000, 1·12, 25·2), (185 000, 1·10, 24·9), (555 000, 1·18, 29·2), and (420 000, 1·22, 38·1). The copolymers are designated PStPIp(III), (PStPIp)$_4$Si(IV), PSt/PIp(V), and PSt/PIp(VI), respectively. Technical white oil, which is a selectively bad solvent for polystyrene, was used for the study. Within experimental error the micelles formed by the two block copolymers were monodisperse in size. The graft copolymer micelles on the other hand showed measurable size distributions ($\bar{V}_w/\bar{V}_n = 1·3$ and 1·7). The second method[43] involved allowing a drop of micellar solution to spread and evaporate on a carbon substrate. In one case it was applied to the PStPIp(I)/DMAc which had been successfully investigated by GPC. An electron micrograph of micelles which were stained in solution with osmium tetroxide prior to isolation is shown in Fig. 4. The osmium

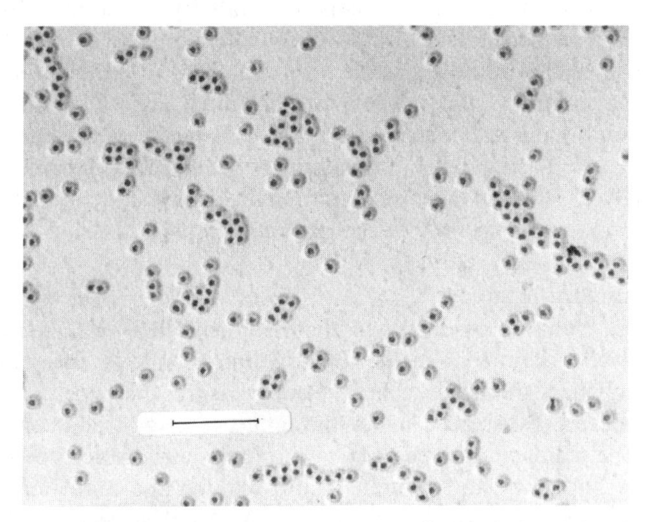

FIG. 4. An electron micrograph of collapsed micelles isolated from a micelle solution of block copolymer PStPIp(I) in DMAc. The micelles were stained with OsO_4 and lightly shadowed in the dry state with C/Pt. Scale mark indicates 200 nm. (From reference 43, with permission Royal Society of Chemistry.)

tetroxide reacted selectively with olefinic bonds of the polyisoprene and served to enhance contrast between the two types of block; thus polyisoprene appears black and polystyrene dark grey. The electron micrograph provided strong evidence in support of the claim that the micelles were essentially monodisperse in size. On final evaporation of the solvent during specimen preparation the micelles were thought to collapse to form particles having a density similar to that of bulk polymer. From shadowing it was concluded that the spherical micelles became approximately disc-shaped during isolation because they tended to flow and wet the carbon substrate. Because of this distortion in shape it was not possible to determine a molecular weight accurately from the micrograph. A molecular weight of several millions was indicated from approximate determinations.

With some modification to allow for the involatility of the solvent medium the latter method was also used to show that the micelles formed by the PStPEp(II) block copolymer in the base lubricating oil had a narrow size distribution as first indicated by GPC.[59]

Using an evaporation technique Horii *et al.*[60] studied micelles formed by some poly(vinyl acetate)-*g*-polystyrene graft copolymers in ethyl acetoacetate, which is a poor solvent for polystyrene. It was found possible using dialysis to replace the ethyl acetoacetate by either methanol or acetonitrile without affecting the colloidal stability of the micelles. The dried micelles were spherical when isolated from methanol or acetonitrile, but were somewhat flattened when prepared from the original solvent. Neither the nature of the solvent nor the initial concentration of the solution was found to influence the average diameter of the spheres. It seems likely that the polymeric content of the micelles remained fixed during dialysis. A broad distribution of micelle sizes was attributed by the authors to the polydispersity of the graft copolymers.

2.3. Ultracentrifugation

Tuzar *et al.*[61] have demonstrated that sedimentation velocity analysis can be a very useful tool in the investigation of micelle formation by block copolymers. With this technique one measures the velocity at which a solute species is displaced under the influence of a strong centrifugal force.[62] The sedimentation velocity of a solute depends on its molecular weight, its buoyancy and friction factor, and on the centrifugal force applied. Detailed studies[61] were made of an ABA polystyrene-*b*-polybutadiene-*b*-polystyrene block copolymer, designated here PStPBdPSt(VII), for which $\bar{M}_w = 1{\cdot}4 \times 10^5\,\mathrm{g\,mol^{-1}}$ in the unassociated state and the polystyrene content by weight equals 52 %. The copolymer was dissolved in

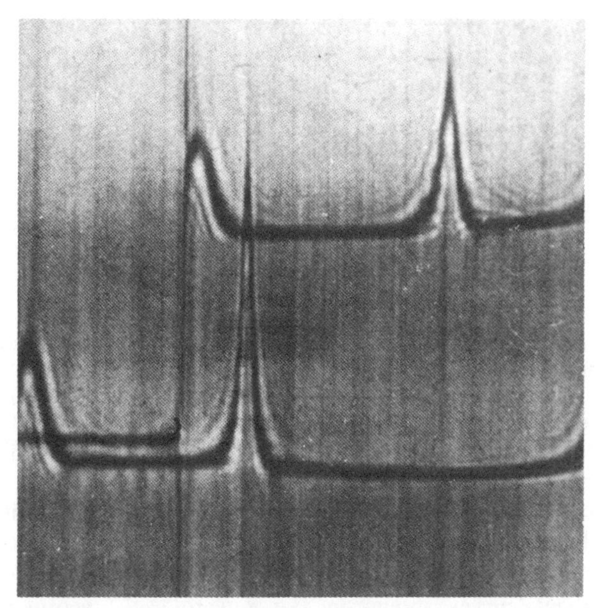

FIG. 5. Gradient curves showing sedimentation of block copolymer PStPBdPSt(VII) in tetrahydrofuran/allyl alcohol (volume fraction of allyl alcohol = 0·46) at 36 °C. For the upper curve, $c = 10^{-3}$ g cm^{-3} and for the lower curve, $c = 4 \times 10^{-3}$ g cm^{-3}. (From reference 61, with permission Hüthig & Wepf Verlag.)

the mixed selective solvent tetrahydrofuran/allyl alcohol. Tetrahydrofuran is a relatively good solvent for polystyrene whilst allyl alcohol is a rather strong precipitant for polybutadiene. The cores of the micelles would have consisted mainly of polybutadiene blocks swollen with a solvent mixture that was enriched with tetrahydrofuran by preferential adsorption. Interpretation of the results was much simplified since mixtures of tetrahydrofuran ($n_D^{25} = 1{\cdot}407$, $\rho^{25} = 0{\cdot}8820$) and allyl alcohol ($n_D^{25} = 1{\cdot}412$, $\rho^{25} = 0{\cdot}8469$) could be regarded as isorefractive and isopycnic. Three types of solution behaviour were observed: (i) with volume fractions of allyl alcohol up to 0·42 at 25 °C (or up to 0·44 at 35 °C) the solutions were clear and gave only a single maximum on the gradient curve corresponding to molecularly dissolved copolymer chains; (ii) on passing to higher volume fractions of allyl alcohol (0·44–0·50 at 25 °C and 0·46–0·52 at 35 °C) the solutions became turbid and a second, faster component appeared corresponding to compact copolymer micelles (see Fig. 5); it was deduced from the shape of the gradient curves that the micelles had a comparatively

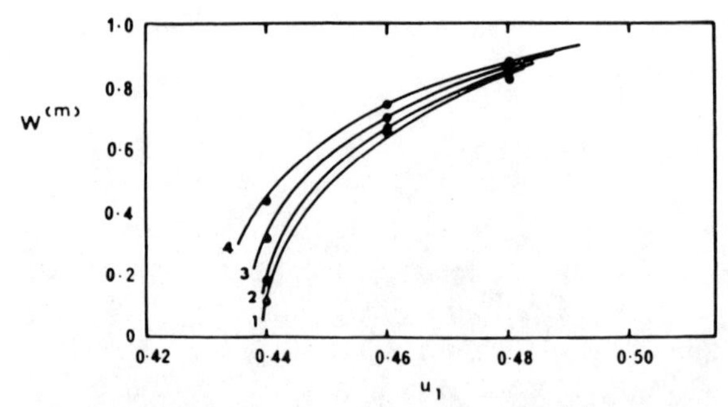

FIG. 6. Dependence of the weight fraction of micelles, $W^{(m)}$, on composition of the solvent mixture for block copolymer PStPBdPSt(VII) in tetrahydrofuran/allyl alcohol at 25 °C; u_1 is the volume fraction of allyl alcohol. Curves 1, 2, 3 and 4 are for copolymer concentrations of 4×10^{-3}, 3×10^{-3}, 2×10^{-3} and 10^{-3} g cm^{-3} respectively. (From reference 61, with permission Hüthig & Wepf Verlag.)

narrow distribution of weights; (iii) when the volume fraction of allyl alcohol in the mixture was 0·52 or more, very rapid sedimentation occurred due to macroscopic precipitation of the copolymer.

The gradient curves shown in Fig. 5 were each recorded after a sedimentation time of 13 min. The observation of two sharp peaks in experiments of this length implies that the rates of dissociation and formation of micelles were relatively slow even when the thermodynamic equilibrium was evenly balanced between micelles and free chains. The concentrations of the free chains and of the micelles were determined from areas under the gradient curves. As shown in Figs 6 and 7 the weight

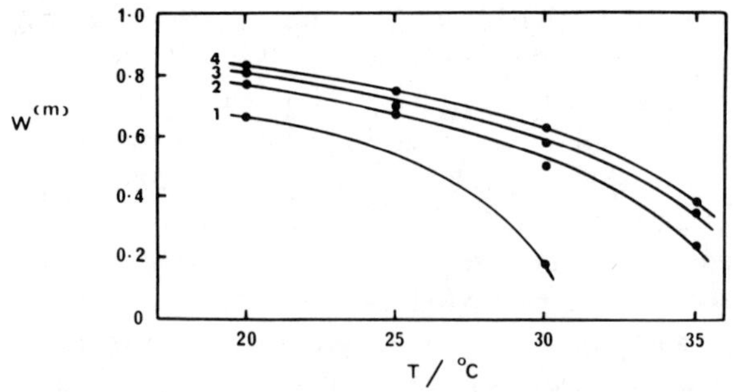

FIG. 7. Dependence of the weight fraction of micelles, $W^{(m)}$, on temperature for block copolymer PStPBdPSt(III) in tetrahydrofuran/allyl alcohol having a volume fraction of allyl alcohol equal to 0·46. Curves 1, 2, 3 and 4 are for copolymer concentrations of 4×10^{-3}, 3×10^{-3}, 2×10^{-3} and 10^{-3} g cm^{-3}, respectively. (From reference 61, with permission Hüthig & Wepf Verlag.)

fractions of micelles, $W^{(m)}$, depended on the composition of the solvent, on the temperature and on the overall concentration of the copolymer.

Following Tuzar *et al.*, Price and Naylor used sedimentation velocity analysis to study the spherical micelles formed by PStPIp(I) in DMAc at 26·5 °C. Measurements were made within the concentration range $(0{\cdot}03\text{–}0{\cdot}21) \times 10^{-2}\,\mathrm{g\,cm^{-3}}$ and the Schlieren patterns were corrected for self-sharpening by Fujita's method. In agreement with GPC they observed a single sharp peak at each concentration indicating that (a) the thermodynamic equilibrium overwhelmingly favoured micelle formation and (b) the micelles had a very narrow size distribution. Substitution of extrapolated values ($c \to 0$) of the sedimentation coefficient and translational diffusion coefficient into Svedberg's relation gave $4{\cdot}1 \times 10^6\,\mathrm{g\,mol^{-1}}$ for the molecular weight of the micelles.

3. DETERMINATION OF THE MOLECULAR WEIGHT, SIZE AND INTERNAL STRUCTURE OF COPOLYMER MICELLES

3.1. Elastic Light Scattering

Elastic light scattering has been the method most widely used for investigating copolymer micelles. The experiment involves measuring the intensity of scattered light (integrated in effect over a narrow range of frequency) as a function of scattering angle. For a system of particles with a spherically symmetric segment distribution, the normalised intensity of scatter in the absence of multiple scattering is given by[63,64]

$$I = F^2(\theta)\left[1 - 4\pi(N/V)\int_0^\infty (1 - \mathrm{g}(r))(\sin kr)(kr)^{-1} r^2\,\mathrm{d}r\right] \qquad (12)$$

In this expression N is the number of particles in volume V, r is the distance from the centre of any reference particle, $\mathrm{g}(r)$ is the radial distribution function, $F^2(\theta)$ is a scattering factor for isolated particles, and $k = 4\pi\lambda^{-1}\sin(\theta/2)$, where λ is the wavelength of the radiation in the medium and θ is the angle between the incident beam and the direction of scattering.

In the limit of infinite dilution the term in square brackets in eqn (12) becomes unity and light scattering measurements when suitably extrapolated to meet this condition provide a means of determining the form of $F^2(\theta)$. The relationship is usually written in the form

$$(Kc/R_\theta)_{c=0} = [MP(\theta)]^{-1} \qquad (13)$$

where M is the molecular weight of the solute, R_θ is the difference between the Rayleigh ratio of the solution and that of the pure solvent, $P(\theta) \equiv F^2(\theta)$, and K is an optical constant containing amongst other quantities the square of the refractive index increment. Since at zero angle $F^2(\theta) = 1$ a double extrapolation of light-scattering results to zero concentration and zero angle gives a measure of the M.W. of the solute. Provided the solute is homogeneous in chemical composition the method yields a $\bar{M}_w$[65]

$$(K_c/R_\theta)_{\substack{c=0\\ \theta=0}} = \bar{M}_w^{-1} \tag{14}$$

In the case of copolymers that are heterogeneous in chemical composition and for which there is a significant difference between the refractive index increments of the polymer blocks, use of eqn (14) yields an apparent M.W.[66,67] rather than the true $\bar{M}_w$. This apparent M.W. depends on the refractive index of the solvent used in the study. Because of the complexities this factor can introduce into structural studies we will only concern ourselves with light scattering results for copolymers which can confidently be treated as homogeneous in chemical composition.

For a solution of a homopolymer at sufficiently low angles eqn (13) takes the form

$$(K_c/R_\theta)_{c=0} = \bar{M}_w^{-1}[1 + (16\pi^2/3\lambda^2)\langle \overline{S^2} \rangle_z \sin^2(\theta/2)] \tag{15}$$

The z-average root-mean-square radius of gyration, $\langle \overline{S^2} \rangle_z^{1/2}$ may be calculated therefore from the limiting slope of the plot $(Kc/R_\theta)_{c=0}$ against $\sin^2(\theta/2)$.

For a block copolymer, even when the chains are homogeneous in chemical composition (and M.W.), the method yields an apparent mean-square radius of gyration rather than the true value. For a block copolymer in which both components have a Gaussian segment distribution, the apparent mean-square radius of gyration[68]

$$\langle S_*^2 \rangle = y\langle S_A^2 \rangle + (1-y)\langle S_B^2 \rangle + y(1-y)l^2 \tag{16}$$

In eqn (16), $\langle S_A^2 \rangle$ and $\langle S_B^2 \rangle$ are the mean-square radii of gyration of the components, A and B, about their centres of gravity, G_A and G_B, l^2 is the mean-square of the distance $G_A G_B$ separating the two centres of gravity and $y = W_A v_A / v$; W_A is the weight fraction of the A component, and v_A and v are the refractive index increments of the A component and the total copolymer, respectively. Only if $v_A = v_B$ will the light scattering method yield the true radius of gyration. As a first approximation the optical properties of micelles formed by a block copolymer can be treated using a concentric sphere model. In that case the apparent mean-square radius of gyration

$$\langle S_*^2 \rangle = \frac{3}{5}\left[\frac{(m_f - m_s)b^5 + (m_c - m_f)a^5}{(m_f - m_s)b^3 + (m_c - m_f)a^3}\right] \tag{17}$$

where a and b signify the radii of the core and overall micelle, respectively, and m_c, m_f and m_s the refractive indices of the swollen core, of the swollen micelle fringe and the pure solvent, respectively.

For finite concentrations of a homogeneous solute the light-scattering equation in the limit of zero angle is

$$(Kc/R_\theta)_{\theta=0} = M^{-1}\left[1 - 4\pi(N/V)\int_0^\infty (1 - g(r))r^2\,dr\right]^{-1} \tag{18}$$

For very dilute solutions the equation reduces to

$$(Kc/R_\theta)_{\theta=0} = M^{-1} + 2A_2c + \cdots \tag{19}$$

where A_2 is the second virial coefficient. For solute particles which are heterogeneous in M.W. but homogeneous in chemical composition the second virial coefficient obtained from light scattering is an average of the form

$$A_2 = \sum_i \sum_j W_i M_i W_j M_j A_{ij} \Big/ \Big(\sum_i W_i M_i\Big)^2 \tag{20}$$

Thus, if micelles and free-chains are both present in solution, A_2 will contain at least three contributions arising from micelle/micelle, micelle/free-chain, and free-chain/free-chain interactions. $(Kc/R_\theta)^{-1}_{\theta=0}$ is sometimes referred to as the apparent $\bar{M}_w$.[13] Only when the three contributions cancel to give $A_2 = 0$ (an unlikely situation) can $(Kc/R_\theta)^{-1}_{\theta=0}$ be set equal to the $\bar{M}_w$ of the mixture of micelles and free-chains. (Note: the word 'apparent' here is being used to indicate quite a different effect to that mentioned in the paragraph immediately following eqn (14).)

Determinations of (Kc/R_θ) over a range of concentration and scattering angle have been made for a number of block and graft copolymers in selectively bad solvents for one of the polymer blocks.[30,32,34,40,42,43,69–75] The most useful results are those for which the type of association process involved was first established by one of the methods discussed earlier. Light-scattering measurements on block copolymer PStPIp(I) in DMAc[43] were made at four temperatures, 8·3, 26·5, 50·5 and 90 °C. Some of the results are shown in Fig. 8 where Kc/R_θ extrapolated to zero angle is plotted

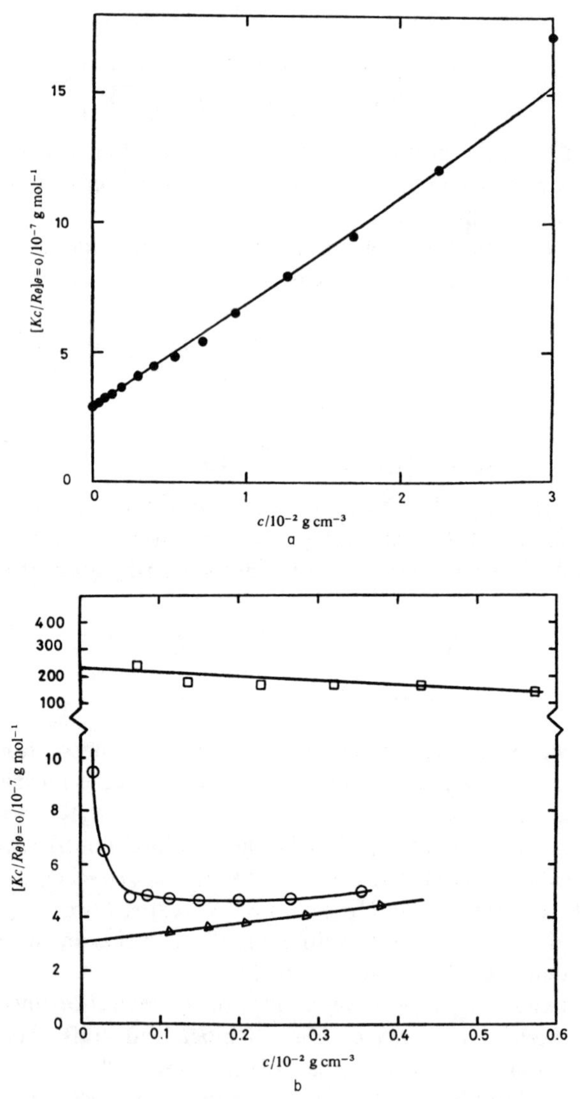

FIG. 8. Plots of $(Kc/R_\theta)_{\theta=0}$ against concentration for block copolymer PStPIp(I) in DMAc at (a) 26·5 °C and (b) 8·3 °C (△), 50·5 °C (○) and 90 °C (□). (From reference 43, with permission Royal Society of Chemistry.)

against concentration. The results obtained at 26·5 °C were consistent with the equilibrium being overwhelmingly in favour of micelle formation down to the lowest concentration, $c = 5 \times 10^{-2}\,\mathrm{g\,dm^{-3}}$, investigated. The intercept yielded $\bar{M}_w = (3{\cdot}42 \pm 0{\cdot}17) \times 10^6\,\mathrm{g\,mol^{-1}}$ (indicating that each micelle contains about 83 chains) and the limiting slope $A_2 = (1{\cdot}93 \pm 0{\cdot}15) \times 10^{-5}\,\mathrm{cm^3\,g^{-2}\,mol}$. The slope of a plot of $(Kc/R_\theta)_{c=0}$ against $\sin^2(\theta/2)$ yielded $\langle S_*^2 \rangle^{1/2} = 19{\cdot}6\,\mathrm{nm}$, which for this system was not expected to be too different from the true radius of gyration.

The equivalent thermodynamic sphere[76] was considered as a possible model for the micelles. For this case

$$A_2 = (16/3)(N\pi R^3/M^2) \tag{21}$$

where R is the equivalent thermodynamic radius. Equation (21) led to $R = 28{\cdot}3\,\mathrm{nm}$ at 26·5 °C. This value was reasonably consistent with $\langle S^2 \rangle^{1/2} \simeq 19{\cdot}6\,\mathrm{nm}$ since, for a uniform sphere $R = 1{\cdot}29\langle S^2 \rangle^{1/2} = 25{\cdot}3\,\mathrm{nm}$. Difficulties were encountered, however, when an attempt was made to explain the magnitude of the third virial coefficient for this system. The results at the other temperatures served to confirm conclusions drawn from GPC results. At $T = 8{\cdot}3\,°\mathrm{C}$ the equilibrium remained overwhelmingly in favour of micelles; $\bar{M}_w = 3{\cdot}26 \times 10^{-6}\,\mathrm{g\,mol^{-1}}$ and $A_2 = 1{\cdot}82 \times 10^{-5}\,\mathrm{cm^3\,g^{-2}\,mol}$. At 50·5 °C the slope of the plot of $(Kc/R_\theta)_{\theta=0}$ against c could only be explained if $\bar{M}_w$ was a function of concentration i.e. if both free-chains and micelles were present in proportions dependent on the concentration. At 90 °C the light-scattering plot was approximately linear and the intercept yielded $\bar{M}_w = 43\,000\,\mathrm{g\,mol^{-1}}$ which was in good agreement with the value established for free-chains in a characterisation study carried out in a good solvent for both components.

A quantitative interpretation of light-scattering results for intermediate regions in which both micelles and free-chains are present in significant proportions is very difficult. Tuzar *et al.*[40] have attempted a solution for their PStPBdPSt(VII)/(tetrahydrofuran/allyl alcohol) system by using the weight fractions they obtained from ultracentrifugation. They were forced, however, to argue that $A_2 = 0$ (through a cancellation of contributions) in order that they might calculate the molecular weight of the micelles, $\bar{M}_w^{(m)}$, from the relation

$$\bar{M}_w^{(m)} = [(Kc/R_\theta)_{\theta=0} - \bar{M}_w^{(f)} W^{(f)}]/(W^{(m)}) \tag{22}$$

where $M_w^{(f)}$ and $W^{(f)}$ are the $\bar{M}_w$ value and weight fraction of free-chains, respectively, and $W^{(m)}$ is the weight fraction of micelles. For a copolymer concentration of $c = 4 \times 10^{-3}\,\mathrm{g\,cm^{-3}}$ and a solvent mixture having a

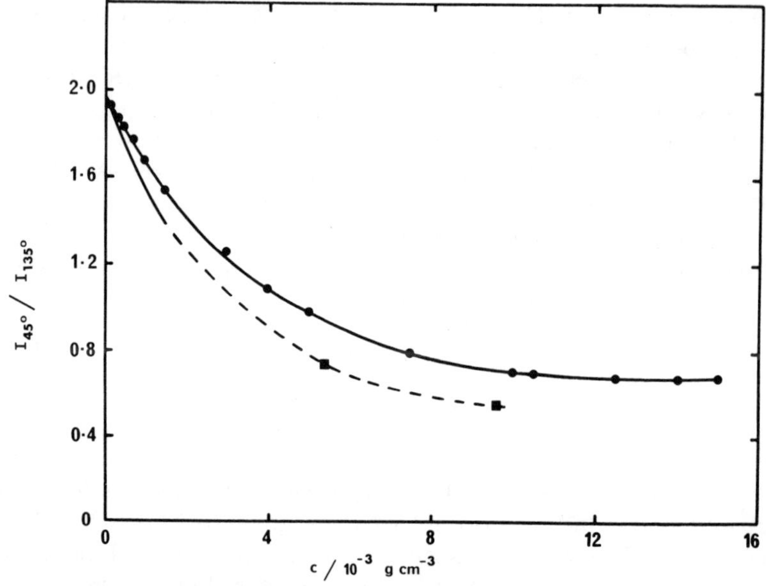

FIG. 9. Plot of dissymmetry ratio against concentration for block copolymer PStPEp(II) in *n*-hexane at 25 °C; the upper curve passes through the experimental points (●) whilst the lower curve indicates the behaviour predicted by the simple hard sphere model. (From reference 77, with permission IPC Business Press, Ltd. ©)

volume fraction of allyl alcohol equal to 0·46, they found that the $\bar{M}_w$ of the micelles decreased from $8{\cdot}5 \times 10^6$ to $6{\cdot}1 \times 10^6\,\mathrm{g\,mol^{-1}}$ on raising the temperature from 20 to 35 °C.

Price *et al.*[77] have used light scattering to study micelle formation by block copolymer PStPEp(II) in *n*-hexane at 25 °C. The equilibrium was overwhelmingly in favour of micelle formation and the micelles had a narrow size distribution over the concentration range studied. Of particular interest was the dissymmetry ratio

$$\left(= \frac{\text{reduced scattering intensity at } 45^\circ}{\text{reduced scattering intensity at } 135^\circ} \right)$$

which decreased with increase in concentration and fell below unity at $c = 4{\cdot}8 \times 10^{-3}\,\mathrm{g\,cm^{-3}}$. The effect of concentration on the dissymmetry ratio (Fig. 9) was predicted reasonably well by again assuming the micelles behaved as hard spheres. On increasing the concentration the hard spheres were considered to pack closer together and to become ordered locally in the same manner as other non-crystalline arrays. The hard sphere diameter ($=150\,\text{nm}$) was determined from A_2 using eqn (21). The first part of the dissymmetry curve ($c < 1{\cdot}5 \times 10^{-3}\,\text{g cm}^{-3}$) was established using an expression derived by Fournet.[78] The dissymmetry ratio at $c = 5{\cdot}4 \times 10^{-3}\,\text{g cm}^{-3}$ was calculated from eqn (12) using values of $g(r)$ determined for a number density of hard spheres $^{n}\rho = 0{\cdot}372\,^{n}\rho_0$ (where $^{n}\rho_0$ is the number density of a closed packed system) by Monte Carlo machine calculations.[79] The dissymmetry ratio at $c = 9{\cdot}6 \times 10^{-3}\,\text{g cm}^{-3}$ was determined from $g(r)$ values derived by Kirkwood and Boggs[80] for what they termed an idealised liquid structure in which the average distance between the centre of a sphere and the centre of its nearest neighbours was 1·15 diameter. The analysis was sufficiently successful to show that the micelles remained colloidally stable even when packed at relatively high number densities. If however the concentration had been increased a little further a transition from a liquid-like to a more highly ordered mesomorphic type structure would eventually have occurred.

Information on the segment distribution of a particular polymer component in a micelle may be obtained by rendering the other component 'invisible' by using an isorefractive solvent. Such measurements have been made by Utiyama *et al.*[73] on polystyrene-*b*-poly(methyl methacrylate) micelles using a mixture of toluene and furfuryl alcohol which masked out the fringe component.

Tanaka *et al.*[42,69] have studied systems in which the core component was rendered invisible. They compared experimental results for polystyrene-*b*-poly(methyl methacrylate) micelles in a mixture of toluene/*p*-cymene with calculated particle scattering functions. Excellent agreement was obtained with a three-phase model consisting of a compact spherical core containing only poly(methyl methacrylate) chains, a mixed phase within which the poly(methyl methacrylate)/polystyrene junctions were located and an outer fringe consisting of polystyrene chains. The polystyrene chains were assumed to obey random flight statistics and to have their ends fixed on an impermeable core surface.

3.2. Small Angle Elastic Scattering of X-rays and Neutrons

The basic theory underlying these two techniques is similar to that for light scattering. The form of the equations remains the same except for expressions involving optical constants which require modification. Whilst differences in polarisabilities are involved in light scattering, X-ray and neutron scattering are dependent on differences in electron densities and collision cross-sections of atoms, respectively. With light scattering the wavelength of the radiation is greater than the molecular dimensions of the particles, whilst it is smaller with X-ray and neutron scattering. Hence equivalent measurements need to be performed at lower angles for X-ray and neutron scattering than for light scattering.

Pleštil and Baldrian[81,82] have demonstrated that small-angle X-ray scattering may be usefully employed to investigate the molecular weight, overall size and phase structure of micelles. Few studies of dilute micellar solutions have yet been made with this technique, however, and these studies are limited to concentrations down to $c = 0{\cdot}5 \times 10^{-2}\,\mathrm{g\,cm^{-3}}$.

The most detailed study that has been reported[82] deals with solutions of a polystyrene-*b*-polybutadiene block copolymer in heptane, which is a selectively bad solvent for polystyrene; the $\bar{M}_n$ of this copolymer and its polystyrene block were 64 700 and 5700 $\mathrm{g\,mol^{-1}}$, respectively. Measurements in heptane were made at four temperatures ranging from 18 to 50 °C. The analysis of results was complicated by the knowledge (gained from ultracentrifugation) that a significant proportion of the chains were in the free-chain form.

Plots of $\log(I/c)$ against θ showed a shoulder appearing on the central peak at about 0·003 rad and a weak maximum within the range 0·0130 to 0·0145 rad. Absolute intensity measurements taken below $\theta = 0{\cdot}003$ rad were extrapolated to zero angle (using Guinier's Equation[83]). The $\bar{M}_w$ of the micelles and the degree of swelling of the individual components were determined; the number of copolymer chains in the micelles varied from 190 (at 18 °C) to 100 (at 50 °C). It was argued that at angles greater than 0·0055 rad the shape of the scattering curves was governed mainly by scattering from the micelle cores. For each of the four temperatures a two-phase concentric sphere model for the micelles was used to calculate the diameter of the micelle cores (22·2–18·1 nm), the overall diameter of the micelles (55·8 – 53·2 nm) and the radius of gyration (19·1 – 18·0 nm) of the micelles, from absolute intensity measurements. Support for the model was provided by showing that an alternative treatment involving direct analysis of relative intensity measurements gave a similar set of values.

Small-angle neutron scattering is potentially a more powerful technique for investigating micelle structure than X-ray scattering owing to the possibility of producing large contrasts by deuteration. Because of the difference in coherent scattering lengths between hydrogen and deuterium, measurable scattering can be generated by dispersing a small concentration of fully deuterated polymer in a hydrogenous matrix or vice versa. The technique offers a means of studying the multiphase structure of micelles and the average dimensions of copolymer chains and their component blocks in a micellar framework. There are no reports to date of its use in studying copolymer micelles in organic solvents. It has been used however[84] in a study of micelles formed from a polystyrene-*g*-poly(oxyethylene) graft copolymer in water. There is also a report[85] of its use in characterising the dimensions of a polystyrene-poly(dimethylsiloxane) block copolymer adsorbed on dispersed polymer particles in the solvents *n*-heptane, Freon 113(1,1,2-trichloro-1,2,2-trifluoroethane) and silicone fluid.

3.3. Osmometry

This technique provides a method for determining the $\bar{M}_n$ of micelles. Regrettably, little use has been made of it to date. In sufficiently dilute solution

$$\pi(RTc)^{-1} = (\bar{M}_n)^{-1} + A_2c + \cdots \tag{23}$$

where π is the osmotic pressure, and for a polydisperse system $A_2 = \sum_i \sum_j A_{ij} W_i W_j$. For situations in which both micelles and free chains are present in significant proportions virial terms pose the same type of problem in M.W. studies as arise in light scattering.

Values of π/c have been reported for several diblock polystyrene-*b*-polyisoprene copolymers in dimethylformamide (which is a bad solvent for polyisoprene) and *n*-hexane (which is a bad solvent for polystyrene). However, in the absence of supporting experimental evidence it is not possible to assess fully the validity of the assumptions which were made concerning the attainment of complete polymer micellisation. In principle osmometry should be a good method for detecting small concentrations of free chains since it is related to the $\bar{M}_n$. In practice difficulties arise when it is required to study very dilute solutions because of the lack of sensitivity of the method.

3.4. Quasi-elastic Light Scattering

Brownian motion of particles in solution gives rise to a spectral distribution in the scattered light. From measurement of the line width of scattered laser light by photon correlation spectroscopy (PCS) the translational diffusion coefficient may be determined.[87] Unlike the traditional method of investigating translational diffusion PCS does not require a macroscopic concentration gradient and therefore can more readily be applied to investigate association processes. The method has been used to determine the translational diffusion coefficients of micelles formed by a number of block copolymers in selectively bad solvents for one of the polymer components.[72,88–90] If the micelles are treated as hydrodynamically equivalent spheres then the translational diffusion coefficient for the limiting case of infinitely dilute solution is given by the Stokes-Einstein relation

$$D_0 = \frac{kT}{6\pi\eta R_d} \tag{24}$$

where R_d is the hydrodynamic radius and η the viscosity of the solvent. For micelles polydisperse in size distribution PCS yields approximately a z-average diffusion coefficient and therefore a z-average reciprocal hydrodynamic radius $(\overline{R_D^{-1}})_z$.

Studies have been made on monodisperse micelles under conditions where the association equilibrium overwhelmingly favoured micelle formation.[72] For the block copolymer PStPIp(I) in DMAc at 26·5°C, $R_D = 228$ Å. Plots of D_c against temperature were found to show fairly abrupt changes in shape in regions where the micelles were beginning to break up to form free chains.

3.5. Ultramicroscopy

Under the most favourable conditions the resolving power of an optical microscope fitted with an oil-immersion objective is limited to one-third the wavelength of the light source. In spite of this limitation an arrangement termed an ultramicroscope, which was devised by Siedentopf and Zsigmondy, makes it possible to monitor the movements of particles much smaller than the resolving power. A strong beam of light is directed horizontally into the liquid containing the suspended particles and the vertically mounted microscope is focused on a region just below the surface. The observer sees bright spots of scattered light moving in a dark field, use being made of the fact that the diffraction image round a particle is considerably larger in size than the particle itself. For organic particles the technique is particularly useful for observing Brownian motion of particles having diameters in the range 0·02–1 μm. With present day light microscopes similar studies may be carried out simply by using a special condenser (either a paraboloid or cardioid type) which ensures that only scattered light enters the objective lens.

Micelles formed by surfactants are generally too small to be observed by ultramicroscopy. Recently it was found however that micelles formed by a number of high M.W. block copolymers could be studied by this direct method.[89] From measurements of the mean-square planar displacement of particles $\overline{l^2}$ in time t, the translational diffusion coefficient may be calculated using the Einstein relation[91,92]

$$D = \overline{l^2}/4t \tag{25}$$

For polydisperse particles the method gives a number-average diffusion coefficient (provided all the particles are visible) and hence a number-average reciprocal hydrodynamic radius $(\overline{R_D^{-1}})_n$ for spherical particles. The ratio $(\overline{R_D^{-1}})_z^{-1}/(\overline{R_D^{-1}})_n^{-1}$ has been used as a measure of polydispersity. The results obtained have supported conclusions drawn from electron microscopy studies.[89] One of the most useful aspects of the technique is that it enables a direct estimate to be made of the number density of micelles in solution. It is also a very sensitive method of detecting colloidal instability.

3.6. Formation of Non-spherical Micelles

It has been argued that under certain conditions spherical micelles formed from block copolymers can aggregate reversibly to form worm-like micelles. In one case of interest, studies were carried out[93,94] on a polystyrene-*b*-polybutadiene-*b*-polystyrene block copolymer in ethyl acetate, which is a selectively bad solvent for polybutadiene; the $\bar{M}_n$ values of the three blocks in this copolymer, which we designate here PStPBdPSt (VIII), were 12 900, 66 000 and 13 700 g mol^{-1} and $\bar{M}_w/\bar{M}_n$ for the overall copolymer was 1·09. At sufficiently low temperatures, solutions of this copolymer were clear but had a distinct bluish cast suggesting the presence of micelles. On raising the temperature a cloud-point was observed for concentrations up to 0·01 g cm^{-3}. At concentrations down to 0·001 g cm^{-3} the visually observed cloud-point could be located to within $\pm 0{\cdot}2°$, but became more diffuse at lower concentrations. On raising the temperature further there was a gradual transition from a cloudy white solution to a clear solution. Solutions could be heated quite slowly through the cloudy region without precipitation. If, on the other hand, the cloudy solutions were allowed to stand for 24 h, the copolymer precipitated from some of them as a gel. The latter quickly disappeared on either heating or cooling. For a given concentration precipitation only occurred over a fairly narrow range of temperature. In Fig. 10 the turbidity/c is plotted as a function of temperature for $c = 2{\cdot}5 \times 10^{-3}$ g cm^{-3}; the cloud-point is indicated by a broken line. The cloud-point moved to higher temperatures with increasing concentration as shown by the upper plot in Fig. 11.

A second polystyrene-*b*-polybutadiene-*b*-polystyrene copolymer, designated PStPBdPSt (IX), having shorter blocks than the first copolymer,

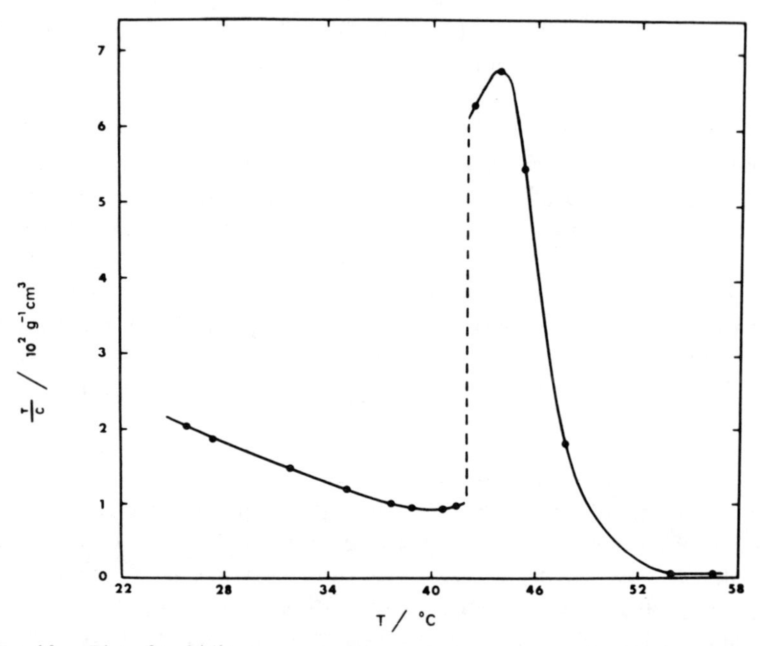

FIG. 10. Plot of turbidity/concentration against temperature for block copolymer PStPBdPSt(VIII) in ethyl acetate at $c = 2{\cdot}5 \times 10^{-3}\,\mathrm{g\,cm^{-3}}$. (From reference 94, with permission Royal Society of Chemistry.)

showed a similar type of cloud-point behaviour (see Fig. 11) but at lower temperatures. On the basis of additional results from light scattering and electron microscopy it was argued[94] that the onset of cloudiness was due to the formation of long worm-like micelles from spherical micelles. An electron micrograph of worm-like micelles isolated from a cloudy solution is shown in Fig. 12; since the specimen was left unshadowed only the OsO_4-stained polybutadiene is visible. Spherical micelles with a polybutadiene core were indicated by electron micrographs (similar to the one in Fig. 4) for solutions having a bluish cast. The spherical micelles had a fairly narrow size distribution. The worm-like micelles had a wide distribution of contour lengths but possessed a narrow size distribution in the radial direction.

The clearing of solutions at higher temperatures was interpreted as being due to the dissociation of worm-like micelles to form copolymer molecules in the free-chain form. One way of viewing the conversion of spherical micelles to worm-like micelles which occurred on either raising the temperature or lowering the concentration was to consider it as a reversible

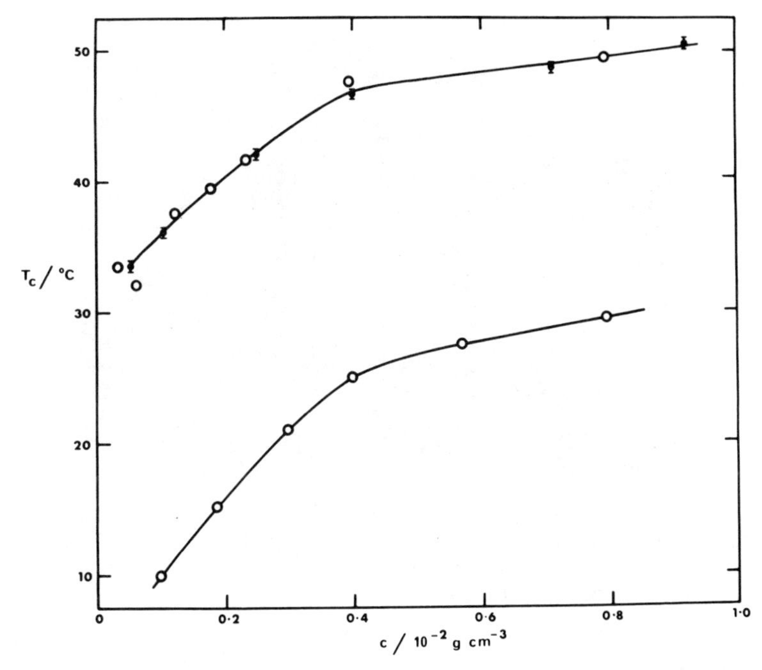

FIG. 11. Dependence of cloud-point on concentration for block copolymers PStPBdPSt(VIII), upper curve, and PStPBdPSt(IX), lower curve, in ethyl acetate; ♦, turbidimetric values, ○, visually observed values. (From reference 94, with permission Royal Society of Chemistry.)

polymerisation. The cloud-point temperature was seen effectively as a 'floor temperature'[95] below which long micelles became unstable. (Note: Kratochvíl[96] has argued correctly that homopolymer impurities might be left insoluble over a narrow temperature range and could easily be mistaken for worm-like micelles. This type of behaviour could immediately be ruled out as the explanation for the results discussed above since precipitation when it occurred involved the bulk of the copolymer sample.)

As yet we do not have a sound statistical theory for dilute copolymer solutions which is able to predict quantitatively the relationship between chain structure and micelle shape and size. As a first requirement any such theory would have to explain the remarkably narrow size distribution which we have seen is exhibited by many micellar systems. Some aspects of the statistical problem have been considered semi-quantitatively by Inoue *et al.*[97] Additional insight may be gained both from consideration of models which have been proposed for amphiphilic micelles,[8] and for block

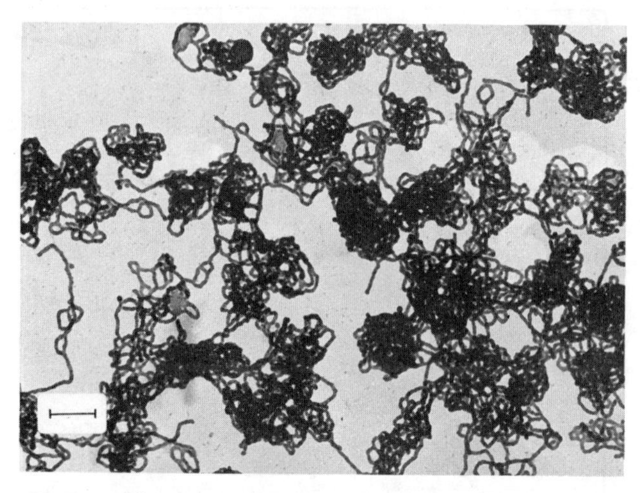

FIG. 12. Electron micrograph of collapsed worm-like micelles isolated from a cloudy micellar solution of block copolymer PStPBdPSt(VIII) in ethyl acetate. Specimen was stained in solution with OsO_4. Scale mark indicates 500 nm. (From reference 94, with permission Royal Society of Chemistry.)

copolymers in the bulk and concentrated solution.[98,99] The factor most likely to be responsible for the change in micelle shape in the PStPBdPSt(VIII and IX)/ethyl acetate systems is the degree of swelling of the polybutadiene core. Equilibrium swelling experiments[94] carried out on a sample of polybutadiene, $\bar{M}_n = 135\,000\ \mathrm{g\,mol^{-1}}$, gave a gel phase containing 57 wt % ethyl acetate at 25 °C and 68 wt % at 50 °C. The cores of the PStPBdPSt micelles must therefore have contained appreciable amounts of solvent since for a given temperature they might be expected to swell rather more than polybutadiene homopolymer. Changes which lead to absorption of more solvent by the core of a spherical micelle increase the core surface area per copolymer chain and consequently reduce the number density of chains emerging from the core. To compensate for this effect the core surface area per chain may be reduced either by forming worm-like micelles from spherical micelles or by increasing the association number of the spherical micelles. In keeping with expectations the formation of worm-like micelles by PStPBdPSt (VIII and IX)/ethyl acetate systems became more favourable on making changes which increased the degree of swelling of the core material (i.e. raising the temperature and lowering the copolymer concentration).

One of the factors likely to influence the size and shape of micelles is chain geometry. With ABA micelles having a core composed of swollen B blocks, each chain must emerge from the core in two places. With AB micelles on the other hand only a single exit per chain is required. It can be expected, however, that worm-like micelles and lamellar micelles will occur fairly generally regardless of the chain geometry, particularly in situations for which the solvent is selectively bad for the longer blocks. Furthermore there is a possibility that as precipitation occurs these forms could develop into essentially bicontinuous structures through the formation of cross-links between micelle cores. For this reason the precipitating particles could quickly become globular in overall shape.

Some support for the formation of elongated micelles is provided by the work of Utiyama *et al.*[73] By elastic light scattering they found that over a limited range of solvent composition an AB polystyrene-*b*-poly(methyl methacrylate) block copolymer in the mixed solvent toluene/furfuryl alcohol formed micelles with elongated polystyrene cores. Further addition of furfuryl alcohol, the non-solvent for polystyrene, led to the formation of spherical micelles. The transient formation of large particles at the onset of micellisation of an ABA polystyrene-*b*-polybutadiene-*b*-polystyrene block copolymer has also been noted by Tuzar *et al.*[100]

Evidence for the formation of worm-like or lamellar micelles has been discussed by Mandema *et al.*[90] They studied in detail an AB polystyrene-*b*-poly(ethylene/propylene) block copolymer with $\bar{M}_w = 48\,000\ \mathrm{g\,mol^{-1}}$, $\bar{M}_w/\bar{M}_n \simeq 1{\cdot}16$ and a polystyrene content of 38%; additional measurements were reported for a second high M.W. sample ($\bar{M}_w = 80\,000\ \mathrm{g\,mol^{-1}}$) of similar composition. Elastic light scattering measurements made on solutions of the first copolymer in decane, which is a selective non-solvent for polystyrene, revealed very similar behaviour to that found for the PStPBdPSt (VIII and IX)/ethyl acetate systems. Supporting evidence was provided by photon correlation spectroscopy. On the other hand viscosity measurements did not give any clear indication of the formation of worm-like or lamellar micelles. More recently[101] they have reported light-scattering evidence for the formation of worm-like or lamellar micelles in selective mixed solvents.

Large changes in light-scattering dissymmetry indicating the formation of extended micelles were observed thirty years ago[102] for amphiphiles in aqueous media. For these systems[103] however it appears that an increase in total amphiphile concentration always favours the formation of extended micelles. Stigter[104] has studied the variation of the intrinsic viscosity of dodecylammonium chloride micelles with changes in micelle M.W. induced by changes in ionic strength. He concluded that the micelles were rod-like, possessing a singly connected core and a continuous surface, but that the rods must possess considerable flexibility. Evidence for flexible rod-like micelles[105] has also been provided by the angular dependence of light scattering from cetyltrimethylammonium bromide micelles at high ionic strength. The results fitted the theoretical curve for a stiff rod but the molecular weight of the micelles estimated from the dimensions of the rod was less than the directly-determined M.W. Whilst the extended amphiphilic micelles have undoubtedly some degree of flexibility it appears nevertheless that they are considerably stiffer than the worm-like micelles observed for block copolymers in organic solvents.

4. VISCOSITY

4.1. Solution Viscosity

It has long been known that the viscosity of simple liquids may be increased by the addition of macromolecular or colloidal particles. For spheres which are large compared with the solvent molecules Einstein[106] showed that

$$\lim_{\phi \to 0} (\eta \eta_0^{-1} - 1) = 2{\cdot}5\phi \tag{26}$$

where ϕ is the volume fraction of spheres, η is the viscosity of the solution and η_0 is the viscosity of the pure solvent. The quantity $(\eta \eta_0^{-1} - 1)$ is termed the specific viscosity, η_{sp}. At finite concentrations where interparticle interactions influence the flow behaviour, experimental results for the specific viscosity, η_{sp}, have been fitted to a series expansion in powers of the volume fraction:[107,108]

$$\eta_{sp} = 2{\cdot}5\phi + \alpha\phi^2 + \beta\phi^3 + \cdots \tag{27}$$

The volume fraction of spheres can be set equal to $(4\pi R^3/3)N_A M^{-1}c$, where c is the concentration of spheres expressed in terms of mass per unit volume of solution, R the radius of the spheres and M their M.W. Hence

$$\lim_{c \to 0} (\eta_{sp}/c) \equiv [\eta] = (10\pi N_A/3)R^3 M^{-1} \tag{28}$$

The quantity $[\eta]$, which is termed the intrinsic viscosity, is a measure of the extra energy dissipated as heat in the flowing system due to the presence of the solute particles. By treating R as an effective hydrodynamic radius, R_η, eqn (28) may be applied to other types of particles including coils and spherical micelles. For a polydisperse solute the relation is

$$[\eta] = (10\pi N_A/3)(\overline{R_\eta^3})_n M_n^{-1} \tag{29}$$

Provided the thermodynamic equilibrium overwhelmingly favours micelle formation down to the lowest concentrations that can be meaningfully studied by viscometry there appears to be no problem in establishing a suitable extrapolation procedure for determining $[\eta]$. In one study[43] the usual plots of η_{sp}/c against c were found to be adequate. From a knowledge of $[\eta]$ and $\bar{M}_n$ the average hydrodynamic radius $(\overline{R^3})_n^{1/3}$ may be determined. Alternatively, eqns (24) and (29) may be combined to give

$$\bar{M}_n = (1{\cdot}54 \times 10^{-2})k^3 N_A \pi^{-2} T^3 \eta^{-3} (\overline{R_\eta^3})_n (\overline{R_D^{-1}})_z^3 [\eta]^{-1} (\overline{D_0})_z^{-3} \quad (30)$$

For micelles which are essentially monodisperse in size distribution it has been argued[72] that the ratio (R_η/R_D) can be set equal to 1·05 without introducing an error greater than about 5% in the calculated M.W.

The temperature dependence of $[\eta]$ was used by Staudinger[109] in his classic studies as a means of distinguishing between true polymer solutions and association colloids. At that time it was believed that $[\eta]$ would be independent of temperature for a polymer in molecular solution, whilst a marked dependence of $[\eta]$ on temperature was expected for association colloids. It has long been realised that this criterion is not valid since, near a θ-point, $[\eta]$ varies strongly with temperature for a polymer in molecular solution.[76,110,111] On the other hand the temperature dependence of $[\eta]$ has been found to be very small for many surfactants in micellar form.

Many studies of η_{sp}/c have been reported for solutions of graft and block copolymers containing significant proportions of both micelles and free-chains. The influences of temperature[72,74,90,112] and solvent composition[32,113,114] have been investigated. In most cases the results were extrapolated to zero concentration. For solutions of fixed concentration η_{sp}/c has, for some systems, been observed to increase sharply on passing from a predominantly micellar to a free-chain solution (see Fig. 13). In other cases a decrease in η_{sp}/c has been observed. Depending on the relative M.W.'s and hydrodynamic radii of the micelles and free-chains both types of behaviour can be expected.

Elias has considered in detail the dependence of η_{sp}/c on c over the range where shifts in the equilibrium position strongly influence the observed behaviour. Model calculations were carried out for both closed[115] and open [116] end-to-end associations of homogeneous polymers. Each type of species, i, was assumed to contribute additively to η_{sp}/c by an amount predicted by the Huggins equation[117]

$$(\eta_{sp}/c)_i = [\eta] + (k_\eta)_i [\eta]_i^2 c_i \quad (31)$$

where $(k_\eta)_i$ is a constant for the ith type of species having concentration c_i

and intrinsic viscosity $[\eta]_i$. Even for closed association the analysis predicted a wide variety of curves for (η_{sp}/c) versus c. The shape of the individual curves (including the possibility of a maximum or a minimum) depended on the particular choice of seven independent parameters. Inclusion of thermodynamic and hydrodynamic cross-interactions in the analysis would have introduced further possibilities. It is very clear from

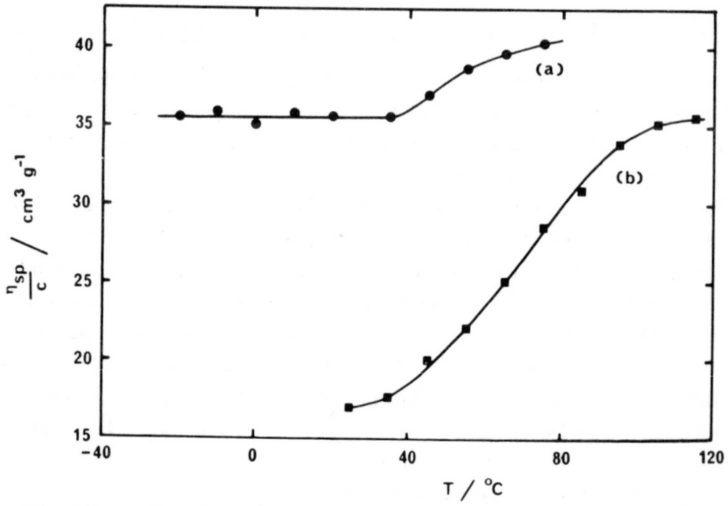

FIG. 13. Plots of η_{sp}/c against temperature for block copolymer PStPIp(III) at $c = 6{\cdot}49 \times 10^{-3}\,\mathrm{g\,cm^{-3}}$ (curve (a)) and graft copolymer PSt/PIp(V) at $c = 4{\cdot}97 \times 10^{-3}\,\mathrm{g\,cm^{-3}}$ (curve (b)) in *n*-decane.

the studies of Elias and coworkers that measurements of η_{sp} do not offer a useful means of investigating quantitatively the nature of an association process.

From a technical standpoint, the viscometric properties of block copolymer solutions are important. One example is their use as viscosity index improvers in lubricating oils. The viscosity index is an arbitrary number which indicates the resistance of a lubricant to viscosity change with temperature.[118] Viscosity index improvers must be soluble in the usual non-polar paraffinic and aromatic lubricating oils in the temperature range −30–200 °C. Polymer concentrations from 1 to 5 wt % are used. Over the years many homopolymers, random copolymers and random block copolymers have been investigated as additives. More recently, presumably to take advantage of the viscometric properties of associating systems, investigations have been made on polybutadiene-*b*-polystyrene[119] and poly(ethylene/propylene)-*b*-polystyrene block copolymers.[120–22] The current trend is to use polymeric additives which are able to serve two or more of the roles: viscosity index improver, dispersant and pour-point depressant.

4.2. Internal Viscosity

In principle micelles may have cores which are liquid-like, glassy or crystalline. The fluidities of the hydrocarbon cores of micelles formed from simple surfactants in aqueous media have been studied by a number of techniques. Comparisons of the mobilities of fluorescence[123] and ESR probe molecules[124,125] solubilised in micelles and dissolved in organic solvents have shown that amphiphiles with ionic head groups and saturated hydrocarbon tails (less than 16 carbon atoms) have liquid-like cores at room temperature.[8] It has been suggested that amphiphilic micelles with very long saturated hydrocarbon chains may have ordered hydrocarbon cores at room temperature, which become liquid-like only on heating to higher temperatures.[8]

Use of NMR offers a method of studying the fluidity of micelle cores without the uncertainties introduced by the perturbing effects of probes.[125] Unfortunately, even in favourable situations it is usually only possible to interpret the results semi-quantitatively.

Heatley and Begum[126] have measured ^{13}C spin-lattice relaxation times for the *cis*-CH_2 peak of micelles formed from a polystyrene-*b*-polybutadiene-*b*-polystyrene block copolymer in ethyl acetate. The results were compared with data obtained for *cis*-polybutadiene homopolymer in the same solvent. It was argued that the mole fraction of polybutadiene segments in the liquid-like cores of the micelles must be greater than 0·6.

There is no direct evidence as yet for colloidally stable copolymer micelles with crystalline or glassy cores. It seems likely, however, that in such systems as PStPEp (II) block copolymer in base lubricating oil, as studied by Price *et al.*,[59] the micelle cores were in a glassy state at 20 °C. The glass transition temperature of the core material in these micelles would be somewhat lower than that for the polystyrene blocks because of the presence of a small percentage of lubricating oil. The occurrence of colloidally stable aggregates with crystalline cores formed by poly(oxyethylene) blocks is suggested by results obtained for an AB polystyrene-*b*-poly(oxyethylene) copolymer in ethylbenzene.[32,127] Further studies are required to characterise the structure and size distribution of particles formed in such systems.

5. SOLUBILISATION OF POLYMERS

Early attempts to fractionate graft copolymers from homopolymer impurities revealed that homopolymers could be held in solution through the solubilising action of copolymers.[26,28,128] When attempts were made to isolate a particular homopolymer by addition of a non-solvent a stable emulsion was often obtained rather than the desired liquid–liquid separation. The behaviour is analogous to the stabilisation of hydrophobic substances by surfactants in aqueous media.[4] Studies have been made of the solubilisation of homopolymers by block and graft copolymers in dilute solution.[129,130] The results have been reviewed by Tuzar and Kratochvíl.[41] An important application of the phenomenon is to be found in dispersion polymerisation.

5.1. Dispersion Polymerisation

In this type of process the monomer is polymerised in a diluent (usually an organic liquid) in which it is soluble, but in which the polymer is insoluble.[131–3] Whilst virtually any type of polymerisation may be used, the majority of studies to date have involved free-radical addition polymerisation. The polymer is prevented from precipitating by the presence of stabilisers. The most effective stabilisers are block and graft copolymers which sterically stabilise[134,135] the polymeric product. The requirement is for a copolymer which contains one component that will become adsorbed onto the polymeric product and act as anchor, and a second component which will remain flexible and extend out into the diluent. Sufficient copolymer must be present in the diluent to maintain adequate coverage as the polymerisation progresses.

After a very short period (less than the time for conversion of 1% of the monomer to polymer) the number of growing particles in free-radical addition polymerisation remains constant. Additional particles do not form because it is argued that before a growing radical can reach any appreciable length it coalesces with one of the already existing particles. During the main part of the reaction the majority of propagation steps occur at particle sites where the ratio of polymer to monomer is high and the viscosity is high. It is claimed that these conditions are responsible for dispersion polymerisations being much faster than solution polymerisations carried out under comparable conditions. As in the case of Trommsdorff effect (i.e. the rapid increase in rate observed during the later stages of bulk free-radical polymerisation) termination is retarded relative to propagation.[136] The main differences between dispersion and conventional emulsion[131] polymerisation are firstly that in emulsion polymerisation the monomer is insoluble in the diluent (an aqueous medium) and is effectively stored in the form of surfactant-stabilised droplets, and secondly that in emulsion polymerisation there is usually no more than one growing radical per particle site whereas in dispersion polymerisation there may be several.

Block and graft copolymers possessing the structural and thermodynamic properties required by a stabiliser will more than likely form micelles in the diluent prior to polymerisation. The rate of dissociation of the micelles will play a major role in governing the effectiveness of the stabiliser. It has already been pointed out that the rate of dissociation of micelles in organic solvents can be extremely slow. The conditions must be chosen carefully therefore to ensure that the copolymer molecules have a sufficient degree of mobility to maintain adequate coverage of the reaction product. Cases have been mentioned in which dispersion polymerisations proceeded satisfactorily at high temperatures, but failed at low temperatures.[137] The addition of a plasticiser, which swells the micelle cores and polymeric product, has been found useful in improving stabilisation. Conditions under which the anchor component is found in an essentially glassy state are obviously to be avoided. In addition it is probably advisable to avoid conditions in which the micelle/free-chain equilibrium of the stabiliser overwhelmingly favours micelle formation.

By carefully controlling the experimental conditions (and in some cases by using a seeding procedure) it is possible to prepare dispersions with narrow size distributions. Particle sizes produced by dispersion polymerisation generally fall in the range 100–10 000 nm.

The main technical applications of polymer dispersions have been in the field of surface coatings. In these applications, however, the total concentrations of polymer involved (20–70 wt %) far exceed any concentrations we have considered in the rest of this review.

6. COLLOIDAL STABILITY

Using the method of dispersion polymerisation Everett and Stageman[139,139] prepared colloidal dispersions of poly(methyl methacrylate), PMMA, and polyacrylonitrile, PAN, in hexane. The two stabilisers used in the investigation were ABA poly(dimethylsiloxane)-*b*-polystyrene-*b*-poly(dimethylsiloxane) block copolymers; for the two copolymers, IX and X, $\bar{M}_w = 2 \times 10^4$ and 3×10^4 g mol^{-1} and the wt % values of the polysiloxane blocks were 42·5 and 49·0, respectively. In Table 1 the results are given for four dispersions;[139] the letters describing the latex type refer to the core monomer whilst the Roman numerals refer to the stabiliser used. The core radii were estimated from transmission electron micrographs (TEMs). The hydrodynamic radii were measured by PCS using very dilute samples in hexane at 30 °C.

TABLE 1
RESULTS OBTAINED[b] BY EVERETT AND STAGEMAN[139] FOR COLLOIDAL DISPERSIONS OF POLY(METHYL METHACRYLATE) AND POLYACRYLONITRILE

Sample no.	*Latex type*	*Average core radius, nm*	*Average hydrodynamic radius, nm*	*Flocculation temperature in propane, °C*	
				Upper	*Lower*
1.	PMMA IX	111 ± 14	123 ± 6	56 ± 1	−82 ± 4
2.	PAN X	85 ± 5	110 ± 5	64 ± 1	−113 ± 4
3.	PAN IX	50 ± 8	70 ± 5	68 ± 1	−84 ± 4
4.	PAN X	40 ± 6	65 ± 5	76 ± 1	−115 ± 4

[a] With permission, Royal Society of Chemistry.

The colloidal particles were transferred successfully to a number of lower liquid alkanes (down to ethane) and to liquid xenon. The dispersions exhibited both upper and lower flocculation temperatures; values extrapolated to zero concentration are given in Table 1 for samples in propane. The flocculation temperatures for small particles and thick adsorbed layers correlated closely with the phase properties of the stabilising component poly(dimethylsiloxane) in the same liquid[140,141] and this result was similar to one obtained earlier by Napper[142] for more complex systems. The range of stability of dispersions was much reduced for larger particles with thinner adsorbed layers.

At either the upper or lower θ-temperatures for a polymer–solvent system the excluded volume effect and the net segment–segment attraction just balance. For a polymer with an infinitely high molecular weight these two θ-temperatures would coincide with the lower critical solution temperature (LCST) and the upper critical solution temperature (UCST), respectively.[76,143,144] At some temperature near the upper θ-temperature of poly(dimethylsiloxane) the surface-attached poly(dimethylsiloxane) blocks must phase-separate from the dispersion medium and collapse on the surface. Below this temperature the excluded volume effect will start to dominate. Hence when the flexible polymer chains surrounding two of the

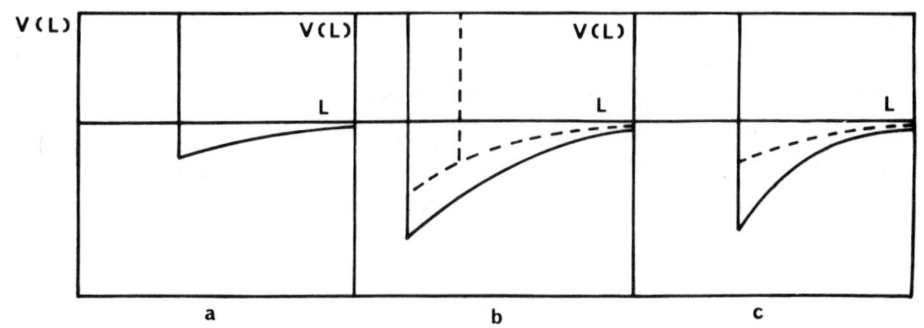

FIG. 14. Potential energy, $V(L)$, diagrams to describe the total interaction between two sterically stabilised particles as a function of separation distance, L. The repulsive potential is taken to be an infinitely steep barrier which prevents very close approach. (a) Stable dispersion; (b) flocculated dispersion, the instability being caused by (i) decrease in range of repulsive potential, (ii) collapse of polymer layer with increase in concentration and consequent increase in depth of the potential energy minimum; (c) flocculated dispersion, the instability being caused by an increase in attractive potential and the consequent increase in depth of the total energy minimum. For cases (b) and (c) the location of the original potential (case (a)) is shown by the dashed lines. (From reference 139, with permission Royal Society of Chemistry.)

particles interpenetrate there will be a repulsive effect arising from a fall in configurational entropy.[134,135,145] If the temperature is lowered to near the lower θ-temperature, the relative importance of segment–segment interactions and adsorption forces will be reasserted and the surface-attached poly(dimethylsiloxane) blocks will again collapse on the surface.

Potential energy diagrams[139] for the total interaction between two sterically stabilised particles as a function of separation distance, L, are obtained schematically in Fig. 14. Flocculation is envisaged as arising either from a decrease in the range of the steric repulsion and/or from an increase in the attractive potential. Everett and Stageman interpreted their results semi-quantitatively in terms of the temperature-dependence of the net attractive force between the colloidal particles and the change in adsorbed layer thickness as the θ-temperature of the stabilising polymer was approached. The effect of particle size was accounted for by the dependence of the attractive potential on particle size.

No detailed studies have been reported to date of the colloidal stability of copolymer micelles in the absence of stabilised homopolymer. It would be particularly interesting if such studies were carried out on spherical micelles under conditions which ensured that the micelle/free-chain equilibrium overwhelmingly favoured micelle formation. The factors requiring investigation are similar to those considered by Everett and Stageman for sterically stabilised particles. Spherical copolymer micelles in certain solvents can be expected to exhibit both upper and lower critical solution temperatures and to behave effectively as large compact macromolecules over a wide range of temperature. In special cases it might well be possible to pass from a lower to an upper flocculation temperature without the micelles dissociating at any stage to form free chains.

REFERENCES

1. McBain, J. W., *Trans. Faraday Soc.*, 1913, **9**, 99.
2. Hartley, G. S., *Aqueous Solutions of the Paraffin-chain Salts*, Hermann et Cie, Paris, 1936.
3. Hartley, G. S., *Kolloid-Z.*, 1939, **88**, 33.
4. Shinoda, K., Nakagawa, T., Tamamushi, B.-I. and Isemura, T., *Colloidal Surfactants*, Academic Press, New York, 1963.
5. Becker, P., *Non-ionic Surfactants*, ed. M. J. Schick, Marcel Dekker, New York, 1967.
6. Fisher, L. R. and Oakenfull, D. G., *Chem. Soc. Revs.*, 1977, **6**, 1.
7. Jones, M. N., *Biological Interfaces*, Elsevier Scientific Publishing, Amsterdam, 1975.
8. Tanford, C., *The Hydrophobic Effect*, Wiley-Interscience, New York, 1973.
9. Frank, H. S. and Evans, M. W., *J. Chem. Phys.*, 1945, **13**, 507.
10. Jones, E. R. and Bury, C. R., *Phil. Mag.*, 1927, **4**, 841.
11. Elias, H-G., *J. Macromol. Sci. Chem.*, *A*, 1973, **7**(3), 601.
12. Ben-Naim, A., *Hydrophobic Interactions*, Plenum Press, New York and London, 1980.
13. Elias, H-G., *Light Scattering from Polymer Solutions*, ed. M. B. Huglin, Academic Press, New York and London, 1972.
14. Hall, D. G. and Pethica, B. A. *Non-ionic surfactants*, ed. M. J. Schick, Marcel Dekker, New York, 1967.
15. Mittal, K. L. and Mukerjee, K. L., *Micellization, Solubilization and Microemulsions*, vol. 1, ed. K. L. Mittal, Plenum Press, New York and London, 1977.
16. Kegeles, G. and Rao, M. S. N., *J. Amer. Chem. Soc.*, 1958, **80**, 5721.
17. Nichol, I. W., Bethune, J. L., Kegeles, C. and Hess, E. L., *The Proteins*, 2nd edn, ed. H. Neurath Academic Press, New York and London, 1964.
18. Lässer, H. R. and Elias, H.-G., *Kolloid-Z.*, 1972, **250**, 46, 58.
19. Ray, A., *Nature*, 1971, **231**, 313.
20. Debye, P. and Prins, J., *Kolloid Sci.*, 1958, **13**, 86.
21. Kertes, A. S. and Gutman, H., *Surface and Colloid Sci.*, vol. 8, ed. E. Matijevic, Wiley, New York, 1976.
22. Kertes, A. S., *Micellization, Solubilization and Microemulsions*, vol. 1, ed. K. L. Mittal, Plenum Press, New York and London, 1977.
23. Steiner, R. F., *Arch. Biochem. Biophys.*, 1952, **39**, 333.
24. Rossotti, F. J. C. and Rossotti, H., *J. Phys. Chem.*, 1961, **65**, 926.

25. MERRETT, F. M., *Ric. Sci.*, 1955, **25**, 279.
26. MERRETT, F. M., *Trans. Faraday Soc.*, 1954, **50**, 759.
27. JONES, M. H., *Can. J. Chem.*, 1956, **34**, 948.
28. HARTLEY, F. D., *J. Polym. Sci.*, 1959, **34**, 397.
29. SCHLICK, S. and LEVY, M., *J. Phys. Chem.*, 1960, **64**, 883.
30. KRAUSE, S., *J. Phys. Chem.*, 1961, **65**, 1618.
31. BRESLER, S. YE., PYRKOV, L. M., FRENKEL, S. YA., LAIUS, L. A. and KLENIN, S. I., *Vysokomol. Soed.*, 1962, **4**, 250.
32. GALLOT, Y., FRANTA, E., REMPP, P. and BENOIT, H., *J. Polym. Sci., C*, 1964, **4**, 473.
33. DONDOS, A., REMPP, P. and BENOIT, H., *J. Chim. Phys.* 1963, **62**, 821.
34. KRAUSE, S., *J. Phys. Chem.*, 1964, **68**, 1948.
35. DONDOS, A., REMPP, P. and BENOIT, H., *J. Polym. Sci., B*, 1966, **4**, 293.
36. MOLAU, G. E. and WITTBRODT, W. M., *Macromolecules*, 1968, **1**, 260.
37. SAKURADA, I., IKADA, Y., UEHARA, H., NISHIZAKI, Y. and HORII, F., *Makromol. Chem.*, 1970, **139**, 183.
38. MOLAU, G. E., *Block Polymers*, ed. S. L. Aggarwal, Plenum Press, New York and London, 1970.
39. PRICE, C. and WOODS, D., *Eur. Polymer J.*, 1973, **9**, 827.
40. TUZAR, Z., PETRUS, V. and KRATOCHVÍL, P., *Makromol. Chem.*, 1974, **175**, 3181.
41. TUZAR, Z. and KRATOCHVÍL, P., *Advances in Colloid and Interface Science*, 1976, **6**, 201.
42. TANAKA, T., KOTAKA, T. and INAGAKI, H., *Bull. Inst. Chem. Res., Kyoto Univ.*, 1977, **55**, 206.
43. BOOTH, C., NAYLOR, DE V. T., PRICE, C., RAJAB, N. S. and STUBBERSFIELD, R. B., *J.C.S. Faraday, 1*, 1978, **74**, 2352.
44. SADRON, C., *Pure Appl. Chem.*, 1962, **4**, 347.
45. KRESHECK, G. C., HAMORI, E., DAVENPORT, G. and SCHERAGA, E. M., *J. Amer. Chem. Soc.*, 1966, **88**, 246.
46. NAKAGAWA, T., *Colloid Polym. Sci.*, 1974, **252**, 56.
47. PRICE, C., HUDD, A. L., BOOTH, C. and WRIGHT, B., *Polymer. Comm.*, 1982, **23**, 650.
48. ALTGELT, K. H., MOORE, J. C. and CANTOW, M. J. R., *Polymer Fractionation*, Academic Press, New York, 1967.
49. GRUBISIC, Z., REMPP, P. and BENOIT, H., *J. Polymer Sci., B*, 1967, **B5**, 753.
50. BOOTH, C., FORGET, J-L., GEORGII, I., LI, W. S. and PRICE, C., *Eur. Polymer J.*, 1980, **16**, 255.
51. BOOTH, C., FORGET, J-L., LALLY, T. P., NAYLOR, T. D. and PRICE, C., *Chromatography of Synthetic and Biological Polymers*, vol. 1, ed. R. Epton, Ellis Horwood, London, 1979.
52. TUNG, L. H., *Separation Sci.*, 1970, **5**, 339.
53. SOLC, K. and ELIAS, H.-G., *J. Polym. Sci., Polym. Phys. Ed.*, 1973, **11**, 137.
54. SIEGEL, B. M., JOHNSON, D. H. and MARK, H., *J. Polym. Sci., A*, 1950, **5**, 111.
55. HEYN, A. N. J., *J. Polym. Sci.*, 1959, **41**, 23.
56. QUAYLE, D. V., *Br. Polym. J.*, 1969, **1**, 15.
57. RICHARDSON, M. J., *Proc. Roy. Soc., A.*, 1964, **50**, 279.

58. Newman, S., *J. Appl. Polym. Sci.*, 1962, **6**, 515.
59. Price, C., Hudd, A. L. and Stubbersfield, R. B., *Polymer*, 1980, **21**, 9.
60. Horrii, F., Ikada, Y. and Sakurada, I., *J. Polym. Sci., Polym. Chem. Ed.*, 1974, **12**, 323.
61. Tuzar, A., Petrus, V. and Kratochvíl, P., *Makromol. Chem.*, 1974, **175**, 3181.
62. Fujita, H., *Foundations of Ultracentrifugal Analysis*, Wiley, New York, 1975.
63. Zernicke, F. and Prins, J., *Z. Phys.*, 1927, **41**, 184.
64. Debye, P. and Menke, H., *Ergebn. Tech. Rontgenk.*, 1931, **2**, 1.
65. Zimm, B. H., *J. Chem. Phys.*, 1948, **16**, 1093.
66. Stockmayer, W. H., Moore, L. D., Fixman, M. and Epstein, B. N., *J. Polym. Sci.*, 1955, **16**, 517.
67. Bushuk, W. and Benoit, H., *Canad. J. Chem.*, 1958, **36**, 1616.
68. Leng, M. and Benoit, H., *J. Chim. Phys.*, 1961, **58**, 480.
69. Tanaka, T., Kotaka, T. and Inagaki, H., *Polym. J.*, 1972, **3**, 327.
70. Tanaka, T., Kotaka, T. and Inagaki, H., *Polym. J.*, 1972, **3**, 338.
71. Tuzar, Z. and Kratochvíl, P., *Makromol. Chem.*, 1972, **160**, 301.
72. Price, C., McAdam, J. D. G., Lally, T. P. and Woods, D., *Polymer*, 1974, **15**, 228.
73. Utiyama, H., Takenaka, K., Mizumori, M., Fukuda, M., Tsunashima, Y. and Kurata, M., *Macromolecules*, 1974, **7**, 515.
74. Krause, S. and Reismiller, P., *J. Polym. Sci., A-2*, 1975, **13**, 663.
75. Price, C. and Woods, D., *Polymer*, 1974, **15**, 389.
76. Flory, P. J., *Principles of Polymer Chemistry*, chapter 12, Cornell University Press, New York, 1953.
77. Price, C., Hudd, A. L. and Wright, B., *Polymer Comm.*, 1982, **23**, 170.
78. Fournet, G., *C.R. Acad. Sci., Paris*, 1949, **228**, 1421.
79. Ree, F. H., Lee, Y-T. and Ree, T., *J. Chem. Phys.*, 1971, **55**, 234.
80. Kirkwood, J. G. and Boggs, E. M., *J. Chem. Phys.*, 1942, **10**, 294.
81. Pleštil, J. and Baldrian, J., *Makromol. Chem.*, 1973, **174**, 183.
82. Pleštil, J. and Baldrian, J., *Makromol. Chem.*, 1975, **176**, 1009.
83. Guinier, A., *Ann. Phys.*, 1939, **12**, 161.
84. Candau, F., Guenet, J-M., Boutillier, J. and Picot, C., *Polymer*, 1979, **20**, 1227.
85. Higgins, J. S., Dawkins, J. V. and Taylor, G., *Polymer*, 1980, **21**, 627.
86. Periard, J. and Riess, G., *Eur. Polym. J.*, 1973, **9**, 687.
87. Pecora, R., *J. Chem. Phys.*, 1964, **40**, 1604.
88. Averbukh, M. Z., Nikonorova, N. I., Rozinoer, Ya. S., Lushchikov, I. I., Shatalov, V. P., Gurari, M. L., Bakeev, N. F. and Kozlov, P. V., *Kolloidnyi Zh.*, 1976, **38**, 419.
89. Price, C., Canham, P. A., Duggleby, M. C., Naylor, T. D., Rajab, N. S. and Stubbersfield, R. B., *Polymer*, 1979, **20**, 615.
90. Mandema, W., Zeldenrust, H. and Emeis, C. A., *Makromol. Chem.*, 1979, **180**, 1521.
91. Einstein, A. *Investigations on the Theory of the Brownian Movement*, ed. R. Fürth and transl. A. D. Cowper, Dover Publications, New York, 1956.
92. Perrin, J., *Ann. Chim. Phys.*, 1909, **18**, 15; *Physik Z.*, 1911, **11**, 461.

93. LALLY, T. P. and PRICE, C., *Polymer*, 1974, **15**, 325.
94. CANHAM, P. A., LALLY, T. P., PRICE, C. and STUBBERSFIELD, R. B., *J.C.S. Faraday 1*, 1980, **76**, 1857.
95. DAINTON, F. S. and IVIN, K. J., *Quart. Rev.*, 1958, **12**, 61.
96. KRATOCHVÍL, P., in a lecture on *Micelle formation by block copolymers* presented at UMIST, Manchester, 1979.
97. INOUE, T., SOEN, T., HASHIMOTO, T. and KAWAI, H., *Block Polymers*, ed. S. L. Aggarwal, Plenum Press, New York and London, 1970.
98. MEIER, D. J., *J. Polym. Sci., Part C*, 1969, **26**, 81.
99. MEIER, D. J. *Block and Graft Copolymers*, ed. J. J. Burke, Syracuse University Press, Syracuse, 1973.
100. TUZAR, Z., SIKORA, A., PETRUS, V. and KRATOCHVÍL, P., *Makromol. Chem.*, 1977, **178**, 2743.
101. MANDEMA, W., EMEIS, C. A. and ZELDENRUST, H., *Makromol. Chem.*, 1979, **180**, 2163.
102. DEBYE, P. and ANACKER, E. W., *J. Phys. and Colloid Chem.* 1951, **55**, 644.
103. ATTWOOD, D., *J. Phys. Chem.*, 1968, **72**, 339.
104. STIGTER, D., *J. Phys. Chem.*, 1966, **70**, 1323.
105. ANACKER, E. W. *Cationic Surfactants*, ed. E. Jungermann, Marcel Dekker, New York, 1970.
106. EINSTEIN, A., *Ann. Phys.*, 1906, **19**, 271.
107. MANLEY, R. ST. J. and MASON, S. G., *Can. J. Chem.*, 1964, **32**, 763.
108. BATCHELOR, G. K. and GREEN, J. T., *J. Fluid Mech.*, 1972, **56**, 401.
109. STAUDINGER, H. *Die hochmolekularen organischen Verbindungen*, Springer, Berlin, 1932.
110. DOTY, P. M., WAGNER, H. and SINGER, S., *J. Colloid Chem.*, 1947, **51**, 32.
111. ELIAS, H.-G., *Makromol. Chem.*, 1967, **103**, 214.
112. PRICE, C. and WOODS, D., *Polymer*, 1973, **14**, 82.
113. CLIMIE, I. E. and WHITE, E. F. T., *J. Polymer Sci.*, 1960, **47**, 149.
114. DEMIN, A. A., RUDKOVSKAYA, G. D., DMITRENKO, L. V., OVSYANNIKOVA, L. A., SOKOLOVA, T. A., SAMSONOV, G. V. and NIKONOVA, I. N., *Vysokomol. Soedin.*, 1974, **16**, 2706.
115. WATTERSON, J. G., LÄSSER, H.-R. and ELIAS, H.-G., *Kolloid-Z.Z. Polymere*, 1972, **250**, 64.
116. WATTERSON, J. G. and ELIAS, H.-G., *Makromol. Chem.*, 1972, **157**, 237.
117. HUGGINS, M. L., *J. Amer. Chem. Soc.*, 1942, **64**, 2716.
118. SMALLHEER, C. V. and KENNEDY SMITH, R., *Lubricant Additives*, Lezius-Hiles, Cleveland, Ohio, 1967.
119. STREETS, N. L., YOCHUM, K. H. and MITACEK, B., SAE, Technical Paper 760267, 1976.
120. ECHERT, R. and ALBRECHT, J., German Patent 2 156 122, 1977.
121. ST. CLAIR, D. J. and CROSSLAND, R. K., US Patent 3 965 019, 1976.
122. ANDERSON, W. S., US Patent 4 036 910, 1977.
123. SHINIZKY, M., DIANOUX, A. C., GITLER, C. and WEBER, G., *Biochemistry* 1971, **10**, 2106.
124. HAMILTON, C. L. and MCCONNELL, H. M., *Structural Chemistry and Molecular Biology*, eds A. Rich and N. Davidson, Freeman and Co., San Francisco, 1968.

125. WAGGONER, A. S., GRIFFITH, O. H. and CHRISTENSON, C. R., *Proc. Nat. Acad. Sci. USA*, 1967, **57**, 1198.
126. HEATLEY, F. and BEGUM, A., *Makromol. Chem.*, 1977, **178**, 1205.
127. FRANTA, E., *J. Chim. Phys.*, 1966, **63**, 595.
128. CERESA, R. J. *Block and Graft Copolymers*, Butterworths, London, 1962.
129. TUZAR, Z. and KRATOCHVÍL, P., *Makromol. Chem.*, 1973, **170**, 177.
130. IKADA, Y., HORRII, F. and SAKURADA, I., *J. Polym. Sci., Polym. Chem.*, 1973, **11**, 27.
131. BLACKLEY, D. C. *Emulsion Polymerisation*, Applied Science Publishers, London, 1975.
132. BARRETT, K. E. J. (Ed.), *Dispersion Polymerization in Organic Media*, Wiley, London, New York, Sydney and Toronto, 1975.
133. BARRETT, K. E. J., *Br. Polym. J.*, 1973, **5**, 259.
134. EVANS, R. and NAPPER, D. H., *Kolloid-Z. u.Z.Polymere*, 1973, **251**, 329 and 409.
135. DOLAN, A. K. and EDWARDS, S. F., *Proc. Roy. Soc., A*, 1974, **337**, 509; 1975, **343**, 427.
136. BARRETT, K. E. J. and THOMAS, H. R., *J. Polym. Sci., A-1*, 1969, **7**, 2621.
137. WAITE, F. A., paper presented to meeting of the *Colloid and Interface Science Group, Faraday Division of the Chemical Society*, Bristol, June, 1974.
138. EVERETT, D. H. and STAGEMAN, J. F., *Colloid Polym. Sci.*, 1977, **255**, 293.
139. EVERETT, D. H. and STAGEMAN, J. F., *Disc. Faraday Soc.*, 1978, **65**, 230.
140. PATTERSON, D., DELMAS, G. and SOMCYNSKY, T., *Polymer*, 1967, **8**, 503.
141. ZEMAN, L., BIROS, J., DELMAS, G. and PATTERSON, D., *J. Phys. Chem.*, 1972, **76**, 1206.
142. NAPPER, D. H., *J. Colloid Interface Sci.*, 1977, **58**, 390.
143. FLORY, P. J., *J. Amer. Chem. Soc.*, 1965, **87**, 1833.
144. FLORY, P. J., *Disc. Faraday Soc.*, 1970, **49**, 7.
145. OSMOND, D. W. J., VINCENT, B. and WAITE, F. A., *Colloid Polym. Sci.*, 1975, **253**, 676.

Chapter 3

MESOMORPHIC PROPERTIES OF BLOCK COPOLYMERS

A. E. Skoulios

Centre de Recherches sur les Macromolécules, Strasbourg, France

SUMMARY

This chapter is concerned with the supramolecular organisation of phase-separated block copolymers that display mesomorphic (liquid-crystalline) order.

The structures of the lamellar, cylindrical and spherical morphologies are described, together with the principles used in identifying the types and deriving their structural parameters from small angle X-ray scattering and electron diffraction data.

The occurrence of mesophases of different types is discussed for a variety of block copolymers in relation to composition, molecular weight, the presence or absence of solvents and the effects of temperature.

1. INTRODUCTION

It has long been known that block copolymers are capable of producing well organised mesomorphic phases. This was found immediately after one of them, namely a polystyrene-*b*-poly(oxyethylene), had been synthesised under good experimental conditions by anionic polymerisation.[1] The solubility properties of this polymer were rather unusual, being surprisingly similar to those exhibited by soaps. The polymer was soluble both in water and in non-polar organic solvents, and furthermore, it was endowed with strong emulsifying power.[2] As, at that time, the lyotropic mesophases of soap[3,4] were being actively investigated by small angle X-ray diffraction,[5] it did not take long to establish with the same experimental technique that block copolymers are also mesomorphic in character and that they show the same type of structures.[2]

The typical properties of soaps are basically due to their amphiphilic character. Their molecules are generally made up of two distinct parts: a long flexible paraffin chain of some ten to twenty carbon atoms, carrying at one end a polar, hydrophilic, frequently ionic, end group. These parts are so dissimilar in chemical nature, and so unlike with regard to their interactions, that they are not soluble in one another; instead, they segregate themselves into separate microdomains, alternatively juxtaposed in space. However, because of the covalent bond linking them together within one single molecule, these parts cannot lead to a true macroscopic phase separation; instead they can cause the formation of interfaces on a molecular scale and force the molecules to stand on well-defined surfaces. Now, it is only natural that block copolymers behave in a similar way; as far as the chemical architecture of the molecules is concerned, block copolymers indeed greatly resemble soaps and amphiphilic compounds, and their blocks are also mutually incompatible.

Intramolecular segregation is certainly the most characteristic, and also the most interesting property of block copolymers. It is responsible for their specific behaviour and their usefulness in the field of technological applications. It leads, as a matter of fact, to the formation of multiphase polymeric materials which can be considered as microcomposites[6] and whose morphology can be controlled to a large extent.

Commonly, segregation occurs in parallel with another characteristic property, namely the tendency of block copolymers to become organised. Depending on the particular chemical and physical experimental conditions, the segregation microdomains may very well be simply interdispersed in a random fashion, so as to form a locally heterogeneous, but otherwise isotropic medium whose morphology would then resemble that of a polymer-in-polymer emulsion. But generally, the microdomains are regularly and periodically distributed in space, producing the same type of mesophases as those presented by soaps and other related amphiphiles.

Since their discovery in 1960, the mesomorphic phases of block copolymers have been studied by many workers. In a first stage of development, the investigations were aiming merely at establishing the generality of the organisation phenomena observed and at describing the main structures encountered. They were carried out primarily with the help of X-ray diffraction and electron microscopy, using as many and varied polymers as possible, both in the presence of preferential solvents for one of the blocks and in the absence of any diluent or swelling agent.[7-10] It was only several years later that the mechanical and technological properties were also considered,[6,7] and that theoretical thermodynamic studies (see Chapter 4) were undertaken to answer questions related to the dimensions of the microdomains, their dependence upon composition or molecular weight and the exact conditions for the microphase separation.

In this chapter, we shall be concerned only with those aspects of the behaviour of block copolymers which are connected with the mesomorphic character. We shall describe the main structures observed and define their geometrical parameters as a function of various factors. We shall also provide an appreciation of the exact conditions necessary for the occurrence of the mesophases.

2. DESCRIPTION OF THE LIQUID-CRYSTALLINE STRUCTURES

Three types of structure have so far been found to characterise the liquid-crystalline organisation of block copolymers: the lamellar, the cylindrical and the spherical structures. For the sake of clarity, these will first be described from a geometrical point of view, reserving the discussion of their properties for the following sections of the chapter.

Undoubtedly, the lamellar structure is altogether the simplest and the most commonly encountered structure of block copolymers. It is formed of a set of two kinds of layers periodically and alternately piled up in space (Fig. 1). One of the layers contains the blocks A of the copolymer, whilst the

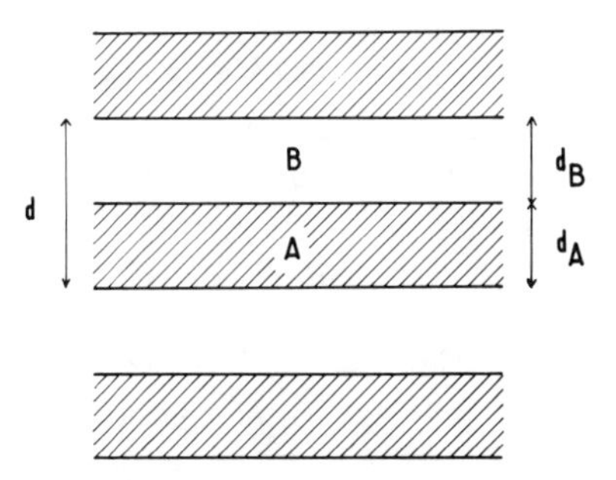

FIG. 1. Schematic representation of the lamellar structure.

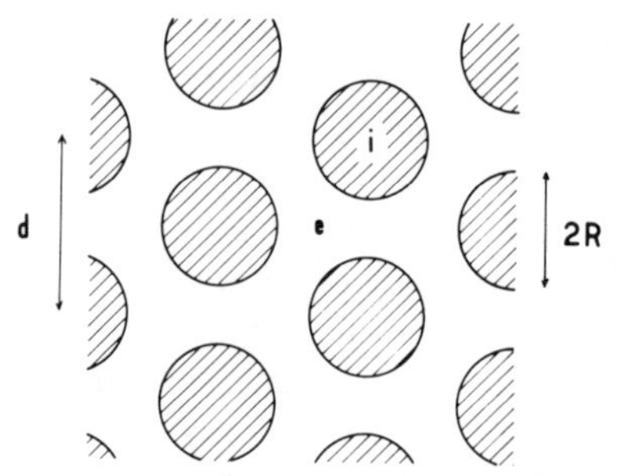

FIG. 2. Schematic representation of the cylindrical structure.

other contains the blocks B. When a solvent is present in the system, it is located according to its affinity with regard to the blocks either in sub-layers A or in sub-layers B;[11] but generally, it is distributed between both sub-layers with a well-defined partition coefficient.[12] From a crystallographic standpoint,[13] this structure corresponds to a one-dimensional crystal lattice formed of an array of equidistant points lying on a straight line.

The cylindrical structure consists of a set of indefinitely long cylindrical microdomains containing one of the blocks, oriented parallel to one another and packed according to a two-dimensional hexagonal lattice; the cylinders are embedded in a continuous matrix formed of the other blocks[11] (Fig. 2). In the same way as for the lamellar structure, any solvent present in the system is assumed to be distributed between the cylinders and the continuous matrix according to its affinity towards the blocks. By analogy with soaps,[4] it is sometimes stated that in the *regular* cylindrical structure the solvent is located predominantly in the continuous matrix, and conversely, that in the *inversed* cylindrical structure the solvent is concentrated inside the cylinders.[9,14] Quite obviously, however, this is merely a matter of convenience.

The last type of structure observed is formed of spherical microdomains containing one of the blocks, embedded in a continuous matrix formed of the other blocks. The spheres can now be assembled according to one of the three possible Bravais cubic lattices:[13] the simple, the body-centred and the face-centred cubic lattices (Fig. 3). Among these, only the last two lattices have so far been established beyond any doubt. The face-centred cubic lattice was observed with Tetronic 904 which is a low M.W. poly(oxyethylene)-*b*-poly(oxypropylene),[15] and the body-centred cubic lattice was observed with Kraton 1107 which is a commercially available polystyrene-*b*-polyisoprene.[16]

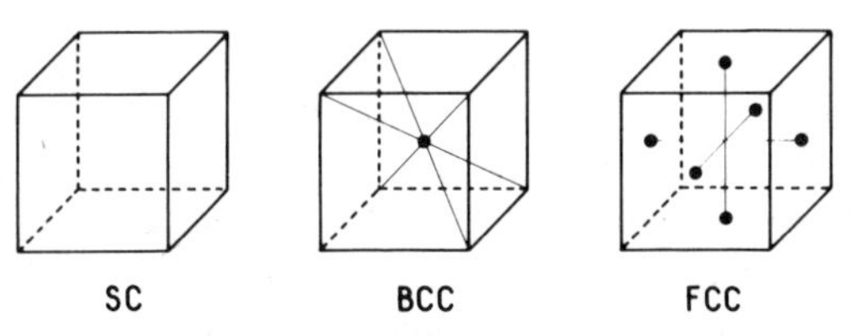

FIG. 3. The three possible cubic arrangements of spheres: simple cubic (SC), body-centred cubic (BCC) and face-centred cubic (FCC).

3. PHYSICAL STATE OF THE BLOCKS

Regardless of whether the structure is in fact lamellar, cylindrical or spherical, block copolymers in the mesomorphic state are all characterised by a common feature: the segregation of the blocks within distinct microdomains, and the localisation of the junction points of the blocks at the interfaces (Fig. 4). The size of the microdomains is not far from the size of the polymer coils themselves; it is thus very large compared to the size of the monomer units and the polymer statistical segments. As a result, the blocks behave quite freely and prove capable of preserving their own properties to a large extent. Depending upon their chemical nature, they can indeed either crystallise,[11,17] or else remain in the amorphous state, acting as a glass or a mobile fluid.[18] They can also swell differently in solvents presenting preferential or specific interactions towards them.[2,12] The rapid expansion of research in the field of block copolymers is clearly due to the free behaviour of the blocks, which lets one imagine a variety of possible applications in manufacturing new engineering materials with enhanced physical and technological properties.[19]

To illustrate this liberty of the blocks, let us recall the case of the first

FIG. 4. Schematic representation of the blocks at the interfaces.

copolymer tested for mesomorphic behaviour, namely polystyrene-*b*-poly(oxyethylene).[11,17] At room temperature and in the absence of solvents, the POE blocks pass into the crystalline state, irrespective of their attachment of the amorphous PSt blocks. The structure of the system is lamellar and corresponds to an alternating pile of crystalline and amorphous layers; in that respect, it is comparable to what is agreed as being a B-smectic liquid-crystal.[20] The POE layers of the lamellar copolymer are similar to those commonly observed with crystalline homopolymers.[2] The chains are oriented perpendicularly to the lamellae and folded back and forth a number of times; on annealing, the thickness of the layers increases and, correspondingly, the number of folds decreases; finally, the degree of crystallinity of the POE blocks (about 80 %) is comparable to that of the related *homo*poly(oxyethylene) crystallised under similar experimental conditions.

In this connection, the case of poly(oxyethylene)-*b*-poly(ε-caprolactone)[22] is also very instructive. Here, the blocks are both structurally regular and, therefore, able to crystallise. Without going into details, let us simply state that those blocks which crystallise first are found to lock the system in the lamellar structure that fits them best and to impose upon the whole their own folding characteristics; correspondingly, those blocks which crystallise last are obliged to make the best of the structure already formed and hence they crystallise rather poorly.

To illustrate the behaviour of block copolymers of which the blocks are unable to crystallise, it is worth citing the exemplary case of polystyrene-*b*-polyisoprene.[18,23] With this polymer indeed, the segregation of the blocks results in the existence of two distinct, readily detectable glass transition temperatures, one characteristic of the PIp, the other of the PSt blocks. To be precise, it should be added that when there is appreciable mixing of the species, there is also a third, intermediate transition. This occurs when the molecular weight of the copolymer is sufficiently low that the thickness of the interfacial regions is no longer small compared to the size of the segregation microdomains.

4. EXPERIMENTAL EVIDENCE OF THE MESOMORPHIC ORDER

The mesomorphic character of block copolymers and their structural resemblance to soaps was first established by small-angle X-ray diffraction (SAXS).[2] The periodic arrangement in space of the segregation micro-

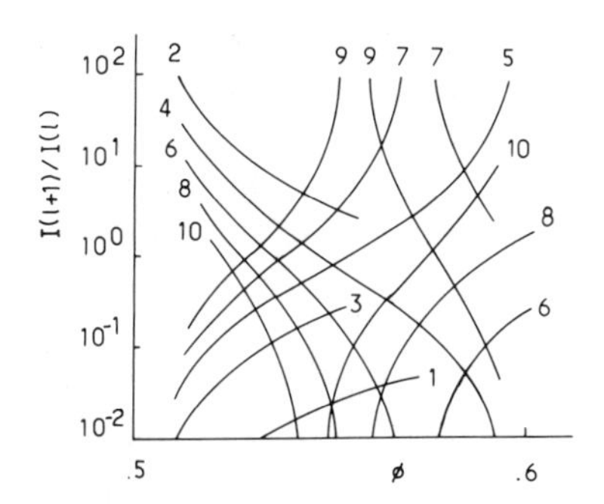

FIG. 5. Variation of the relative intensity of the Bragg reflections (taken two-by-two) as a function of the volume fraction of one of the blocks in a lamellar copolymer.

domains gives rise to well-defined, and often very sharp, Bragg reflections. The angular position and the intensity of these reflections are related to the crystallographic structure of each sample considered.

The sharpness of the diffraction lines provides valuable information about the perfection of the crystal lattices involved and their coherent extension in space.[13] Usually, the ordering of block copolymers is found to be well developed and to extend over quite a number of unit cells. But sometimes, especially when the sample is not allowed to reach equilibrium, the organisation turns out to be less than perfect. This is because the structural units, that is the lamellae, the cylinders or the spheres, have not had enough time during their formation all to attain exactly the same shape and size. As a result, there are fluctuations of the crystal lattice both on a local and on a macroscopic scale.

The number of Bragg reflections observed is also important (Fig. 5). Usually, the X-ray patterns of block copolymers contain no more than a few diffraction lines; occasionally, however, they may contain up to fifteen lines, indicating a high quality of ordering, along with a great sharpness of interfaces. The structures can then be described precisely and commented upon in terms of classical crystal lattices.

As mentioned above, three types of structure have been proposed so far to describe the mesomorphic ordering of block copolymers: the lamellar, the cylindrical and the spherical type of structure. The first corresponds to a one-dimensional crystal lattice characterised by SAXS patterns which contain a series of equidistant diffraction lines. The reciprocal Bragg spacings ($2 \sin \theta/\lambda$; 2θ being the Bragg diffraction angle and λ the wavelength of the X-radiation) of these lines are in the ratio 1:2:3:4:5....

The reciprocal spacing of the first order of diffraction (Bragg reflection 001) is equal to $1/d$ (see Fig. 1). The cylindrical type of structure corresponds to a two-dimensional hexagonal crystal lattice; it is characterised by X-ray patterns which contain a series of diffraction lines with reciprocal spacings in the ratio $1:\sqrt{3}:\sqrt{4}:\sqrt{7}:\sqrt{9}$....

The reciprocal spacing of the first order of diffraction (Bragg reflection 100) is equal to $2/\sqrt{3}d$ (see Fig. 2). The spherical type of structure corresponds to one of the three possible three-dimensional Bravais cubic lattices.[13] The face-centred cubic lattice is easy to identify; it is characterised by X-ray patterns which contain a set of reflections with reciprocal spacings in the ratio $\sqrt{3}:\sqrt{4}:\sqrt{8}:\sqrt{11}:\sqrt{12}$....

This is definitely not the case with the other two cubic lattices. The corresponding X-ray patterns are usually far too poor;[24] they contain only a few reflections which, of course, cannot help decide whether the reciprocal spacings are in the ratio $1:\sqrt{2}:\sqrt{3}:\sqrt{4}:\sqrt{5}:\sqrt{6}:\sqrt{8}$... as required for a simple cubic lattice, or in the ratio $1:\sqrt{2}:\sqrt{3}:\sqrt{4}:\sqrt{5}:\sqrt{6}:\sqrt{7}:\sqrt{8}$... as necessary for a body-centred cubic lattice. However, save for this ambiguous case, small-angle X-ray diffraction provides much useful information about the symmetry of the system and about the way in which the structural elements are packed together. It also provides quantitative information about the distances among the structural elements. The reciprocal spacing of the first diffraction line (Bragg reflection 100 or 110 or still 111) is equal to $1/a$ or $\sqrt{2}/a$ or $\sqrt{3}/a$ depending on whether the unit cell is simple or body-centred or face-centred (see Fig. 3).

First identified with the help of SAXS, all these structures were fully confirmed by electron microscopy.[25-31] It should be added, however, that this confirmation was achieved only with copolymers of which one of the block types could be visualised by staining. Such was the case notably with styrene/diene block copolymers for which staining is currently achieved by a technique due to Kato:[32] the sample is exposed to OsO_4 vapour which reacts only with the double bonds of the diene blocks; the diene microdomains can thus be stained alone and gain sufficient contrast with respect to the styrene microdomains. Another requirement for transmission electron microscopy to be used profitably is that the sample should be sufficiently thin: its thickness must actually be of the order of only a few unit cells, that is of about 10^2–10^3 Å. Sectioning is effected using ultramicrotomy, either from the solvent-cast films or from bulk material (see reference 7).

Although conclusions drawn from single observations with electron microscopy are generally unreliable, the systematic examination of carefully prepared samples of block copolymers did actually allow an accurate checking of the structures found by X-ray diffraction. Such is the case particularly with the lamellar and the cylindrical structures. As the exact sectioning direction of the sample with respect to its structural features cannot be guaranteed with the usual non-oriented specimens, selections have to be used from a large number of micrographs to illustrate the maximum of morphological configurations. The lamellar structure appears typically as a set of black-and-white, parallel and equidistant streaks. The cylindrical structure, for its part, will usually appear as a collection of white (or black), circular, hexagonally-packed dots on a black (or white) background; but it can also appear as a set of parallel streaks, if the samples happen to be cut parallel to the axis of the cylinders (see, for example, Fig. 1 of Chapter 1).

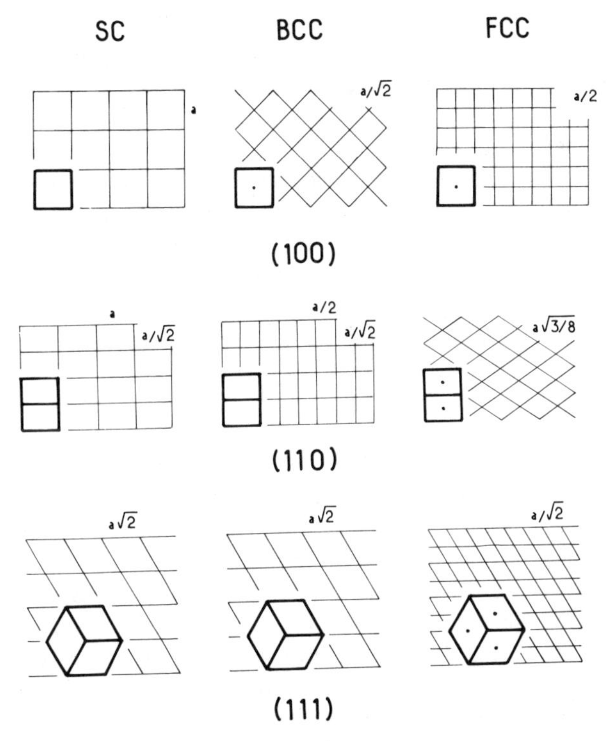

FIG. 6. Simple, body-centred and face-centred cubic arrangements of spheres viewed along the (100), (110) and (111) directions.

It has also been claimed that spherical structures have been fully identified by electron microscopy.[8] However, the actual situation is not as straightforward as it appears at first sight. To distinguish properly the three possible cubic lattices by electron microscopy, one has in fact to proceed to an accurate and detailed quantitative analysis of many micrographs viewed along different directions with respect to the cubic unit cell (Fig. 6). Special attention must be paid, not only to comparing the experimental and the calculated values of such parameters as the ratio between the diameter of the spots and their nearest distance apart,[16] but also to comparing the dimensions of the square lattice with those of the hexagonal one, viewed in the same specimen along the (100) and (111) directions of the cubic system.

5. QUANTITATIVE CHARACTERISATION

To characterise quantitatively the liquid-crystalline structure of block copolymers, one can be satisfied with specifying the value of only those geometrical parameters which are given directly by SAXS. These are, successively, the total thickness, d, of the layers in the case of the lamellar structure (Fig. 1), the distance, d, between the axes of two neighbouring cylinders in the case of the cylindrical structure (Fig. 2), and the edge, a, of the cubic cell in the case of the spherical structure (Fig. 3). However, by simply assuming that the molar volumes of the constituents are strictly additive—and this is probably a reasonable assumption—it is possible to calculate some further geometrical parameters which of course can prove very useful in describing the structure in more detail, and in discussing aspects of the chain conformation.

In addition to the overall thickness, d, of the layers in the case of the lamellar structure, it is thus of interest to calculate the respective thicknesses, d_A and d_B, of the sub-layers containing blocks A and blocks B (Fig. 1), and also the molecular area, S, which represents the average area occupied at the interface by every covalent junction point between the blocks. To perform this calculation, one has to know the value of several parameters: the weight content, X_A, of the copolymer in blocks A; the number-average molecular weight, M_A, of blocks A; the specific volumes v_A, v_B and v_S of blocks A and B, and of the solvent that may be present in the system; the weight fraction, $1-c$, of the solvent in the mixture; its partition coefficient p_A, that is the weight fraction of solvent incorporated preferentially in sub-layers A; and, finally, the number, ν, of blocks B which are covalently linked to each block A. From simple geometrical considerations it is, indeed, clear that one can write:[17,33]

$$d_A = d \Bigg/ \left[1 + \frac{c(1-X_A)v_B + (1-c)(1-p_A)v_S}{cX_A v_A + (1-c)p_A v_S} \right]$$

$$d_B = d - d_A$$

$$S = \frac{2M_A v_A}{\mathfrak{N} d\nu} \left[1 + \frac{1-X_A}{X_A}\frac{v_B}{v_A} + \frac{1-c}{c}\frac{v_S}{v_A}\frac{1}{X_A} \right]$$

where $\mathfrak{N}$ is Avogadro's number ($\mathfrak{N} = 6{\cdot}02 \times 10^{23}$).

Likewise, in the case of the cylindrical structure it is of interest to calculate the radius, R, of the cylinders (Fig. 2) and the molecular area, S:[17,33]

$$R = d \sqrt{(\sqrt{3}/2\pi) \Bigg/ \left[1 + \frac{c(1-X_i)v_e + (1-c)(1-p_i)v_S}{cX_i v_i + (1-c)p_i v_S} \right]}$$

$$S = \frac{2M_i v_i}{\mathfrak{N} d\nu} \sqrt{(2\pi/\sqrt{3})\left(1 + \frac{1-X_i}{X_i}\frac{v_e}{v_i} + \frac{1-c}{c}\frac{v_S}{v_i}\frac{1}{X_i} \right)\left(1 + \frac{1-c}{c}\frac{v_S}{v_i}\frac{p_i}{X_i} \right)}$$

In these expressions, subscripts i and e designate, respectively, the *internal* and the *external* blocks, that is those blocks which are located inside and outside the cylinders. As for ν, it represents here the number of *external* blocks attached to each *internal* one. Now, the problem is to determine which block is located inside the cylinders, and which goes into the continuous matrix. Obviously, the direct proof of the correct location of the blocks can only be given by electron microscopy, provided of course that this technique can be applied under the circumstances and that the two types of block can be properly differentiated from each other.[31] In certain favourable cases, however, X-ray diffraction can also be used to clarify the situation. Indeed, close inspection of the intensity of the diffraction lines—at least, when the X-ray patterns contain many reflections—should give information about the distribution of the electron density within the specimen and, consequently, about the value of the size of the cylinders. On comparing this value with that calculated from the equation above, using successively the two possible assumptions (blocks A either inside or outside the cylinders), one can decide without difficulty which assumption is correct. Similarly it is clear that a detailed analysis of the type of structure occurring and of the variation of its parameters as a function of the chemical composition, the solvent content of the system and the M.W. of the blocks, should also contribute to solving the problem.[14]

Finally, regarding the spherical structure, it is of interest to calculate, in addition to the edge, a, of the cubic cell which is given directly by X-rays, the radius, R, of the spheres and the molecular area, S, of the polymer chains:[15,24]

$$R = a\sqrt[3]{(3/4\pi n)\Big/\left[1 + \frac{c(1 - X_i)v_e + (1 - c)(1 - p_i)v_S}{cX_iv_i + (1 - c)p_iv_S}\right]}$$

$$S = \frac{3M_i}{\mathfrak{N}av}\left(v_i + \frac{1 - c}{c}v_S\frac{p_i}{X_i}\right)\sqrt[3]{\frac{4\pi}{3}\left(1 + \frac{c(1 - X_i)v_e + (1 - c)(1 - p_i)v_S}{cX_iv_i + (1 - c)p_iv_s}\right)}$$

In these expressions, n is the number of spheres contained in each cubic cell, that is 1, 2 or 4 depending on whether the cubic lattice is simple, body-centred, or face-centred.[13]

In addition to the information on the size of the unit cell (that is on the thickness of the lamellae) on the distance between the cylinders, and on the edge of the cubic cell (which is related only to the angular position of the Bragg reflections) X-rays can give further useful information from the distribution of intensities of the diffraction lines. This is particularly the case with the lamellar structure when the diffraction patterns are sufficiently rich. Without entering into undue details of X-ray crystallographic techniques, let us say that, for a lamellar system in which ϕ is the volume fraction of blocks A and X the width of the electron density profile across the interface between sublayers,* the intensity of the $\mathbf{l}^{\text{th}}$ order of diffraction is given to a first approximation by:[34,35]

$$\mathrm{I}(\mathbf{l}) \propto \left[\left(\frac{\sin \mathbf{l}\pi\phi}{\mathbf{l}\pi\phi}\right)\left(\frac{\sin \mathbf{l}\pi\phi(X/d_A)}{\mathbf{l}\pi\phi(X/d_A)}\right)\right]^2$$

From the knowledge of how $\mathrm{I}(\mathbf{l})$ varies with $\mathbf{l}$, it is possible to get an estimate of ϕ or X.

From the practical standpoint, it is often convenient to avoid lengthy experiments, such as measuring the intensity of each diffraction line, and tedious calculations, such as correcting the intensities for Lorentz factor, polarisation, slit collimation, etc.[13] In the determination of the value of parameter ϕ, it is also attractive to avoid taking into account the diffuseness of the interfaces. To do so, one needs only to take advantage of the extremely steep dependence of $\mathrm{I}(\mathbf{I})$ on $\mathbf{l}$ and ϕ, especially when ϕ is about $1/2$

* Actually, depending on the degree of incompatibility of the two types of block, the interfaces between the sub-layers A and B are never infinitely sharp. The electron density varies, perhaps rapidly, but always continuously from its constant value in the pure A layers to its value in the pure B layers.

and $\mathbf{l}$ rather large. Indeed, minute changes in the values of ϕ can affect dramatically the relative intensities of the reflections. Actually, one only has to compare visually the intensities of the diffraction lines taken successively two by two, and to state whether these are of the same or of different orders of magnitude. For example, if the 6th, 7th and 8th orders of diffraction are of comparable intensities, $I(6) \simeq I(7) \simeq I(8)$, then ϕ has a value of approximately 0·465. This method of evaluating ϕ, and hence d_A/d, has been applied successfully in the case of styrene/isoprene block copolymers. The partition coefficient of a set of solvents was thus studied as a function of their affinities towards the PSt and the PIp domains[12] and the degree of solubilisation of PSt *homo*polymers was analysed as a function of the molecular weight.[36]

In an attempt to estimate the diffuseness of the interfaces, a suitably selected lamellar styrene/isoprene block copolymer was analysed recently, using the same crystallographic technique.[34] In that particular sample, the PSt and PIp sub-layers were of exactly the same thickness ($d_A = d_B \simeq 450$ Å; $\phi = 1/2$). As a result, the only diffraction lines which could be observed were those of odd $\mathbf{l}$, and their intensity was a monotonically decreasing function of $\mathbf{l}$. A simple visual inspection of the X-ray diagrams showed then that the diffraction lines could be seen up to the fifteenth order ($\mathbf{l} = 15$), indicating that $\mathbf{l}\pi\phi(X/d_A) < \pi$ and $X < 50$ Å. A more elaborate analysis of the X-ray intensity distribution of styrene/isoprene block copolymers[37] has just brought the proof that X is actually of only 20 Å, this value being almost independent of the M.W.

6. OCCURRENCE OF THE MESOPHASES

At this point, one is entitled to ask the following questions: Can a given block copolymer adopt all the three possible types of structure that have been described so far? And if it can, what are the exact conditions that must be fulfilled in order for a given structure to occur? The answer is affirmative; in fact, everything depends on the relative lengths of the blocks, on the solvent content of the system, on the overall M.W. and on temperature.

Let us first consider the role of the relative lengths of the blocks. As has been observed repeatedly with styrene/diene block copolymers, the structure of the mesophases depends very clearly upon the volume fraction of the blocks.[38] By covering a wide range of compositions, the following sequence of structures was established: for volume fractions of polystyrene smaller than 15%, $\phi_{PSt} < 0{\cdot}15$, the structure corresponds to PSt spheres embedded in a polydiene continuous matrix; for $0{\cdot}15 < \phi_{PSt} < 0{\cdot}40$ the structure corresponds to PSt cylinders in a polydiene matrix; for $0{\cdot}40 < \phi_{PSt} < 0{\cdot}60$, the structure is lamellar; for $0{\cdot}60 < \phi_{PSt} < 0{\cdot}85$, the structure is again cylindrical, but *inversed*, with a PSt matrix; and finally, for $\phi_{PSt} > 0{\cdot}85$, the *inversed* spherical structure corresponds to polydiene spheres embedded in a continuous polystyrene matrix.

The same sequence of structures was also established as a function of the solvent content of the system. Thus, the structural behaviour of a homologous series of poly(methyl methacrylate)/poly(hexyl methacrylate) diblock copolymers with a constant PHMA block and PMMA blocks of varied lengths was studied in the presence of acetonitrile, used as preferential solvent for the PMMA blocks.[39] The conclusion drawn was that, depending upon the relative lengths of the blocks, the dry system actually displays the whole range of structures, namely the *inversed* cylindrical, the lamellar, the *regular* cylindrical and, finally, the *regular* spherical structure. It was also shown that, in the swollen state, a given member of the homologous series is often capable of exhibiting more than one kind of structure depending on its solvent content.

Such a behaviour is easy to understand. For the lamellar structure, indeed, the interface between the segregation microdomains is planar and the space offered to the blocks on either side of the interface is of comparable importance; it is only logical, therefore, that the lamellar structure should occur when the two blocks are of roughly the same volume. In the same way, it is also clear why the structure should be cylindrical, or even spherical, when one of the blocks is more bulky than the other; the space offered on the convex side of the interface is larger than the corresponding space on the concave side. The presence of a solvent should simply result in an increase in the apparent volume of the soluble blocks. The occurrence of one or the other structure as a function of the relative lengths of the blocks can not only be analysed intuitively, but also discussed more deeply in terms of thermodynamics.[40]

Let us now consider the role of the M.W. It seems reasonable to expect the establishment of mesomorphic order only with block copolymers of sufficiently high M.W. and this has been confirmed in practice. Indeed, a systematic small-angle X-ray diffraction and transmission electron microscopy study was performed recently[41] with a set of lamellar styrene/isoprene block copolymers containing 50 wt % of one of the monomers. The conclusion drawn was that, below a M.W. of about 2×10^4 for the diblock and of 4×10^4 for the triblock copolymers, the system is in the disordered state; the structure corresponds to a simple interdispersion of modular PSt and PIp microdomains. Above the same M.W. threshold the system undergoes a sharp transformation; its structure becomes lamellar and develops over quite a number of layers: suddenly, the corresponding X-ray diffraction patterns show many Bragg reflections which are very sharp. The transformation as a function of M.W. is so striking that one is tempted to speak of a true phase transition. This is easy to understand both intuitively (the segregation of the blocks and, hopefully, the ordering of the microdomains thereupon, should occur when the blocks are sufficiently large) and theoretically, on thermodynamic grounds.[40]

The same type of sudden order–disorder transformation also occurs as a function of temperature. The melting of a set of styrene/isoprene block copolymers, that is, the transition from the mesomorphic to the disordered state, was studied very recently[41] by SAXS. The conclusion drawn was that the disordering of the lamellar structure on heating goes through an irreversible stage where the thickness of the lamellae increases considerably as a function of time. It turned out that the thickening effect is the faster, the higher the temperature and the lower the M.W. Quite surprisingly, however, and contrary to any classical thermodynamic expectation, it was finally found that the *melting* temperature (around 180 °C) does not depend significantly on M.W., nor on the type of structure of the mesophase.

7. VARIATION OF THE STRUCTURAL PARAMETERS

As mentioned in Section 5, SAXS gives a means of measuring the thickness of the layers, the diameter of the cylinders or the spheres, the molecular area at the interfaces, and therefore, of characterising quantitatively the mesomorphic structure of block copolymers. Now the actual value of the structural parameters depends upon many factors and it is of interest to consider here some of them, at least from a general standpoint.

Let us first analyse the role of adding a solvent to the polymer. Located in those microdomains which contain the more soluble blocks, the solvent clearly contributes to increasing the volume of the corresponding microphases. However, it does not necessarily contribute to increasing their size. As we shall see, this is due to the random conformation of the blocks and to their expansion on swelling parallel to the interface.

To illustrate the swelling behaviour of block copolymers, let us briefly recall the results of an X-ray study of the lamellar structure of a styrene/isoprene diblock copolymer swollen in a series of different solvents.[12] This work describes, first, how the partition coefficient of the solvents may be estimated from a mere visual inspection of the intensities of the diffraction lines. It then shows that, on swelling, the geometrical parameters of the structure may vary differently depending upon the value of the partition coefficient. Thus, the total thickness, d, increases considerably when the solvent used is N,N-dimethylformamide which has a partition coefficient of $p = 0{\cdot}75$, favourable to PSt; in contrast, it is almost independent of the degree of swelling when the solvent used is not particularly selective with respect to one of the blocks. On the other hand, the molecular area S increases the most with solvents which are less selective towards the blocks; the swelling proceeds parallel to the interfaces when the solvent used does not preferentially dissolve one of the blocks, whereas it proceeds normally to the lamellae when the solvent used is preferential for one of the blocks. As a result, one particular sub-layer increases in thickness only when it incorporates a significant proportion of the solvent; otherwise, it may even become thinner as it spreads to cover a larger surface without increasing appreciably in volume.

Apart from the solvent content, the length of the blocks also drastically affects the structural parameters. The majority of the studies devoted to this phenomenon were carried out using mixtures of block copolymer and solvent.[8,24,33] These studies showed that the microdomains, and also the molecular area of the molecules at the interfaces, increased in size with the length of the blocks. However, they were difficult to analyse quantitatively, because of the necessity of taking into account the particular nature of each solvent used and its actual partition coefficient.

Regarding the influence of M.W. on the structural parameters of block copolymers in the absence of solvent, several studies have been performed in the last decade with styrene/diene block copolymers but the results were obtained from incomplete experimental data, and the conclusions were neither very coherent nor convincing. Some stated that the size of the structural elements is proportional to the square root of the M.W.,[42,43] others claimed[44] that it is proportional to the M.W. raised to power 0·55–0·60, still others merely announced that it increases with the length of the blocks in a monotonic fashion.[45] In order to try to clarify the situation, it was decided in a recent investigation[41] to go over the whole work again and to study systematically a homologous series of lamellar styrene/isoprene diblock copolymers. It thus became clear that the lamellar thickness increases in reality more steeply than predicted ($d \propto M^{0{\cdot}79}$).

Let us consider, lastly, the dependence of the structural elements upon temperature. For brevity, it is worth stating immediately that this dependence is extremely weak. The repeat distances of the structural elements are almost insensitive to a change of temperature, indicating that the conformation of the polymer chains (in the amorphous state) is not temperature-dependent.[7,46] However, after a careful scrutiny of the experimental data obtained with lamellar styrene/isoprene block copolymers,[34,35] it became clear that up to about 100 °C the lamellar thickness increases slightly ($\simeq 5 \times 10^{-4}$ per °C) with temperature, while beyond this limit it decreases again. After determining the respective thickness of PSt and PIp sub-layers from the distribution of the intensities of the X-ray reflections, it appeared that, in the first stage only the PIp sub-layers increased in thickness, the glassy PSt ones remaining unchanged; and in the second stage, both sub-layers decreased simultaneously in thickness. This peculiar and weak dependence of the spacing on temperature is, therefore, the mere result of the thermal expansion coefficient of the species, which is much smaller for the glassy state.

REFERENCES

1. Richards, D. H. and Szwarc, M., *Trans. Faraday Soc.*, 1959, **55**, 1644.
2. Skoulios, A., Finaz, G. and Parrod, J., *Comptes Rendus Acad. Sci., Paris*, 1960, **251**, 739.
3. Ekwall, P., in *Advances in Liquid Crystals*, vol. 1, ed. G. H. Brown, Academic Press, New York, 1975, p. 1.
4. Skoulios, A., *Ann. de Phys.*, 1978, **3**, 421.
5. Luzzati, V., Mustacchi, H. and Skoulios, A., *Disc. Faraday Soc.*, 1958, **25**, 43; Skoulios, A., *Adv. Coll. Interface Sci.*, 1967, **1**, 79.
6. Aggarwal, S. L., Livigni, R. A., Marker, L. F. and Dudek, T. J., in: *Block and Graft Copolymers*, eds. J. J. Burke and V. Weiss, Syracuse University Press, Syracuse, NY, 1973, p. 157.
7. Folkes, M. J. and Keller, A., in: *Physics of Glassy Polymers*, ed. R. N. Haward, Applied Science Publishers, London, 1973, p. 748.
8. Gallot, B., in: *Liquid-Crystalline Order in Polymers*, ed. A. Blumstein, Academic Press, New York, 1978, p. 191.
9. Gallot, B., in: *Advances in Polymer Science: Structure and Properties of Polymers*, vol. 29, ed. H.-J. Cantow *et al.*, Springer-Verlag, Berlin, 1979, p. 85.
10. Skoulios, A., in: *Advances in Liquid Crystals*, vol. 1, ed. G. H. Brown, Academic Press, New York, 1975, p. 169.
11. Skoulios, A. and Finaz, G., *Comptes Rendus Acad. Sci., Paris*, 1961, **252**, 3467.
12. Ionescu, M. L. and Skoulios, A., *Makromol. Chem.*, 1976, **177**, 257.
13. Guinier, A., *X-Ray Crystallographic Technology*, Hilger and Watts, London, 1952.
14. Ailhaud, H., Gallot, Y. and Skoulios, A., *Kolloid Z. Z. Polym.*, 1971, **248**, 889.

15. TSOULADZÉ, G. and SKOULIOS, A., *J. Chim. Phys. Physico-chimie Biol.*, 1963, **60**, 626.
16. PEDEMONTE, E., TURTURRO, A., BIANCHI, U. and DEVETTA, P., *Polymer*, 1973, **14**, 145.
17. SKOULIOS, A. and FINAZ, G., *J. Chim. Phys. Physico-chimie Biol.*, 1962, **59**, 473.
18. FEDORS, R. F., *J. Polym. Sci., C*, 1969, **26**, 189.
19. AGGARWAL, S. L., *Polymer*, 1976, **17**, 938.
20. SACKMAN, H. and DEMUS, D., *Mol. Cryst. Liq. Cryst.*, 1973, **21**, 239.
21. MANDELKERN, L., *Crystallization of Polymers*, McGraw-Hill, New York, 1964.
22. PERRET, R. and SKOULIOS, A., *Makromol. Chem.*, 1972, **162**, 147; *ibid.*, 1972, **162**, 163.
23. MEYER, G. C. and WIDMAIER, J. M., *J. Polym. Sci., Polym. Phys. Ed.*, in press.
24. GROSIUS, P., GALLOT, Y. and SKOULIOS, *A.*, *Europ. Polym. J.*, 1970, **6**, 355; *Makromol. Chem.*, 1970, **132**, 35.
25. VANZO, E., *J. Polym. Sci., A1*, 1966, **4**, 1727.
26. HENDUS, H., ILLERS, K. and RÖPTE, E., *Kolloid Z. Z. Polym.*, 1967, **216**, 110.
27. FISCHER, E., *J. Macromol. Sci.-Chem.*, 1968, **2**, 1285.
28. DLUGOSZ, J., KELLER, A. and PEDEMONTE, E., *Kolloid Z. Z. Polym.*, 1970, **242**, 1125.
29. DLUGOSZ, J., FOLKES, M. J. and KELLER, A., *J. Polym. Sci.*, 1973, **11**, 929.
30. GALLOT, B., MAYER, R. and SADRON, C., *Comptes Rendus Acad. Sci., Paris*, 1966, **263**, 42.
31. DOUY, A. and GALLOT, B., *Comptes Rendus Acad. Sci., Paris*, 1971, **C272**, 440.
32. KATO, K., *J. Polym. Sci., B*, 1966, **4**, 35.
33. GROSIUS, P., GALLOT, Y. and SKOULIOS, A., *Makromol. Chem.*, 1969, **127**, 94.
34. TERRISSE, J., *PhD Thesis*, University of Strasbourg, 1973.
35. SKOULIOS, A. in: *Block and Graft Copolymers*, eds. J. J. Burke and V. Weiss, Syracuse University Press, Syracuse, NY, 1973, p. 157.
36. PTASZYNSKI, B., TERRISSE, J. and SKOULIOS, A., *Makromol. Chem.*, 1975, **176**, 3483.
37. TODO, A., UKIUNO, H., MIYOSHI, K., HASHIMOTO, T. and KAWAI, H., *Polym. Eng. Sci.*, 1977, **17**, 587.
38. MOLAU, G. E., in: *Block Polymers*, ed. S. L. Aggarwal, Plenum Press, New York, 1970, p. 102.
39. AILHAUD, H., GALLOT, Y. and SKOULIOS, A., *Makromol. Chem.*, 1972, **151**, 1.
40. LEIBLER, L., *Macromolecules*, 1980, **13**, 1602.
41. HADZIIOANNOU, G. and SKOULIOS, A., *Macromolecules*, in press.
42. BRADFORD, E. and VANZO, E., *J. Polym. Sci., A1*, 1968, **6**, 1661.
43. INOUE, T., SOEN, T., HASHIMOTO, T. and KAWAI, H., *Block Polymers*, ed. S. L. Aggarwal, Plenum Press, New York, 1970, p. 53.
44. KRÖMER, H., HOFFMANN, M. and KÄMPF, G., *Ber. Bunsenges Physik. Chem.*, 1970, **74**, 859.
45. DOUY, A. and GALLOT, B., *Makromol. Chem.*, 1972, **156**, 81.
46. GROSIUS, P., GALLOT, Y. and SKOULIOS, A., *Comptes Rendus Acad. Sci., Paris*, 1970, **C270**, 1381.

Chapter 4

MICRODOMAIN STRUCTURE AND THE INTERFACE IN BLOCK COPOLYMERS

EUGENE HELFAND

and

Z. R. WASSERMAN

Bell Laboratories, Murray Hill, New Jersey, USA

SUMMARY

A statistical thermodynamic theory is presented which may be used to determine the size and shape of microdomains in amorphous block copolymer systems. The discussion is mostly in physical terms. A FORTRAN program is included to facilitate calculations. Particular emphasis in the exposition is placed on the nature and role of the interface between domains, and its relation to homopolymer interfaces. A theoretical phase diagram as a function of block copolymer composition is presented. Other theories, particularly that of Meier, are reviewed. The concluding section points to problems which appear to require further attention.

1. INTRODUCTION

Like a child contemplating the results of tying the cat's tail to that of the dog, scientists perhaps find a certain mischievous delight in considering the effect of joining two immiscible polymer blocks into one macromolecule. The immiscible units attempt to separate, but by virtue of their connectivity they can never get very far away from each other. The result is that they either segregate into microdomains or remain homogeneously mixed with each other. Putting this thermodynamically, the system's enthalpy can be lowered by phase separation. However, the connectivity of the blocks puts restrictions on how the molecules must then arrange themselves, and this results in a loss of entropy. Thus the question is whether the net free energy, H-TS, is lower in the microdomain or in the homogeneous state. In the event that a domain structure does result the further questions must be asked: (1) what is the thermodynamically most stable geometric form of the microdomains? and (2) what size will they be? The key to answering these questions theoretically is calculation and minimisation of the free energy as a function of domain size and shape. It would be nice, though, if the theory also enabled us to develop some intuitive sense of what the ingredients of the free energy function are, and that will be our primary aim in this chapter.

The discussion will be limited to block copolymers which form amorphous domains. After defining essential terms, we will introduce the narrow interphase approximation (NIA) and discuss the physical effects which contribute to the free energy. Polymeric interfaces play a major role, and these are examined in Section 4. In Section 5 an effort is made to present the flavour of a general mean-field theory of inhomogeneous polymers and its application to block copolymers[1-4] and interfaces.[4-7] The theory of these earlier sections is presented in the context of diblock copolymers which form lamellar microdomains. Little more is needed to generalise to triblock geometries. In Section 8 a FORTRAN program is presented which makes it quite easy to use the present theory for prediction of domain properties. In Section 9 some comments are made on the interpretation of the important scattering experiments which probe domain interfacial structure. Section 10 examines the validity of the narrow interphase approximation. In Section 11 the phase diagram of domain morphology is discussed, as well as the question of domain stability. A brief but critical examination of other theories, particularly that of Meier,[8-10] is presented in Section 12. The concluding remarks deal primarily with those questions that remain most pressing.

2. DEFINITIONS AND MODEL

Let us establish several definitions which will be used throughout this chapter. We will be considering two types of polymer units, these being labelled with an index K equal to A or B. The macromolecules will be modelled as Gaussian random walks. The degree of polymerisation of a block will be called Z_K, and the effective length of a single segment (Kuhn statistical length) is b_K. Thus the blocks, if unperturbed, would have mean squared end-to-end distances of $Z_K b_K^2$. The bulk density of pure polymer K, in monomer units per unit volume, will be called ρ_{0K}. The actual density at point $\mathbf{r}$ will be termed $\rho_K(\mathbf{r})$, or in reduced units

$$\tilde{\rho}_K(\mathbf{r}) \equiv \rho_K(\mathbf{r})/\rho_{0K} \tag{1}$$

This is also the volume fraction if there is zero volume of mixing, i.e.,

$$\tilde{\rho}_A(\mathbf{r}) + \tilde{\rho}_B(\mathbf{r}) = 1 \tag{2}$$

The symbol κ will be used for compressibility.

A most important parameter in any discussion of blends is the interaction parameter, α (or χ, discussed below), defined in terms of a particular model of the mixing free energy. It is assumed that in a homogeneous blend of A and B the contact free-energy density of mixing may be written as

$$\alpha k_B T \tilde{\rho}_A \tilde{\rho}_B / (\tilde{\rho}_A + \tilde{\rho}_B) \tag{3}$$

where k_B is Boltzmann's constant and T is temperature. The interaction parameter α has dimensions of m^{-3}.

In the theoretical discussion of blends and block copolymers, considerable algebraic and conceptual simplification is introduced by assuming that the polymer pair is symmetric. By this is meant that some of the properties of pure polymers A and B are identical, viz. the densities $\rho_{0A} = \rho_{0B} = \rho_0$ and the segment lengths $b_A = b_B = b$. However, we still allow the block degrees of polymerisation Z_A and Z_B to be general. For a symmetric polymer it is possible to define a dimensionless interaction parameter χ as $\chi = \alpha/\rho_0$. Physically, $\chi k_B T$ is the free-energy change per segment of A in taking a chain of A from pure A surroundings and transferring it to pure B surroundings. At times the discussion below will be limited to the case of symmetric A and B for simplicity of exposition, but it should be emphasised that in all cases the more general results are available.

The abbreviation for three polymeric materials which occur frequently in the discussion will be: PSt or S (polystyrene); PBd or B (polybutadiene); PIp or I (polyisoprene).

3. FREE ENERGY OF BLOCK COPOLYMERS IN A MICRODOMAIN STRUCTURE

In this section we will consider a block copolymer system and examine an expression for its free energy when it forms a microdomain structure.[1-4] As

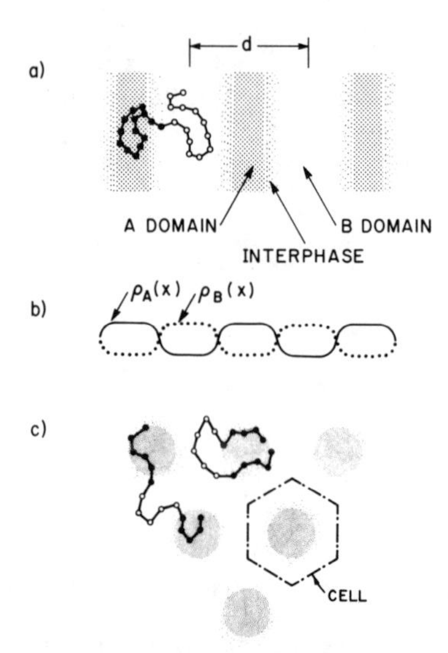

FIG. 1. (a) Schematic diagram of lamellar microdomains in a block copolymer system. (b) The density profiles indicate that the interface is not sharp, but that interphase regions exist. (c) Schematic diagram of a cross-section of cylindrical (or spherical) microdomains; the unit cell is illustrated.

we indicated earlier, this is at the heart of a statistical thermodynamic theory of block copolymers. To begin with let us deal specifically with the case of a diblock copolymer, AB, which forms lamellar microdomains.[3a] The system is depicted schematically in Fig. 1, where we see the alternating layers of A and B with a periodicity distance, d. Given this distance, the widths of the individual domains, d_K, follow from the fact that these widths must be proportional to the volume fractions of the respective materials:

$$\frac{d_K}{d} = \frac{Z_K/\rho_{0K}}{Z_A/\rho_{0A} + Z_B/\rho_{0B}} \tag{4}$$

An associated quantity is the density of block copolymer molecules, N/V, given by

$$\frac{N}{V} = \frac{1}{Z_A/\rho_{0A} + Z_B/\rho_{0B}} \tag{5}$$

In Fig. 1 we also indicate that the interface between domains is not sharp; i.e. the density of A falls off as the density of B builds up. This occurs over a characteristic distance, a_I (discussed more extensively in Section 4), which for a symmetric A/B pair of polymers is given by

$$a_I = (2/6^{1/2})b/\chi \tag{6}$$

frequently of the order of a nanometer or more. Usually this thickness is much less than the width of the domains, so that the domains can be regarded as mostly pure. Under the condition that $a_I/d_K \ll 1$, a number of simplifying assumptions can be made in the theory of block copolymers, and this has been called[3a] the 'narrow interphase approximation' (NIA). Its validity will be discussed in Section 10, and throughout the present section it will be assumed to pertain. In terms of physical parameters, experience indicates that the NIA is quite good for $\chi Z_K \gtrsim 20$.

First let us write an expression for the free energy difference per molecule between a diblock copolymer system in a microdomain morphology and in a homogeneous state:

$$\frac{F}{Nk_BT} = \frac{2\gamma}{k_BT}\left(\frac{Z_A}{\rho_{0A}} + \frac{Z_B}{\rho_{0B}}\right)\frac{1}{d} + \log\left(\frac{d}{2a_J}\right) + \frac{F_{LD}(d_A/b_AZ_A^{1/2}) + F_{LD}(d_B/b_BZ_B^{1/2})}{Nk_BT} - \alpha\frac{(Z_A/\rho_{0A})(Z_B/\rho_{0B})}{Z_A/\rho_{0A} + Z_B/\rho_{0B}} \quad (7)$$

Now we shall discuss each of the terms and identify its physical origin.

In the NIA the interphase has a structure much like that between semi-infinite homopolymer phases of A and B. Thus the free-energy per molecule associated with these interphase regions is $\gamma\Sigma$ where Σ is the surface area per molecule. Simple geometric considerations show that Σ is given by 2 (for the two faces per period) times the volume per molecule (given by eqn (5)) divided by the periodicity distance d, from which the first term of eqn (7) follows. This term is plotted as a function of d and labelled SURFACE in Fig. 2. It is seen to decrease as the size of the domains grows, i.e. the surface-to-volume ratio decreases. The surface term is the usual force to make phases as large as possible. But in block copolymers such domain growth is thermodynamically inhibited, so our next task is to examine the origins of this opposition and put it on a quantitative basis.

In the NIA one can distinguish two sources of opposition to domain growth. The first is related to a restriction on the placement of the block copolymer molecules. Each must have its joint in the interfacial region. For the joint to wander significantly from the interphase, e.g. into the A phase, would involve pulling B units into an unfavourable environment. Using the density profiles appropriate to a homopolymer interface (eqn (12)) one can calculate a measure of the freedom of placement of a joint in the interface; let us call it a_J. It is only slightly different from a_I, and most particularly a_J is a constant independent of the M.W. of the blocks and the periodicity distance d in the NIA. From Fig. 1 we can see that the fraction of the system

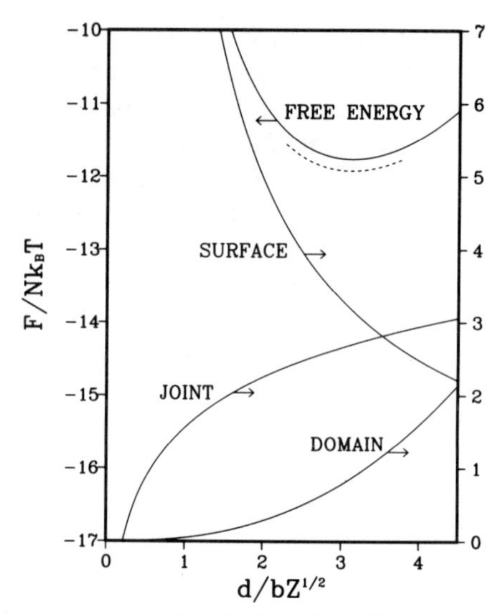

FIG. 2. Free energy per molecule of a lamellar diblock copolymer system as a function of periodicity distance, d, in reduced units. (The blocks are symmetric with $\chi Z = 37$.) The components of the free energy labelled SURFACE, JOINT and DOMAIN are the first three terms of eqn (7), described in the text. The dashed line is the result of a calculation without the NIA. (From reference 3, with permission American Chemical Society.)

available to the block joint is $2a_J/d$, so that the loss of entropy associated with the joint placement restriction is proportional to the logarithm of this ratio, $k_B \log(2a_J/d)$. This is to be multiplied by $-T$ to get the free-energy increase, plotted in Fig. 2 as JOINT.

There is another important factor which limits the growth of domains, and this is associated with the requirement that throughout the K domain the density be maintained at the bulk density of K, ρ_{0K}. A more precise statement is that in the NIA the density pattern across the interface must be the same as across an A/B homopolymer interface. On the scale of distances defined by the domain size the interphase is very narrow in the NIA and may be regarded as a plane. (This sort of two-distance-scale concept is used in various physical problems, such as the treatment of the density across a shock wave.) All the blocks may be considered as having one end, the joint, located in these 'plane' boundaries. Now imagine that the domain is very large, say greater in thickness than $b_K Z_K^{1/2}$. It is obvious that unperturbed molecules usually would not reach into the centre of the domain to fill it up. Actually, it can be shown by the methods of Section 5 that for all domain thicknesses the unperturbed tendency is to have higher density near the boundary, where the blocks always have one end, and less density near the domain centre. But the drive to maintain constant density in a condensed system is a very strong one. Therefore there is a preferential rejection of those configurations which do not fill the domain centre uniformly, and a preferential weighting of the rarer configurations which do reach further into the centre. Any such reweighting of the chain conformational statistics results in a loss of conformational entropy, which can be calculated by methods outlined in Section 5. The corresponding free-energy gain for each domain, a function only of the combined parameter $d_K/b_K Z_K^{1/2}$, is called F_{LD} in eqn (7) (*LD* for lamellar domain). This term, labelled DOMAIN, is plotted as a function of d in Fig. 2. The free energy gain obviously becomes more and more severe as the domain sizes increase, rising roughly according to $d^{2 \cdot 5}$. Thus the necessity of maintaining uniform density with one end of the block anchored in the domain boundary places a severe restriction on how much the domains can grow.

The final term of eqn (7) is independent of domain size and merely serves to set the zero of free energy as a homogeneous mixture of the blocks (if such blocks were stable).

To summarise, the microphases tend to grow in order to reduce the surface-to-volume ratio and the associated interfacial free energy. However, as the domain's volume grows relative to the interphase's volume, the localisation of the joint becomes more severe. Also there must be more biasing toward rarer configurations in order to fill the centre of the domain. The opposing effects produce a free energy which possesses a minimum, and the corresponding value of the periodicity distance, d, is the most stable domain size (cf. Fig. 2). With a functional fit to the free energy term F_{LD} the task of minimisation can be carried out algebraically. This is important since it makes the determination of d quite simple, as will be seen in Section 8 where a straightforward FORTRAN program is presented to perform the algebra.

This has been a brief introduction to the factors which determine block copolymer microdomain size. With the overall picture in mind it is worthwhile examining the various effects in just a bit more detail.

4. POLYMERIC INTERFACES

It was evident in the previous section that an understanding of polymeric interfaces is quite important in the consideration of block copolymer systems, and this we will try to convey briefly here. A longer, but still highly qualitative, account of the subject has been published recently,[7] or one can turn to the original papers[4–6] for details. In this section we will focus on the interface between two homopolymer phases, A and B, and assume high molecular weight for each.

Let us try to get an estimate of how thick a polymer interphase is likely to be. To begin with consider a sharp interface between symmetric polymers A and B, i.e. assume that the density of A abruptly falls from ρ_0 to 0 at a plane $x = 0$, while the B density rises there just as abruptly. Consider, further, a

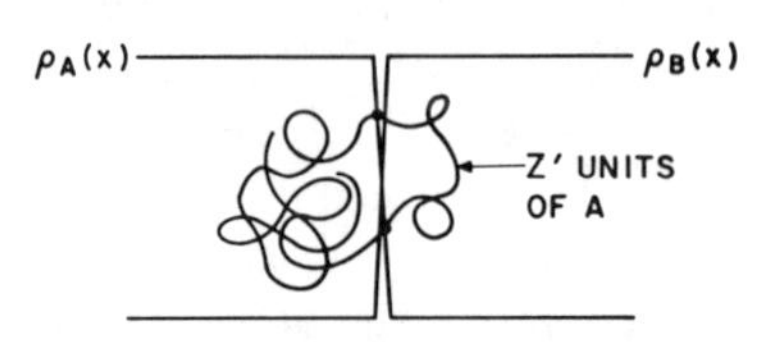

FIG. 3. An imagined, sharp interface between A and B, and the configuration of a typical section of A chain which penetrates into the B phase. (From Helfand, E., *Accts. Chem. Res.*, 1975, **8**, 295; with permission American Chemical Society.)

fluctuation from this situation wherein part of an A molecule wanders across the border into B, winds about for Z' units, and then returns to the A phase, as in Fig. 3. The first question is: what is the typical value of Z' in such spontaneously appearing fluctuations? The energy of such a fluctuation is $\chi k_B T$ for each unit, according to the definition of χ in Section 2, or $\chi k_B T Z'$ in all. Typically, fluctuations have energy $O(k_B T)$ so that $\chi k_B T Z' \approx k_B T$, or

$$Z' \approx 1/\chi \tag{8}$$

The characteristic depth of penetration of Z' units is $O(bZ'^{1/2})$. Using eqn (8) for Z' we get as an estimate of the spontaneous interpenetration of the phases, i.e. interphase thickness a_I

$$a_I \approx b/\chi^{1/2} \tag{9}$$

One can next go on to an estimate of the interfacial free energy, i.e. interfacial tension. To approximate the interfacial energy observe that mixing occurs in a region of volume $a_I S_I$, where S_I is the interfacial area. In this region there are $\rho_0 a_I S_I$ units, each with a typical energy some fraction of $\chi k_B T$, so that the total energy of mixing is

$$\gamma_E S_I \approx \rho_0 a_I S_I \chi k_B T$$

or

$$\gamma_E \approx \chi^{1/2} \rho_0 b k_B T \tag{10}$$

In addition there is a loss of conformational entropy associated with the fact that a polymer molecule which approaches the interface must turn back into its own phase. It amounts to $O(k_B)$ for each loop in the interface. This can be shown in various ways. One way is to say that one of the Z' links loses a finite fraction of its orientational freedom. Another is to say that each loses a fraction c/Z' of its freedom because the turning is spread out. Thus the loop loses conformational entropy

$$k_B Z' \log(1 - c/Z')$$

which is $O(k_B)$ for large Z'. In an interphase of volume $a_I S_I$ there are $\rho_0 a_I S_I / Z'$ loops so the total gain of conformational free energy is

$$\gamma_C S_I \approx k_B T \rho_0 a_I S_I / Z'$$

or, substituting

$$\gamma_C \approx \chi^{1/2} \rho_0 b k_B T \tag{11}$$

dimensionally just like γ_E.

The above argument clearly has a large inconsistency. We began by assuming that the interphase was sharp and then found that there was actually a considerable degree of interpenetration. A self-consistent theory[4-7] will be discussed in the following section, but it will be convenient to present the germane results here. One finds that the density pattern of A and B across a symmetric interface can be depicted as in Fig. 4, viz.,

$$\tilde{\rho}_A(x) = \tfrac{1}{2}[1 - \tanh(2x/a_I)] \tag{12}$$

$$a_I = (2/6^{1/2}) b / \chi^{1/2} \tag{13}$$

Here a_I is defined in terms of a wedge-shaped fit to the density pattern, as illustrated, with the slope of the wedge taken as the true value of $d\tilde{\rho}/dx$ at

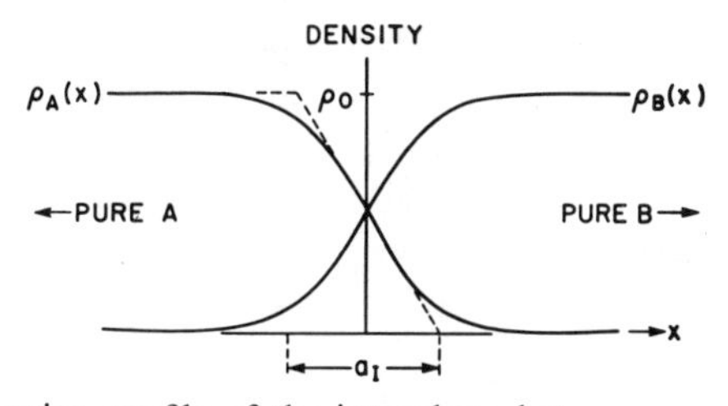

FIG. 4. Concentration profile of the interphase between symmetric homopolymers A and B. The interfacial thickness, a_I, is taken as that of a wedged-shaped fit, indicated by the dotted line. (From Helfand, E., *Accts. Chem. Res.*, 1975, **8**, 295; with permission American Chemical Society.)

$x = 0$. Other significant measures of the interphase width will be discussed in Section 9. The theory finds that the interfacial tension is

$$\gamma = 6^{1/2}\chi^{1/2}\rho_0 b k_B T \tag{14}$$

(The temperature dependence is not evident since χ is temperature-dependent.) Except for the numerical constants, a_I and γ agree with the earlier estimates.

For an unsymmetric interface the formulas for the density profile and interfacial tension[6] are rather more complicated. For example, γ is given by

$$\gamma = k_B T \alpha^{1/2}\left[\frac{\beta_A + \beta_B}{2} + \frac{1}{6}\frac{(\beta_A - \beta_B)^2}{\beta_A + \beta_B}\right] \tag{15}$$

$$\beta_K^2 \equiv \rho_{0K} b_K^2/6 \tag{16}$$

Interfacial tensions calculated in this way compare well with experiments.[6] The greatest uncertainty is in the value of the interaction parameter, α.

In eqn (7), for the block copolymer free energy one may use experimental interfacial tensions, except for the fact that they are generally unknown for the common block copolymer pairs. (It is startling that the interfacial tension between PSt and PBd has never been measured, in spite of the importance of this pair in block copolymers and in impact modification.) In all our calculations[2,3] γ has been estimated by eqn (15) with literature values of ρ_{0K}, b_K, and a value of α due to Rounds and McIntyre[11] based on oligomer miscibility.

5. THEORY OF INHOMOGENEOUS POLYMERS

In this section we shall give a sketchy indication of the structure of the theory[1-6] that may be used to calculate density patterns and modifications of conformational statistics in regions of polymer inhomogeneity such as interfaces and microdomains. The information about the conformational statistics of sections of polymer chains is embodied in a function $Q_K(\mathbf{r}, t; \mathbf{r}_0)$ which is proportional to the probability that a type K chain of t segments has one end at $\mathbf{r}_0$ and the other end at $\mathbf{r}$. Under field-free conditions Q_K is the well known Gaussian distribution; i.e. it satisfies the diffusion equation

$$\frac{\partial Q_K}{\partial t} = \frac{b_K^2}{6}\nabla^2 Q_K \tag{17}$$

with the initial condition

$$Q_K(\mathbf{r}, 0; \mathbf{r}_0) = \delta(\mathbf{r} - \mathbf{r}_0) \tag{18}$$

If the polymer is in an external field $U_K(\mathbf{r})$ per segment of K, then Q_K satisfies a modification of this equation.

$$\frac{\partial Q_K}{\partial t} = \frac{b_K^2}{6} \nabla^2 Q_K - \frac{U_K(\mathbf{r})}{k_B T} Q_K \tag{19}$$

(Q_K is not normalised.) To understand this formula, consider the solution when the diffusional displacement term is not there; i.e. $b = 0$. Then

$$Q_K = \exp\left[-\frac{U_K(\mathbf{r})t}{k_B T} \right] \delta(\mathbf{r} - \mathbf{r}_0) \tag{20}$$

The effect of the U_K has been to decrease the probability of the polymer being at the point $\mathbf{r}$ by a Boltzmann factor, $U_K(\mathbf{r})t$ being the energy of a t segment chain at $\mathbf{r}$. Thus the full eqn (19) describes the decrease of relative probability of those chains which do wander (by which is meant wandering in chain space, not in time) away from regions of high potential. The Kuhn parameter b_K determines the 'rate' at which such wandering can occur.

We are not interested in the statistics of a chain in an external field, but rather in the field generated by the other molecules. To treat this problem in a self-consistent field manner we replace the external potential by the chemical potential difference between a unit of the chain at $\mathbf{r}$ and that unit in pure K. We consider two effects which may give rise to such a chemical potential difference. One is the presence of another type of chain near $\mathbf{r}$. This leads to the free-energy difference of eqn (3), or a chemical potential difference

$$\frac{\alpha k_B T}{\rho_{0K}} \left[\frac{\tilde{\rho}_{K'}(\mathbf{r})}{\tilde{\rho}_A(\mathbf{r}) + \tilde{\rho}_B(\mathbf{r})} \right]^2 \tag{21}$$

where K' is the species opposite from K. The other effect is that the overall density at $\mathbf{r}$ may be different from that which the AB mixture would have at equilibrium. Assuming zero volume change on mixing the densities of A and B at $\mathbf{r}$ should satisfy

$$\tilde{\rho}_A(\mathbf{r}) + \tilde{\rho}_B(\mathbf{r}) = 1 \tag{22}$$

Deviations from this condition give rise to a free-energy density of

$$(1/2\kappa)[\tilde{\rho}_A(\mathbf{r}) + \tilde{\rho}_B(\mathbf{r}) - 1]^2 \tag{23}$$

or a chemical potential for component K of

$$(1/\kappa\rho_{0K})[\tilde{\rho}_A(\mathbf{r}) + \tilde{\rho}_B(\mathbf{r}) - 1] \tag{24}$$

Replacement of U_K by the chemical potential in eqn (19) yields the equation for $Q_A(\mathbf{r}, t; \mathbf{r}_0)$

$$\frac{\partial Q_A}{\partial t} = \frac{b_A^2}{6}\nabla^2 Q_A - \left\{\frac{\alpha}{\rho_{0A}}[\tilde{\rho}_B(\mathbf{r})]^2 + \frac{1}{\kappa k_B T \rho_{0A}}[\tilde{\rho}_A(\mathbf{r}) + \tilde{\rho}_B(\mathbf{r}) - 1]\right\} Q_A \tag{25}$$

This gives the conformational function Q_K in terms of the density pattern, but to close the equation one must express the density in terms of the Q_K. For instance in a mixed homopolymer system

$$\tilde{\rho}_K(\mathbf{r}) = (1/Z_K)\int_0^{Z_K} \mathrm{d}t \int \mathrm{d}\mathbf{r}_1\, \mathrm{d}\mathbf{r}_0 Q_K(\mathbf{r}_1, Z_K - t; \mathbf{r}) Q_K(\mathbf{r}, t; \mathbf{r}_0) \tag{26}$$

which can be easily understood. The density at $\mathbf{r}$ is decomposed into those chains that have one end at $\mathbf{r}_0$, have a segment t units down the chain at $\mathbf{r}$, and have their other end at $\mathbf{r}_1$. The total density is obtained by integrating over all units t and over all locations for the chain ends, and finally normalising. For a diblock copolymer $\mathbf{r}_0$ might represent the joint, and another factor $Q_{K'}(\mathbf{r}_0, Z_{K'}; \mathbf{r}_2)$, with an integration over $\mathbf{r}_2$, must be included. The reader who seeks more detail, such as the expressions for the free energy in terms of the Q_K, is referred to the literature.[1-6]

One topic worth pursuing further physically is the term which maintains the overall reduced density constant. The compressibility, κ, of a condensed phase is very small. Therefore, if the term with κ in the denominator is to be of the same magnitude as all other terms in eqn (25), the density deviation, $\tilde{\rho}_A(\mathbf{r}) + \tilde{\rho}_B(\mathbf{r}) - 1$, must be small also. For instance, in an interface one must have

$$\tilde{\rho}_A(\mathbf{r}) + \tilde{\rho}_B(\mathbf{r}) - 1 = \mathrm{O}(\chi\kappa\rho_0 k_B T) \approx 10^{-3} \tag{27}$$

Substituting this back into the free-energy expression, eqn (23), we see that the free energy directly associated with these density deviations is very small. Indirectly there is a free energy contribution, however. The density deviation adjusts itself, as just noted, to have finite impact, like all other terms, on the conformational statistics, and these statistics are reflected in the conformational entropy. This is what we were referring to in Section 3 when we said that in a block copolymer microdomain there is an extra probability weighting attached to those conformations which reach into the centres of the domains to fill them uniformly. Such a redistribution of probabilities in order to satisfy extra constraints (in this case, uniform density) always leads to a loss of entropy. Again we refer the reader to the original literature[2-4] for a discussion of how to treat the chemical potential expression, eqn (24), in the limit as $\kappa \to 0$. In this limit eqn (24) becomes an indeterminate form, akin to a set of Lagrange multipliers to keep the density constant.

For the problem of an A/B interface the eqns (25) and (26) may be solved analytically for high M.W., leading to the density pattern eqn (12) and interfacial tension eqn (14).[5,6] For the problem of a polymer block starting at the wall of a domain and uniformly filling that domain, numerical solution is necessary. Using that numerical solution the domain free energy, F_D, may be calculated for lamellar,[3a] cylindrical[3c] or spherical[3b] morphologies. Having once performed this calculation and analytically fitted the results, further use of the theory is quite simple, as we shall see in Section 8.

6. TRIBLOCK COPOLYMERS

The formalism outlined above is as applicable to triblock (or multiblock) copolymers as to diblocks. F_{LD} was a function which expressed the loss of conformational entropy involved in uniformly filling a domain with the condition that the blocks have one end anchored at the domain wall. When the blocks are centre blocks of triblock copolymer molecules, then the appropriate condition is that both ends be anchored at the wall.

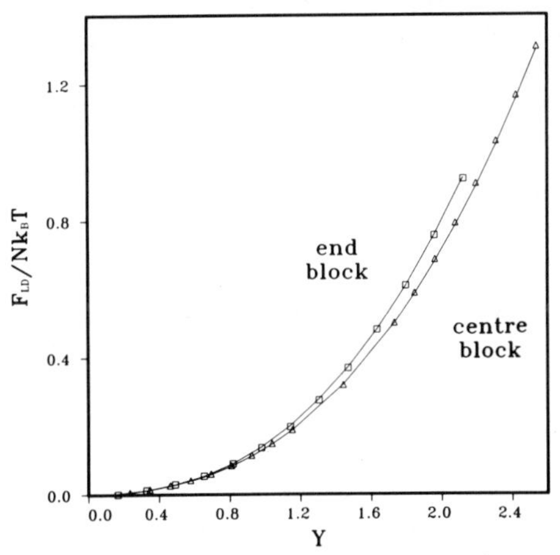

FIG. 5. Plot of the lamellar domain free energy terms (free energy associated with keeping the density in the domain uniform) versus reduced variable Y defined in the text. Squares are calculated points for an end block, while triangles are for a centre block (as in the centre of a triblock copolymer). These free energies do not differ greatly. The algebraic fits adhere well to the calculated points.

Another way of looking at the difference between centre and end block domains is the following:

Compare a domain filled with centre blocks of degree of polymerisation Z, with the same domain filled with twice as many end blocks of degree of polymerisation $\frac{1}{2}Z$. Case two can be obtained from case one by cutting each centre block in half. The difference is not great, as one can see in Fig. 5. Here is a plot of $F_{LD}(Y)$ for end blocks, with $Y \equiv d/bZ^{1/2}$. This is contrasted with the function $F_{LD}^{(c)}(Y)$ for centre blocks, with $Y \equiv d/[b(\frac{1}{2}Z)^{1/2}]$. The conclusion is that as a rule of thumb the thermodynamic properties of triblock copolymers of M.W.'s $M_A - M_B - M_A$ do not differ greatly from those of diblock copolymers of M.W.'s $M_A - (\frac{1}{2}M_B)$. However, since we shall show that it is very easy to perform calculations on either diblock or triblock copolymers there is little need to make this assumption in quantitative work.

7. CYLINDRICAL AND SPHERICAL GEOMETRIES

The consideration of cylindrical or spherical morphologies introduces little new, conceptually or mathematically. Solutions to the equations must be found which possess the appropriate symmetry. In Fig. 1 we have drawn the unit cell for cylindrical symmetry (or this may be regarded as a cross-section of the cell for close-packed cubic symmetry). A minor, simplifying approximation we have made is to treat the hexagonal cross-section of the unit cell as a circle of equal area (similarly the dodecahedron of the spherical morphology is sphericalised).

We have indicated that the domain free-energy term of lamellae, F_{LD}, is a function only of the one reduced size variable, $d/bZ^{1/2}$. This is likewise true of the domain free energy of the inner cylindrical (spherical) region. However, the outer ring of the unit cell, which is the part usually termed the matrix, requires two parameters for its characterisation. These may be chosen as a reduced thickness of the annulus, and the ratio of the inner to the outer diameter, termed ξ.

8. CALCULATION OF THE DOMAIN SIZE, SPACING AND FREE ENERGY

In Section 3 a discussion was presented of the free energy for lamellae of diblock copolymers as a function of the periodicity distance, d. Similar expressions exist for triblocks and other morphologies. In Fig. 2 the total free energy was plotted, and because of the opposing forces of the constituent terms it possesses a minimum, which is the predicted value for *d*. The location of the minimum can be found analytically using algebraic fits to the domain free energies. Rather than write out equations for *F* and its derivatives, which can be found elsewhere,[3] we present here a FORTRAN program (see Appendix) which embodies these formulas and calculates the system's properties. Let us discuss how to use the program, how to modify it, and what output it produces:

The program is set up to calculate the properties of di- or triblock copolymers in lamellar, cylindrical or spherical domains. Parameters have been explicitly included to perform calculations for PSt/PBd or PSt/PIp pairs. In the line 10 DATA statement room has been left to insert the parameters of two more polymer types. To do so replace the zeros with: RHO—the density of pure polymer (mol monomer/m^3); XM—monomer M.W. (kg); B—Kuhn statistical length (m); POLYMR—a two letter designation of the polymer name. The mixing parameter, α, has been assigned the value given by Rounds and McIntyre.[11] For other polymer pairs or to revise, change line 60 to the value of ALPHA (m^{-3}) desired, possibly temperature-dependent. If one wishes to use an experimental rather than theoretical interfacial tension change line 70 (and its continuation line) to the value of $\gamma/k_B T \equiv$ GAMOKT (m^{-2}) desired, possibly temperature-dependent. (Then the value of α is irrelevant.)

The program adopts the naming convention for cylinders and spheres that one has domains of A imbedded in a matrix of B. Thus an SBS which forms PSt domains would be termed ABA, but if it formed PBd domains it would be called BAB. For lamellae, which is A and which is B is arbitrary. The input then consists of (read in according to FORMAT 998): PA—the two-letter designation of polymer A (PS, PI, PB, or other assigned names); PB—two letter designation of polymer B; DIMENSION—1, 2, or 3 to specify lamellae, cylinders, or spheres, respectively; J—type of of block copolymer with 1 = diblock, 2 = triblock BAB and 3 = triblock ABA; MOL.WT.A—M.W. A blocks (kg); MOL.WT.B—M.W. B blocks (kg); TEMP—temperature (K). The temperature may be that at which the structure is set, e.g. by one component becoming glassy. We have used $T = 363{\cdot}18$ K for most of our reported results. The only other input is a blank line to terminate the program.

As output one obtains a statement of the domain structure and a labelled list of properties. These include the M.W.'s (MOL.WT.) and degrees of polymerisation (Z) of the blocks; diameter or lamellar thickness of the A domain; lattice spacing of the array (D); temperature (T); parameter ξ (XI) explained in Section 7; free energy per molecule divided by $k_B T$ (F/NKT); number of molecules (N) in a square metre of lamellae, a metre of cylinder, or a spherical domain; and Σ (SIGMA) the interfacial area per block joint.

Calculations have been performed comparing the theory to all reports of block copolymer domain sizes which we have been able to find, and for which the required parameters are known. They are reported elsewhere.[3] In general the results are quite good.

9. MEASUREMENTS OF INTERPHASE THICKNESS

Measurements of interphase thickness have been made by Hashimoto *et al.*[12] They use X-ray scattering from well-ordered lamellar and spherical domains. For scattering vector $q = 2\pi n/d$ peaks are observed in the scattering pattern. Here n is the order of the reflection, and the peaks are distinct through about nine orders. For larger q, of the order of the reciprocal of the interphase thickness, the scattering function for a single domain is attenuated by a function $|\tilde{h}(q)|^2$, related to the interfacial density profile (for a symmetric interface) by

$$\tilde{h}(q) = \int_{-\infty}^{\infty} \mathrm{d}z\, e^{iqz}\, \mathrm{d}\tilde{\rho}_A(z)/\mathrm{d}z \tag{28}$$

$$= \frac{\pi a_I q/4}{\sinh(\pi a_I q/4)} \tag{29}$$

using eqn (12) for the density. Usually one will have only enough detail of $\tilde{h}(q)$ to fit accurately one parameter. Perhaps the parameter describes predominantly the small q behaviour, as in the expansion

$$\tilde{h}(q) = 1 - \tfrac{1}{2}\sigma^2 q^2 + \cdots \tag{30}$$

This makes σ^2 the second moment

$$\sigma^2 = \int_{-\infty}^{\infty} \mathrm{d}z\, z^2\, \mathrm{d}\tilde{\rho}_A(z)/\mathrm{d}z \tag{31}$$

or, using eqn (12),

$$\sigma = \left(\frac{\pi}{4 \cdot 6^{1/2}}\right) a_I \tag{32}$$

There has been some ambiguity in definitions of interfacial thickness, which we will attempt to clarify here and in a subsequent publication.[13] Hashimoto *et al.*[12] introduced a measure of thickness which, on the basis of the small q behaviour, is $\Delta R = 2 \cdot 3^{1/2} \sigma$ (ΔR is called t in reference 12). Thus

$$a_I = (2/\pi)\Delta R = 0 \cdot 637\, \Delta R \tag{33}$$

ΔR has previously been compared directly with a_I, without a calculation of the proportionality constant. Averaging the 12 ΔR values reported[12] one finds an experimental a_I of 1·2 nm for PSt/PIp. This is to be contrasted with a calculated a_I of 1·4 nm at 25 °C or 1·6 nm at 90 °C according to the theory (eqn (4.6) of reference 6 has been used, but not the unsymmetric density pattern). There may be slight variations in the experimental estimate if the full $\tilde{h}(q)$ is fitted, rather than its expanded form.

10. VALIDITY OF THE NARROW INTERPHASE APPROXIMATION

The NIA assumes that the characteristic width of the interphase, a_I, is much less than the characteristic sizes of the domains. Since a_I is independent of the degrees of polymerisation, Z_K, while the domain size grows with Z_K, the approximation is valid in the limit of large Z_K. One way of gauging the effectiveness of the NIA is to solve the equations, such as eqn (25), without the NIA[2] and compare the results.[2,3a] In Fig. 6 we see the density profile for a symmetric diblock copolymer with $Z_A = Z_B = Z$. The plot goes from the centre of the A lamellar domain to the centre of B. For $\chi Z = 37$, a very typical value, the NIA seems quite appropriate, and indeed the non-NIA calculated density profile is close to that of a homopolymer interphase. The free-energy calculation is also very close to the NIA, as shown by the dashed curve in Fig. 2. Next consider $\chi Z = 10$, corresponding to the M.W. of a PSt block in SB of about 10^4, and close to the limit of stability in the NIA (cf. Section 11). We see in the second part of Fig. 6 that A penetrates noticably all the way to the centre of the B domain. Still the results of the NIA are reasonable with respect to density profile, predicted domain size, and free energy.[3a]

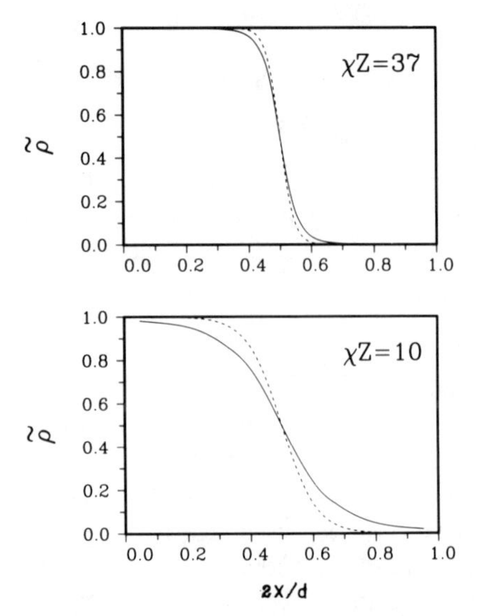

FIG. 6. Density profiles in lamellar domains calculated with (dashed) and without (solid) the NIA. The density is indicated from the centre of one domain ($2x/d = 0$) to the centre of the adjacent domain ($2x/d = 1$). Plots are presented for two values of χZ, as indicated. (From reference 2, with permission American Chemical Society.)

One further comment about the NIA is relevant. There has been some misunderstanding of the use of a zero derivative boundary condition for the domain Q function near the boundary.[3] This is not a reflection boundary condition, but rather a joining condition between the domain and interface solutions.

11. PHASE DIAGRAM AND STABILITY OF DOMAIN STRUCTURES

As a matter of thermodynamic equilibrium the system should choose to assume the geometry which corresponds to the lowest free energy. The most important parameter in this respect is the weight fraction of the components. The total M.W. of the polymer has much less of an impact. This can be seen in Fig. 7 which is a phase diagram showing the regions of absolute stability (lowest free energy) of the various geometries in an SB diblock copolymer according to the theory with the NIA.[13] The variables are weight fraction of PSt and total M.W. The results for triblock

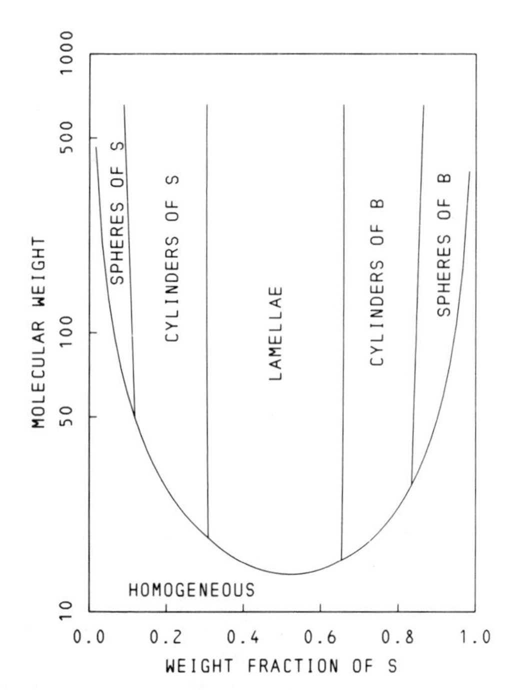

FIG. 7. Phase diagram calculated with the NIA for geometry and stability of SB microdomains in the (weight fraction polystyrene)–(molecular weight) plane. The temperature is fixed at 90 °C. The near-vertical lines mark the transition for absolute stability of one type of geometric array of domains to another. The lower curve indicates the border between microdomains and homogeneous phase as the lowest free energy form. (Molecular weight is given in 000's.)

copolymers do not differ greatly, as is to be expected from the discussion of Section 6.

Gallot[14] has summarised the domain structure observed by the Orleans and other groups[15] on SB (the PBd is mostly 1,2). He reports body-centred cubic spheres up to about 20 % PSt, cylinders between 20 % and about 40 %, lamellae from 40 % to 64 %, PBd cylinders between 64 % and 82 %, a strange orthorhombic array between 82 % and 84 % and body-centred cubic above 84 %. The Bayer group reports similar results.[16] Their PBd blocks are mostly 1,4. They are a bit fuzzier about the phase boundaries, tending to find mixtures of geometries. It appears that they assume their spherical arrays are close-packed, but it is not clear to what extent this was verified. Comparing these observations with the calculations it seems that the theoretical phase lines are too dispersed from the centre. (It should be pointed out that our spherical calculations are for close-packed arrays.) The actual free-energy differences between cylindrical and spherical arrays are quite small, which may account for the difficulty that the theory has in accurately placing the phase boundary, and may also explain the tendency to find both phases coexistent in the experiments.

In Fig. 7 we indicate, too, the phase boundary for the microdomain-to-homogeneous transition; i.e. the line where F goes from negative to positive in the NIA theory. For a given polymer composition and M.W., the temperature at which this transition occurs can also be determined. One of the best studied samples in this respect is an SBS triblock of M.W. 7000–43 000–7000 which forms cylinders of PSt at low temperature. Between 140 and 150 °C the sample undergoes an apparent 'transition' in the rheological behaviour, as has been observed by Chung *et al.*,[17] and Gouinlock and Porter.[18] A change of structure in this range has been noted in electron microscope observations.[19] However, no dramatic transition is apparent in Roe and Chang's[20] X-ray scattering; i.e. evidence of the domain pattern seems to be present above 150 °C.

The theory is also a bit ambiguous on the question of a phase transition. If one uses the NIA, a phase transition at a temperature of 147 °C is found.* However, at this temperature the interphase is becoming broad, so the NIA may be misleading. Some work has been done on solving the equations without the NIA (cf. Section 10), and in tentative calculations no transition has appeared. What does happen is that the penetration of one phase by the other becomes more and more severe. As the temperature continues to rise the domain compositions become almost identical (i.e. one observes only a small density ripple) but still a periodic, inhomogeneous solution has lower free energy than a homogeneous one. At this point, though, the mean field theory which underlies the entire scheme may be called into question, since fluctuations are not being treated properly. The problem is one which should receive further study. It is encouraging, however, that the theory clearly indicates a dramatic change in structural features near the right temperature.

Leibler[21] has attempted to construct a theory of fluctuations about the homogeneous state, and from this to determine the nature of the phase transition. He finds that there should be a first-order phase transition to an inhomogeneous state with body-centred cubic symmetry as temperature is lowered. An exception is for a 50 % composition of a symmetric diblock

* The previous theoretical report[3b] of a transition at 141 °C was an estimate based on spherical rather than the more stable cylindrical geometry.

copolymer, in which case the transition is second order. (Note that our calculations, described in the last paragraph, were for such a composition.)

12. OTHER THEORIES

Quite a few authors have advanced theories of the domain structure in block copolymers. In this section we will point out differences from and similarities to our own. One methodological difference may be regarded as important. Our formulation begins with a general theory of inhomogeneous polymer systems[4] (as embodied in eqn (25)), and works toward the phenomenological description by means of well-defined approximations whose validity can be tested. All the other theories make *ad hoc* assumptions about what the important phenomena are and how they can be quantified as contributions to the free energy.

For the purposes of this section it seems worthwhile to divide block copolymer theories into three groups. One is of those theories,[22] all advanced in the early days, which have omitted important effects now recognised. We will say no more about this category, except to hope that none of the authors disagrees with our classification. The second set of theories (Meier;[8-10] Inoue, Soen, Hashimoto and Kawai;[23] Leary and Williams;[24] and Krigbaum, Yazgan, Tolbert and Boehm[25]) are related to the work of Meier, who was the first to have identified the important elements of a statistical thermodynamic theory of block copolymers. The final class of theory is our own, outlined earlier.

The theory of Meier has evolved from its original formulations[8] and, in particular, additional attention has been devoted to the consequences of there being an interphase region. The importance of this region was emphasised in the theory of Leary and Williams,[24] as well as in our work.[3] Meier begins with a picture of the system as existing in an array of pure microdomains separated by an interphase of thickness λ. He arbitrarily assumes that the density varies across the interphase according to $\sin^2(\pi x/2\lambda)$. He calculates the entropy of mixing in this region (with only a minor numerical error, in our view, due to the assumed density variation, cf. eqn (12)). He also estimates the entropy loss associated with restrictions on placement of the joint near the interface. (Again the error introduced by assuming that the joint is uniformly distributed in the interphase is minor.) Then Meier calculates a term which he attributes[10] to 'the loss in entropy arising from the restriction of the block segments to their respective domain regions and with the perturbations of chain dimensions that occur when the block M.W.'s are unequal.' It is our feeling that this term is quite defective because it is based on random walk statistics modified only by the presence of domain boundaries, instead of a random walk calculation which keeps the density constant, as in eqn (25) above. Efforts to make the density more uniform by homogeneous stretching of the molecules are of some value, but how much it is hard to say. Missing from the Meier theory is any term which accounts for the conformational entropy loss involved in turning the chain back into its domain at the interface, which should depend on λ. In our view this prevents the interface from developing proper structure. A measure of the deficiency of the theory is its inability to provide what we regard as an accurate description of the interface between two homopolymers. On the one hand it appears to predict A–B mixing in only a narrow region, of the order of the range of intermolecular forces and independent of χ. On the other hand it produces spurious overall density deviations a distance of order $bZ^{1/2}$ from the interface.

In one sense the theories of Meier and of Leary and Williams are more ambitious than our NIA. They attempt to go beyond the separation of the terms of our NIA free energy, eqn (7), to make the surface and joint placement terms depend on an interface thickness parameter, to be determined by free energy minimisation. Efforts to do this within the context of our NIA are in progress.

In addition Meier has probed theoretically the effect of added solvent.[9] Some work along these lines has also been done by Hong and Noolandi[26] combining the formalisms presented in this paper for the block copolymer and elsewhere for polymer/solvent systems.[27]

13. CONCLUDING REMARKS

The theory expounded here is a statistical thermodynamic one; i.e. it describes equilibrium or constrained equilibrium situations. It is by no means certain that equilibrium has been achieved in many of the experiments reported. In fact, it has evidently not in samples which exhibit poorly formed domain geometries and differences of domain structure depending on the casting solvent. To some extent equilibrium can be approached by annealing, or preparation of the sample in flow fields which aid in producing appropriate order. However, there is still a problem associated with the fact that as the temperature is lowered, or solvent removed, in sample preparation, materials such as PSt will become glassy so that the domain size and geometry will be 'frozen in.' In general it may be difficult for systems to adjust domain size, especially when the domains are spherical. This involves movement of blocks across foreign domains with large-scale reorganisation, for which there may be no convenient path and only small driving force. Very high M.W. blocks may be especially sluggish.

A matter which we also wish to discuss, in conclusion, is the lack of measurements of required parameters. We urge further work to determine the mixing energy parameters of the common block copolymer materials, especially the temperature-dependence. Precise determinations of the Kuhn statistical length are also required, and in some cases pure densities have not been reported.

We applaud the careful work of Hashimoto, Kawai *et al.*[12] on block copolymer interfaces, and especially their efforts to interpret critically the experiments in a contemporary theoretical context. The predictions of Meier's theory and ours differ on interfacial behaviour, so that such measurements provide a crucial test. We believe that work on homopolymer interfaces is also important as a probe of the general theoretical foundations and as a possible means of measuring the interaction parameter. We hope that the implications of Meier's theory for homopolymer interfaces will be worked out.

The problem of stability of the domain structure on heating or adding solvent now appears to be not as simple as it was once thought to be. Further theoretical work is called for, as well as experiments on polymers more compatible than PSt and PBd.

Theoretical work is needed on the effects of adding homopolymer to block copolymers, and on the dramatic role of block copolymers as compatibilising agents.

There are many questions which remain to be answered about amorphous block copolymer systems. Would that as much were known theoretically, though, about systems in which one of the components forms crystalline domains.

REFERENCES

1. HELFAND, E., *Polym. Prepr.*, 1973, **14**, 70.
2. HELFAND, E., *Macromolecules*, 1975, **8**, 552.
3. HELFAND, E. and WASSERMAN, Z. R., (*a*) *Macromolecules*, 1976, **9**, 879; (*b*) *Macromolecules*, 1978, **11**, 960; (*c*) *Macromolecules*, 1980, **13**, 994; (*d*) *Polym. Eng. Sci.*, 1977, **17**, 535.
4. HELFAND, E., *J. Chem. Phys.*, 1975, **62**, 999.
5. HELFAND, E. and TAGAMI, Y., *J. Polym. Sci., B*, 1971, **9**, 741; *J. Chem. Phys.*, 1971, **56**, 3592.

6. HELFAND, E. and SAPSE, A. M., *J. Chem. Phys.*, 1975, **62**, 1327.
7. HELFAND, E., in: *Polymer Compatibility and Incompatibility*, ed. K. Solc, MMI Press, Midland, Michigan, to be published.
8. MEIER, D. J., *J. Polym. Sci., C*, 1969, **26**, 81.
9. MEIER, D. J., *Polym. Prepr.*, 1970, **11**, 400; *Block and Graft Copolymers*, ed. J. J. Burke and V. Weiss, Syracuse University Press, Syracuse, NY, 1973.
10. MEIER, D. J., *Prepr. Polymer Colloquium*, Society Polymer Science, Kyoto, 1977; *Polym. Prepr.*, 1974, **15**, 171.
11. ROUNDS, N. A., Thermodynamics and phase equilibria of polystyrene–polydiene binary mixtures. *Doctoral Dissertation*, University of Akron, 1971; MCINTYRE, D. and ROUNDS, N. A., private communication.
12. HASHIMOTO, T., SHIBAYAMA, M. and KAWAI, H., *Macromolecules*, 1980, **13**, to be published; HASHIMOTO, T., FUJIMURA, M. and KAWAI, H., *Macromolecules*, 1980, **13**, to be published.
13. HELFAND, E. and WASSERMAN, Z. R., in preparation.
14. GALLOT, B. R. M., *Adv. Polym. Sci.*, 1978, **29**, 85.
15. Reports of experimental observations, contained in Chapter 3, this book.
16. HOFFMAN, M., KÄMPF, G., KRÖMER, H. and PAMPUS, G., *Adv. Chem. Ser.*, 1971, **91**, 351.
17. CHUNG, C. I. and GALE, J. C., *J. Polym. Sci., Polym. Phys. Ed.*, 1976, **14**, 1149; CHUNG, C. I. and LIN, M. I., *ibid.*, 1978, **16**, 545.
18. GOUINLOCK, E. V. and PORTER, R. S., *Polym. Eng. Sci.*, 1977, **17**, 573.
19. CHUNG, C. I., GRIESBACH, H. L. and YOUNG, L., *J. Polym. Sci., Polym. Phys.* Ed., 1980, **18**, 1237.
20. ROE, R.-J. and CHANG, J. C., *Bull. Am. Phys. Soc.*, 1979, **24**, 256.
21. LEIBLER, L., *Macromolecules*, 1981, **14**, 727.
22. KRAUSE, S., *J. Polym. Sci., A-2*, 1969, **7**, 249; *Macromolecules*, 1970, **3**, 84. BIANCHI, U., PEDEMONTE, E. and TURTURRO, A., *J. Polym. Sci., B*, 1969, **7**, 785; *Polymer*, 1970, **11**, 268. MARKER, L., *Polym. Prepr.*, 1969, **10**, 524. KRÖMER, H., HOFFMANN, M. and KÄMPF, G., *Ber. Bunsenges. Physik. Chem.*, 1970, **74**, 859 (cf. reference 16).
23. INOUE, T., SOEN, T., HASHIMOTO, T. and KAWAI, H., *J. Polym. Sci., A-2*, 1969, **7**, 1283.
24. LEARY, D. and WILLIAMS, M., *J. Polym. Sci., B*, 1970, **8**, 335; *J. Polym. Sci., Polym. Phys. Ed.*, 1973, **11**, 345; 1974, **12**, 265.
25. KRIGBAUM, W., YAZGAN, S. and TOLBERT, W., *J. Polym. Sci., Polym. Phys. Ed.*, 1973, **11**, 551; BOEHM, R. and KRIGBAUM, W., *J. Polym. Sci., C*, 1976, **54**, 153.
26. HONG, K. M. and NOOLANDI, J., *Macromolecules*, 1980, **13**, 1602.
27. HELFAND, E. and SAPSE, A. M., *J. Polym. Sci., C*, 1976, **54**, 289.

APPENDIX

```
C          PROGRAM TO DETERMINE DOMAIN SIZE
C      IN A LAMELLAR, CYLINDRICAL OR SPHERICAL
C             BLOCK COPOLYMER SYSTEM
C                   SI UNITS
```

```
C
      INTEGER BLANK,PA,PB,DIMEN,DOMTYP(4,5),UNIT1(3),UNIT2(3),UNIT3(3)
      REAL RHO(5),BETA(5),XM(5),B(5),MWA,MWB
      INTEGER POLYMR(5)
      DATA AVOGAD/6.02252E23/
      DATA BLANK/1H /,UNIT1/3H(/M,3H(/M,1H /,UNIT2/3H**2,1H),1H /,
     $ UNIT3/1H),2*1H /
      DATA DOMTYP/3H   ,3HLAM,3HELL,3HAR ,3HCYL,3HIND,3HRIC,
     $ 2HAL,3H  S,3HPHE,3HRIC,2HAL/
 10   DATA RHO/10100.,13600.,16500.,2*0./,
     $     XM/.10414,.06811,.05409,2*0./,
     $     B/.68E-9,.63E-9,.63E-9,2*0./,
     $     POLYMR/2HPS,2HPI,2HPB,2*1H /
      DATA NPOL/3/
      FUN(X)=-C1+X+X**3*(C3*(C2*X**2+C4)**C5+C7*(C6*X**2+C8)**C9)
C  INPUT SECTION
C  PA AND PB ARE THE NAMES OF THE POLYMER COMPONENTS, A
C   REFERS TO THE ONE WITH CYLINDRICAL OR SPHERICAL DOMAINS
C  DIMEN IS THE DIMENSIONALITY OF THE SYSTEM
C  J INDICATES THE TYPE OF BLOCK COPOLYMER
C    1 = DIBLOCK COPOLYMER, A CYLINDERS OR SPHERES
C    2 = TRIBLOCK COPOLYMER BAB
C    3 = TRIBLOCK COPOLYMER ABA
C  MWA AND MWB ARE THE MOLECULAR WTS. OF THE A AND B BLOCKS (KG)
C  T IS THE TEMPERATURE
C  BLANK LINE AS INPUT TERMINATES THE RUN
C
 20   WRITE(6,999)
      READ (5,998) PA,PB,DIMEN,J,MWA,MWB,T
      IF(PA.EQ.BLANK) STOP
      IF(DIMEN.GE.1.AND.DIMEN.LE.3) GO TO 30
      WRITE(6,990)
      GO TO 20
 30   IF (J.GE.1 .AND. J .LE. 3) GO TO 40
      WRITE(6,997)
      GO TO 20
 40   IA=0
      IB=0
      DO 50 I=1,NPOL
      IF(PA.EQ.POLYMR(I)) IA=I
      IF(PB.EQ.POLYMR(I)) IB=I
 50   CONTINUE
      IF(IA.NE.0.AND.IB.NE.0) GO TO 60
      WRITE(6,996) (POLYMR(I),I=1,NPOL)
      GO TO 20
C  PARAMETER ASSIGNMENT AND CALCULATION
 60   ALPHA=-900.+750000./T
      ZA=MWA/XM(IA)
      Z1=ZA
      IF (J .EQ. 2) Z1=Z1*.5
      ZB=MWB/XM(IB)
      Z2=ZB
      IF (J .EQ. 3) Z2=Z2*.5
      BETA(IA)=SQRT(RHO(IA)/6.)*B(IA)
      BETA(IB)=SQRT(RHO(IB)/6.)*B(IB)
      BETAAV=.5*(BETA(IA)+BETA(IB))
 70   GAMOKT=SQRT(ALPHA)*(BETAAV+(BETA(IA)-BETA(IB))
     $ **2/(12.*BETAAV))
      ZORA=Z1/RHO(IA)
      ZORB=Z2/RHO(IB)
      ZORS=ZORA+ZORB
      RXI=((ZORB/ZORA)+1.)**(1./DIMEN)-1.
      XI=1./RXI
      BTG=AMAX1(BETA(IA),BETA(IB))
      BTS=AMIN1(BETA(IA),BETA(IB))
      Y=(BTS/BTG)**2
      AJ=(BTG/SQRT(ALPHA))*(1.+.4630151*Y+.1077812*Y**2
     $     -(.2452727*Y+.0412496*Y**2)*ALOG(Y))
C  PARAMETERS OF THE DOMAIN FREE ENERGY
      GO TO (80,110,140),DIMEN
 80   IF(J.NE.2) GO TO 90
      ETAA1=.061
      ETAA2=.838
      ETAA3=2.64
      GO TO 100
```

```
90  ETAA1=.0766
    ETAA2=.531
    ETAA3=2.57
    IF(J.NE.3) GO TO 100
    ETAB1=.061
    ETAB2=.838
    ETAB3=2.64
    GO TO 170
100 ETAB1=.0766
    ETAB2=.531
    ETAB3=2.57
    GO TO 170
110 IF(J.NE.2) GO TO 120
    ETAA1=.0166
    ETAA2=1.051
    ETAA3=2.86
    GO TO 130
120 ETAA1=.0274
    ETAA2=.202
    ETAA3=2.605
    IF(J.NE.3) GO TO 130
    ETAB1=.0659+.0357*RXI
    ETAB2=.757-.132*RXI
    IF(ETAB2.LT.0.) ETAB2=0.
    ETAB3=2.59-.147*RXI
    GO TO 170
130 ETAB1=.0792+.0412*RXI
    ETAB2=.539-.206*RXI
    IF(ETAB2.LT.0.) ETAB2=0.
    ETAB3=2.53-.152*RXI
    GO TO 170
140 IF(J.NE.2) GO TO 150
    ETAA1=.00810
    ETAA2=1.07
    ETAA3=2.87
    GO TO 160
150 ETAA1=.0123
    ETAA2=.384
    ETAA3=2.68
    IF(J.NE.3) GO TO 160
    ETAB1=.0705+.0578*RXI
    ETAB2=.596+.599*RXI
    ETAB3=2.55-.132*RXI
    GO TO 170
160 ETAB1=.0852+.0744*RXI
    ETAB2=.348+.220*RXI
    ETAB3=2.48-.123*RXI
170 C1=DIMEN*ZORA*GAMOKT
    C2=6./(B(IA)**2*Z1)
    C3=ETAA1*ETAA3*C2
    C4=ETAA2**2
    C5=ETAA3*.5-1.
    C6=(6./Z2)*(RXI/B(IB))**2
    C7=ETAB1*ETAB3*C6
    C8=ETAB2**2
    C9=ETAB3*.5-1.
C   CALCULATION OF THE FREE ENERGY MINIMUM
    A=C1
    EPS=1.E-15
    AS=.99*A
    FS=FUN(AS)
    F=FUN(A)
180 ANEW=A-F*(A-AS)/(F-FS)
    AS=A
    A=ANEW
    IF(A.NE.AS) GO TO 190
    WRITE(6,995)
    EPS=EPS*10.
    IF(EPS.LE.1.E-8) GO TO 190
    WRITE(6,994)
    GO TO 20
190 FS=F
    F=FUN(A)
    IF(ABS(F).GT.EPS) GO TO 180
C   CALCULATION OF SYSTEM'S PROPERTIES
    SIGMA=DIMEN*ZORA/(A*AVOGAD)
    T1=(1.+RXI)**DIMEN
```

```
      IF(J.EQ.2) T1=SQRT(T1)
      IF(J.EQ.3) T1=SQRT(T1*(T1-1.))
      F=C1/A+ALOG(A*T1/(DIMEN*AJ))
     $ +ETAA1*(C2*A**2+C4)**(.5*ETAA3)-ETAA1*ETAA2**ETAA3
     $ +ETAB1*(C6*A**2+C8)**(.5*ETAB3)-ETAB1*ETAB2**ETAB3
     $ -ALPHA*ZORA*ZORB/ZORS
      DA=2.*A
      GO TO (200,210,220),DIMEN
200   D=2.0
      XN=2.0
      GO TO 230
210   D=1.904626
      XN=3.14159
      GO TO 230
220   D=1.80940
      XN=4.188790
230   D=D*(ZORS/ZORA)**(1./DIMEN)*A
      XN=XN*A**DIMEN*AVOGAD*RHO(IA)/ZA
C     OUTPUT
      IF(J.NE.1) GO TO 260
      GO TO (240,250,250),DIMEN
240   WRITE(6,993)POLYMR(IA),POLYMR(IB),(DOMTYP(J,DIMEN),J=1,4)
      GO TO 320
250   WRITE(6,993)POLYMR(IA),POLYMR(IB),(DOMTYP(J,DIMEN),
     $ J=1,4),POLYMR(IA)
      GO TO 320
260   IF(J.NE.2) GO TO 270
      IOUT=IB
      IIN=IA
      GO TO 280
270   IOUT=IA
      IIN=IB
280   GO TO (290,300,300),DIMEN
290   WRITE(6,992) POLYMR(IOUT),POLYMR(IIN),POLYMR(IOUT),
     $ (DOMTYP(J,DIMEN),J=1,4)
      GO TO 310
300   WRITE(6,992) POLYMR(IOUT),POLYMR(IIN),POLYMR(IOUT),
     $ (DOMTYP(J,DIMEN),J=1,4),POLYMR(IA)
310   F=2.*F
      XN=.5*XN
320   WRITE(6,991) POLYMR(IA),MWA,POLYMR(IA),ZA,POLYMR(IA),DA
      WRITE(6,989) POLYMR(IB),MWB,POLYMR(IB),ZB,D
      WRITE(6,988) T,XI,F
      WRITE(6,987) UNIT1(DIMEN),UNIT2(DIMEN),UNIT3(DIMEN),XN,SIGMA
      GO TO 20
999   FORMAT(//41H PA,PB,DIMENSION,J,MOL. WT. A,MOL. WT. B,4HTEMP)
998   FORMAT(2(A2,1X),2I1,2F6.2,F10.4)
997   FORMAT(19H J MUST = 1, 2 OR 3/)
996   FORMAT(35H0ILLEGAL POLYMER NAME - THE LEGAL
     $ 11H NAMES ARE /20(3XA2))
995   FORMAT(19H0ACCURACY DECREASED)
994   FORMAT(12H NO SOLUTION)
993   FORMAT(/12X19H DIBLOCK COPOLYMER ,A2,1H-,A2,
     $ 6H WITH 4A3,8HDOMAINS,A2)
992   FORMAT(/12X20H TRIBLOCK COPOLYMER ,A2,1H-,A2,1H-,
     $ A2,6H WITH 4A3,8HDOMAINS ,A2)
991   FORMAT(9H0MOL.WT.-A2,5H (KG)F9.2,6X2HZ-A2,F10.2,6X
     $ 13HDOMAIN DIAM.-,A2,4H (M)1PE11.4)
990   FORMAT(33H DIMENSIONALITY MUST BE 1, 2 OR 3)
989   FORMAT(9H MOL.WT.-A2,5H (KG)F9.2,6X2HZ-A2,F10.2,6X
     $ 19HLATTICE SPACING (M)1PE11.4)
988   FORMAT(6H T (K)F19.2,6X2HXIF12.3,6X5HF/NKT1PE25.4)
987   FORMAT(3H N 2A3,A1,1PE15.4,26X13HSIGMA (/M**2)1PE17.4)
      END
```

(Note: In order to make the solution of eqn (25) more tractable for spherical domains the unit cell of the lattice has been sphericalised. This precludes calculation of any free energy difference between different lattices. However, the relation between domain radius and spacing does depend on lattice type. The above program is for a close-packed geometry. To modify for a body-centred cubic, on the line labelled 220 change 1·809 40 to 1·758 88.)

Chapter 5

HETEROCHAIN BLOCK COPOLYMERS: SYNTHESIS AND GENERAL PROPERTIES

I. GOODMAN

School of Polymer Science,
University of Bradford, UK

SUMMARY

Polymers with heteroatom-containing groups as recurring features of the backbone chain structure have occupied a prominent position throughout the history of development of block copolymers and some of the technologically most important block copolymers are members of this class.

Following an outline of the general strategies available for the construction of block copolymers, this chapter reviews the extensive progress of recent years in the evolution of methods for the synthesis of heterochain block copolymers. Developments in precursor oligomer synthesis and in methods for the investigation of blockiness in imperfectly sequenced copolymers are also noted.

The second part of the chapter considers some aspects of structure–property relationships in polymers containing heterochain blocks. Particular attention is given to the phase state and crystallisation tendency of some individual types of blocks in varied molecular environments. The tendency of regularly structured heterochain blocks to undergo phase separation by crystallisation proves to be acutely sensitive to the nature of the other component(s) of structure but such phase separation is often apparent at remarkably small average block sizes and, when present, has a dominating influence upon the morphology and properties of the block copolymers.

1. INTRODUCTION AND HISTORICAL SURVEY

This chapter is concerned with block copolymers having, in at least one type of the constituent blocks, atoms of an element or elements other than carbon as recurrent structural features of the main chain. Substances conforming to this general description were the earliest block copolymers to be studied, and the class includes some of the present day technologically most important block macromolecular materials.

The history of synthetic heterochain block copolymers began effectively in the early 1940s with the introduction of Bayer's 'diisocyanate polyaddition procedure' as a method for the linking together of molecules of preformed aliphatic polyesters, polyesteramides and similar substances to obtain elastomers.[1–6] At that time ideas of elastomer structure and properties were dominated by experience with natural rubber, and there was a prevalent belief that covalent crosslinking was an indispensable prerequisite for serviceability in such materials. The thinking of Bayer and his coworkers was therefore directed largely towards cast elastomers containing network structures, though it is now apparent that at least some of their interesting products were of linear block polymeric character. The work of that period led eventually to the now well known linear polyester- and polyetherurethane thermoplastic block copolymer elastomers and to the elastic ('spandex') fibres in both of which groups the recovery properties and restraint to viscous flow arise from physical constitutional factors rather than from a system of covalent crosslinks.

A second fertile theme in heterochain block copolymers developed from work in the late 1940s by Coleman[7] whose interest in modifying the properties of poly(ethylene terephthalate) fibres was influenced by knowledge (new at that time) of the molecular structure of silk fibroin which had been recognised to be partially of block character. On the basis of this model, poly(oxyethylene) blocks were introduced into the polyester chain structure yielding fibre-forming materials of improved hydrophilicity and dyestuff-receptivity. The melting temperatures of the copolymers, unlike those of random copolymers with similar weight percentages of modifying units, were not greatly reduced from that of the homopolyester, leading to the conclusion (which has significance for block copolymers generally) that this property is determined by the molar rather than the weight fraction of modifying unit. Coleman's investigation was restricted to essentially hard copolymers containing limited amounts of combined polyether units. The composition range was later widened by Charch and Shivers[8] who found that the incorporation of larger amounts (40–70 wt %) of polyether unit gave elastomeric products with good tensile recovery and stress decay characteristics. From this pattern of a linear chain structure composed of high melting readily crystallisable hard segments intercalated with low melting soft segments there has developed the important group of thermoplastic polyetherester and block copolyester elastomers which appeared as commercial materials in the 1970s.

The third class of novel heterochain block copolymers, inspired by interest in non-ionic detergents, was evolved in the early 1950s by Lundsted and coworkers[9-11] who found that substances composed of a central poly(oxypropylene) unit, itself having a sufficient M.W. (>900) to be insoluble in water as a homopolymer and flanked by poly(oxyethylene) blocks, were water-soluble liquids or solids, according to the lengths of the end blocks, with valuable surfactant and detergent properties.

Subsequent developments, which have greatly expanded the range of heterochain block copolymers, form the main subject of this chapter and will be mentioned here only in outline. The discovery by Szwarc of methods for achieving sequential polymerisations by homogeneous termination-free ('living') reactions of vinyl monomers has led not only to the well-established polyvinyl–polydiene block polymer elastomers but also to procedures of wide potential for combining polyvinyl species with heterochain block species through a variety of strategies of synthesis.[12-15] The late 1950s also saw the start of work on polysiloxane-containing block copolymers[16-19] of which those with poly(oxyalkylene) co-units quickly gained industrial importance as stabilising surfactants for the control of cell growth and form in the manufacture of flexible polyurethane foams.[20]

Until well into the 1960s interest in block copolymers as bulk materials was centred particularly upon those of elastomeric character but investigation has since developed on plastics materials of block or segmented character, especially those containing aromatic polycarbonate or polysulphone sequences.[21-25] The 1970s brought an extension and intensification of work in these directions, together with the introduction of new themes of study—both for carbochain and heterochain block polymers—in, for example, polyionic materials,[26-34] copolymers containing blocks of naturally occurring macromolecules,[35-40] and the use of block copolymers as catalysts.[41-43]

These utilitarian developments have naturally been accompanied by a large body of fundamental investigation, partly concerned with the evolution of new methods of synthesis and partly with the exploration of that inner texture of block copolymers which confers their distinctive properties and differentiates them from random copolymers. The phase-separated character of amorphous polyvinyl–polydiene block polymers has been recognised and investigated intensively since the later 1960s and is discussed in detail in Chapters 1, 3 and 4. For heterochain block copolymers, which are of more polar character and often semi-crystalline, the elucidation of supramolecular organisation is a more complex undertaking which calls, for its resolution, on the whole armamentarium of modern polymer physics. Advances in this field have been a notable feature of the later 1970s.

Detailed accounts of earlier work on heterochain block copolymers are given in two standard monographs published during the 1970s.[44,45] The present chapter will necessarily cover some of the ground in order to establish the context for consideration of later developments which are surveyed up to the end of 1980. The approach throughout will be illustrative rather than encyclopaedic, firstly with regard to synthesis and, in the later part of the chapter, by means of a comparative survey of some selected block copolymer properties.

1.1. Nomenclature and Symbolism

The development of any science depends upon the ability of its practitioners to communicate in precise terms and to retrieve information from original and secondary sources without loss or distortion of meaning. These requirements would ordinarily presuppose the existence of a pertinent *lingua franca*. Even for relatively simple homopolymers, however, nomenclature and classification presents a complex and incompletely resolved problem,[46,47] and no generally accepted system yet exists for the naming of block copolymers so as to describe precisely their chemical composition and molecular structural organisation. Some examples of the difficulties are given below:

(a) No firm level of definition exists as to the molecularity appropriate for a component of copolymer structure to be recognised as a 'block'. Conventionally the term is often applied to any structural component which is of sufficient size to display in the copolymer some characteristic feature(s) of the corresponding free homopolymer, yet this view would exclude the case where polymeric blocks of one component are joined together by a small linking group which may modify considerably the properties of the major component without displaying its own homopolymer character. Such copolymers are sometimes termed segmented polymers but this term is also frequently used as synonymous with block polymers; another common designation for these materials is as chain-extended or simply extended polymers.

(b) Many block copolymers are synthesised by the direct or indirect combination of preformed, relatively low M.W. polymeric precursors furnished with reactive end-groups. Such precursors are known variously—and sometimes with different nuances of meaning—as prepolymers, telechelic polymers, oligomers, macromers, segmers or as end-capped or tipped polymers. For uniformity, such precursors will be described as oligomers throughout this chapter.

(c) A typical procedure for the synthesis of some classes of block copolymers commences with the living polymerisation of a monomer, e.g. styrene, initiated bidirectionally by a dianion, e.g. the tetrameric dianion derived from α-methylstyrene (**A**). When all of the styrene has been polymerised, the reactivity of the anionic chain ends is modified by treatment with a reagent **B** so as to permit sequential polymerisation with a second monomer **C**; the ensuing living block copolymer is finally stabilised by treatment with a reagent **D**. The product, which has the general structure (I) where **A**′, **B**′, **C**′, **D**′ represent the residues of the original active ingredients, will often be described as poly-**C**-*b*-polystyrene-*b*-poly-**C** even though this expression pays no regard to the foreign structure fragments resulting from the various steps of chemical reaction, or to the inversion of chain direction that occurs at the centre of the $(\mathbf{A}')_4$ unit.

$$\mathbf{D}'(\text{poly-}\mathbf{C}')\text{—}\mathbf{B}'\text{—}(\text{polystyrene})\text{—}(\mathbf{A}')_4\text{—}(\text{polystyrene})\text{—}\mathbf{B}'\text{—}(\text{poly-}\mathbf{C}')\mathbf{D}'$$

(I)

(d) The polyetheresters already mentioned typify the class of random block copolymers which present a different problem in designation. An important group of these materials is usually prepared by the co-condensation of dimethyl terephthalate with a mixture of tetramethylene glycol (1,4-butanediol) and the oligomeric diol (II)

$$HO[(CH_2)_4O]_xH$$

(II)

which is itself a polyether. The relative proportions of the two diols can be varied at will, and the molecules of the products (which, strictly, are (polyether)ester–polyester block copolymers) are composed of blocks of units (III) and (IV) placed randomly along the chain and with the values of *y* and *z* related to the diol proportions used.

$$[\!-\!O(CH_2)_4O\,.\,CO\!-\!C_6H_4\!-\!CO\!-\!]_y$$

(III)

$$[\!-\!O[(CH_2)_4O]_x\,.\,CO\!-\!C_6H_4\!-\!CO\!-\!]_z$$

(IV)

The structure of (III) will be easily recognised from its common verbal designation as poly(tetramethylene terephthalate), though perhaps less readily from the systematic name poly(oxy-1,4-butanediyloxycarbonyl-1,4-phenylenecarbonyl) now recommended by IUPAC. What, however, of (IV)? The precursor oligomer (II) is named variously in the literature as polytetrahydrofuran, polytetramethylene oxide, polytetramethylene glycol, polyoxytetramethylene glycol, poly(oxytetramethylene) glycol, polytetramethylene ether glycol and polybutylene oxide glycol; the writer is aware of a case where four of these names appear in a single published paper! It will be obvious that certain of these terms are structurally inexplicit and stoichiometrically inaccurate as literal equivalents of formula (II). Additionally, they are not readily adaptable to the precise naming of (IV), a particular problem arising from the 'extra' oxygen atom (with respect to the repeating unit) remaining from the left-hand hydroxyl group of (II).

An important function of nomenclature is for the storage and retrieval of information. It is therefore of interest to examine, with the help of this example, the practice of *Chemical Abstracts*—the prime chemical search tool of the Western world—which, except for a small number of macromolecular materials of well-defined structure, indexes polymers on the basis of their *supposed* monomers or precursors rather than as substances in their own right. Until 1971,[48] the substances in question were indexed under 'terephthalic acid, polyester with 1,4-butanediol and polytetramethylene glycol'. Subsequently,[49] partial use has been made of IUPAC-recommended structure-based nomenclature, leading to a cycle of expressions of the type '1,4-benzenedicarboxylic acid, polymer with 1,4-butanediol and α-hydro-ω-hydroxypoly(oxy-1,4-butanediyl)', permuted to commence with each reactant in turn. It will be noted that these terms presuppose what is not the case in practice, that terephthalic acid (1,4-benzenedicarboxylic acid) is an immediate precursor of the polymers.

However, further search will show the polymers located under a cycle of names of the type 'poly(oxy-1,4-butanediyl),α-hydro-ω-hydroxy, polymer with 1,4-butanediol and dimethyl 1,4-benzenedicarboxylate', and will reveal that the assigned *Chemical Abstracts Registry* numbers, which are intended to be unique with respect to molecular structure, are different for the phantom acid-based polymer [37282-12-5] and real ester-based polymer [9078-71-1]. Clearly the full recovery of information (and the avoidance of mis-information) on an unfamiliar block copolymer calls for a combination of skills in lexicography and chemical guesswork that it is not necessarily possessed by all polymer scientists.

Faced with these difficulties, the terminology used in this chapter will be pragmatic and semi-structural, unavoidably containing some stoichiometric imprecisions, but aiming at a reasonable middle course between the extremes of inscrutable rigour and misleading jargon. To avoid excessively cumbrous formulae and names, considerable use will be made of the abbreviations for structural units listed near the beginning of this volume and of the additional symbols ϕ for the phenylene group (*para*-orientation unless otherwise indicated), **P** as a generalised representation of a polymer molecule, **M** for a monomer and **R** for other reactants. Different varieties of such generalised species will be denoted by superscript letters ($\mathbf{P}^A$, $\mathbf{P}^B$, etc.), and reactive end groups, where the nature of these does not need to be specified precisely, by ◆ or ●. Where appropriate, block copolymers will be indicated by *-b-* between the names of the components, and random copolymers by *-co-*. Poly(alkylene terephthalates) will be abbreviated in the familiar way as 2G.T, 4G.T, etc., according to the numbers of carbon atoms in the alkylene moiety, and linear aliphatic polyesters as 2G.6, 4G.6, etc., the final cipher being the carbon number of the residue of a linear alkanedicarboxylic acid; the oxy(methyl-1,2-ethanediyl)oxy group (—$OCHMeCH_2O$—), as in poly(1,2-propylene terephthalate), will be abbreviated as 1,2-P. The simple alkyl groups from methyl to butyl will be shown in formulae as Me, Et, Pr and Bu, and generalised alkylene and arylene groups as Alk and Ar, respectively.

1.2. Principles of Block Copolymer Synthesis

The following brief account of the major approaches to heterochain block copolymer synthesis will serve as an introduction to the developments described in Section 2.

1.2.1. Adventitious and Imperfect Block Copolymer Syntheses

Copolymers containing some measure of ordered sequences of the component units (blockiness) result from certain ostensibly random copolymerisation reactions, particularly of the condensation and step-polymerisation type, as a result of differential reactivities amongst the functional groups present. Thus, if a bifunctional reactant $\mathbf{R}^A$ (e.g. a dicarbonyl chloride) reacts with a stoichiometrically equivalent mixture of two co-reactants $\mathbf{R}^B$ and $\mathbf{R}^{B'}$ which are of similar chemical type (e.g. diamines) but different reactivities such that $\mathbf{R}^A$–$\mathbf{R}^B$ unions are formed more readily than $\mathbf{R}^A$–$\mathbf{R}^{B'}$, there may be an enrichment of the first-mentioned unions in the early stages of combination leading to partial block structure. An analogous result may occur in the reaction of $\mathbf{R}^A$ with bifunctional $\mathbf{R}^B$ which has combining groups of different reactivities; in this case the blockiness can take the form of a preference for head-to-tail combinations of the units rather than their random head-to-tail/head-to-head presentation along the chain.

Another type of adventitious block polymer formation occurs in the partial inter-randomisation in the melt of mixtures of homopolymers such as polyesters and polyamides which can undergo ester–ester, amide–amide or ester–amide interchange reactions.

These procedures suffer from the deficiencies (1) of relative structural randomness and lack of well-defined sequential specificity in the products, (2) of non-equilibrium character which can lead to continued group interchange and eventual complete randomisation on further thermal treatment.

1.2.2. Synthesis by Sequential Polymerisation

This class of reactions is represented most prominently by the living anionic polymerisations in which a first monomer $\mathbf{M}^A$ is converted to its corresponding polymer $\mathbf{P}^A$ having propagatively reactive groups at one or both ends of its chain according to whether a mono- or a bifunctional initiator is employed. The living polymer then serves as initiator for the polymerisation of a second monomer $\mathbf{M}^B$, yielding a diblock copolymer (V) or a triblock copolymer (VI) depending on the functionality of $\mathbf{P}^A$ (eqn. (1)).

$$n\,\mathbf{M}^A \xrightarrow{\text{monofunctional initiator}} \mathbf{P}^A\blacklozenge \xrightarrow{n\,\mathbf{M}^B} \mathbf{P}^A\text{–}\mathbf{P}^B \quad \text{(V)}$$

$$n\,\mathbf{M}^A \xrightarrow{\text{bifunctional initiator}} \blacklozenge\mathbf{P}^A\blacklozenge \xrightarrow{2n\,\mathbf{M}^B} \mathbf{P}^B\text{–}\mathbf{P}^A\text{–}\mathbf{P}^B \quad \text{(VI)} \qquad (1)$$

The concept is open to considerable elaboration. Thus, the diblock polymer (V), whilst still reactive, may be coupled with a suitable bifunctional reagent to form a triblock copolymer (VII) having $\mathbf{P}^{A}$ units at the outsides of the molecule rather than in the centre as in (VI) (eqn (2)). The reactive forms of (V) and (VI) may also be used to initiate the polymerisation of a further monomer $\mathbf{M}^{C}$ to give products (VIII) and (IX) with three species of blocks per molecule.

$$2\,\mathbf{P}^{A}\text{–}\mathbf{P}^{B}\blacklozenge + \bullet\mathbf{R}\bullet \longrightarrow \underset{\text{(VII)}}{\mathbf{P}^{A}\text{–}\mathbf{P}^{B}\text{–}\mathbf{R}\text{–}\mathbf{P}^{B}\text{–}\mathbf{P}^{A}} \qquad (2)$$

$$\underset{\text{(VIII)}}{\mathbf{P}^{A}\text{–}\mathbf{P}^{B}\text{–}\mathbf{P}^{C}} \qquad \underset{\text{(IX)}}{\mathbf{P}^{C}\text{–}\mathbf{P}^{B}\text{–}\mathbf{P}^{A}\text{–}\mathbf{P}^{B}\text{–}\mathbf{P}^{C}}$$

Success in such continued sequential reaction depends critically upon the real natures of the substances involved, on the reactivity of the end groups existing at each stage or formed by intermediate refunctionalisation, on the efficiency of the reactions employed, and upon the avoidance of interfering side reactions whether extrinsic (e.g. due to terminating impurities) or intrinsic (e.g. due to the occurrence of transfer reactions). In favourable cases the consecutive stages can be achieved with high precision leading to polymers of accurately known structures composed of blocks of predetermined sizes and with narrow M.W. distributions ($\bar{M}_w/\bar{M}_n$ close to unity).

Of sequential polymerisations proceeding by anionic propagation, the most important from the standpoint of this chapter are those using styrene, α-methylstyrene, 1,3-butadiene or isoprene as sources of carbochain blocks, and a limited range of ring-fissile monomers selected from the classes of cyclic ethers and sulphides, lactones, lactams, siloxanes and oxazolidin-2,5-diones as sources of heterochain blocks. Recent work has opened the way to the use of living cationic systems as the basis of heterochain block copolymer synthesis with monomers such as tetrahydrofuran (oxolane) and vinyl alkyl ethers. However, despite its conceptual elegance, sequential polymerisation is subject to considerable limitations with respect to the range of monomers that can be used and to the order in which they can be assembled within the polymer chain. Furthermore, the techniques required are often complex and not readily adaptable to large-scale working.

1.2.3. Random Block Polymerisations

This method of synthesis, which is indicated schematically by eqn (3), comprises essentially a reaction of step-copolymerisation between reactants at least one of which is polymeric. Commonly met systems of this class involve combination through the end group reactant pairs HO + OCN, HO + MeOOC, and NH_2 + ClCO.

$$x \blacklozenge \mathbf{P} \blacklozenge + y \blacklozenge \mathbf{R}^{\mathrm{A}} \blacklozenge + (x+y) \bullet \mathbf{R}^{\mathrm{B}} \bullet \longrightarrow [\text{—}(\mathbf{P}\text{–}\mathbf{R}^{\mathrm{B}})_x\text{—}(\mathbf{R}^{\mathrm{A}}\text{–}\mathbf{R}^{\mathrm{B}})_y\text{—}]_n \quad \text{(X)} \qquad (3)$$

In the simplest procedure (used, for example, for the synthesis of the block polyetheresters already mentioned), a mixture of two reactants of similar chemical type, one being of low M.W. and the other oligomeric, is combined with a common co-reactant $\mathbf{R}^{\mathrm{B}}$ used in stoichiometric equivalence to the total of reactive groups in **P** and $\mathbf{R}^{\mathrm{A}}$. The product (X) is a multiblock copolymer composed of two types of units, one of which includes the structural residue of the starting oligomer; their relative proportions are determined by the values of x and y which can be varied at will, but their placement along the chain is statistical. The following relationships are important in relation to the bulk properties of such copolymers:

(a) Since **P** is oligomeric and $\mathbf{R}^{\mathrm{A}}$ of low M.W., there will be a marked disparity between the weight and molar fractions of the units $\mathbf{P}$–$\mathbf{R}^{\mathrm{B}}$ and $\mathbf{R}^{\mathrm{A}}$–$\mathbf{R}^{\mathrm{B}}$ in the product (X). This is illustrated by Fig. 1 where curves b and d show the relationships for block copolymers in which $\mathbf{R}^{\mathrm{A}}$–$\mathbf{R}^{\mathrm{B}}$ has a M.W. of 200 and $\mathbf{P}$–$\mathbf{R}^{\mathrm{B}}$ has M.W.'s of 1000 and 5000, respectively. In curves a and c, the same M.W.'s are used for $\mathbf{P}$–$\mathbf{R}^{\mathrm{B}}$ but that of $\mathbf{R}^{\mathrm{A}}$–$\mathbf{R}^{\mathrm{B}}$ is taken as 350. This span of values covers that occurring in most practically important block copolymers. It is evident that quite large weight (and hence volume) fractions of the $\mathbf{P}$–$\mathbf{R}^{\mathrm{B}}$ component are present even at relatively low molar fractions.

(b) In random block copolymers (X) made by conventional step-polymerisation reactions, both types of blocks will be polydisperse. Both x and y will thus ordinarily have a distribution of values arising from the statistics of reaction, with a range of block M.W.'s modified further in the case of x by the distribution existing in the starting oligomer $\blacklozenge\mathbf{P}\blacklozenge$. The distribution in values of y is generally unknown but for purposes of comparison, as will be used later, it is helpful to discuss the molecularity of the $\mathbf{R}^{\mathrm{A}}$–$\mathbf{R}^{\mathrm{B}}$ sequences in terms of a notional average segment length (ASL) calculated by assuming the copolymer molecules to be of equal size and made up by interpolating equisized $\mathbf{P}$–$\mathbf{R}^{\mathrm{B}}$ units uniformly into the poly-($\mathbf{R}^{\mathrm{A}}$–$\mathbf{R}^{\mathrm{B}}$)

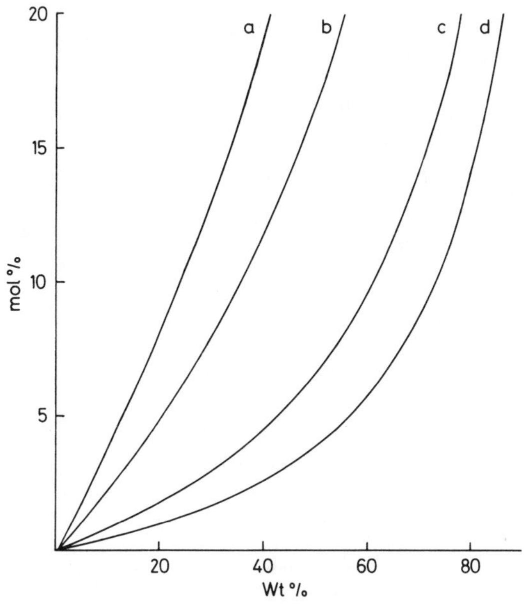

FIG. 1. Mol versus wt % for **P**–$\mathbf{R}^{B}$ blocks in $[—(\mathbf{P}–\mathbf{R}^{B})_x—(\mathbf{R}^{A}–\mathbf{R}^{B})_y—]_n$ multi-block copolymers.

Curve	*Unit M.W.'s*	
	P–$\mathbf{R}^{B}$	$\mathbf{R}^{A}$–$\mathbf{R}^{B}$
a	1000	350
b	1000	200
c	5000	350
d	5000	200

chain. The average segment length is then given by:

$$\text{ASL} = \frac{\text{total number of } \mathbf{R}^{A}\text{–}\mathbf{R}^{B} \text{ units}}{(\text{total number of } \mathbf{P}\text{–}\mathbf{R}^{B} \text{ units}) + 1}$$

Figure 2 shows the relationship for some practically important M.W.'s of $\mathbf{R}^{A}$–$\mathbf{R}^{B}$ and **P**–$\mathbf{R}^{B}$. It can be seen that increase in the weight fraction of the oligomeric component **P**–$\mathbf{R}^{B}$ leads to a rapid fall in the average segment lengths of the $\mathbf{R}^{A}$–$\mathbf{R}^{B}$ units.

Where the chemistry of synthesis permits, as in the formation of block copolyurethanes by the reaction of a diisocyanate with a

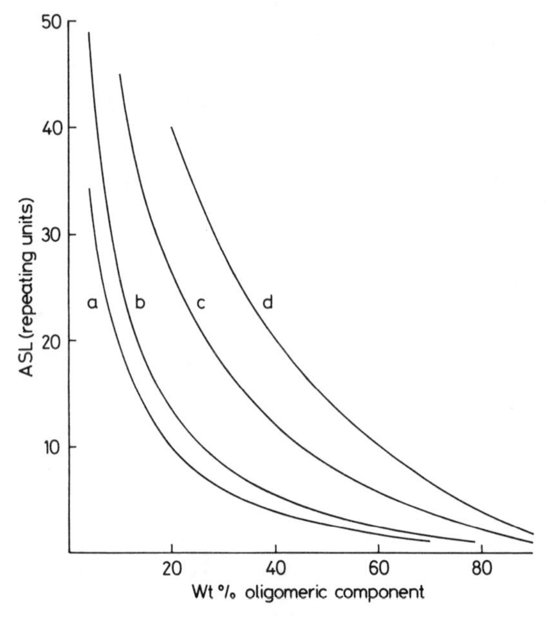

FIG. 2. Average segment length (repeating units) for $\mathbf{R^A}$–$\mathbf{R^B}$ blocks in $[\text{—}(\mathbf{P}\text{–}\mathbf{R^B})_x\text{—}(\mathbf{R^A}\text{–}\mathbf{R^B})_y\text{—}]_n$ random multiblock copolymers in relation to the weight proportion of $\mathbf{P}$–$\mathbf{R^B}$.

Curve	*Unit M.W.'s*	
	P–$\mathbf{R^B}$	$\mathbf{R^A}$–$\mathbf{R^B}$
a	1000	350
b	1000	250
c	2500	250
d	5000	250

short-chain diol and an oligomeric polyether- or polyesterdiol which takes place at temperatures sufficiently low that group interchange is avoided, a higher order of block organisation can be achieved by using a two-stage procedure as shown in eqns (4(a)) and (4(b)). In this example, three molar proportions of the oligomer **P** are first combined with four of the co-reactant $\mathbf{R^B}$ to give a prepolymer (XI) which is then reacted with $\mathbf{R^A}$ and the necessary additional quantity of $\mathbf{R^B}$ to give a product (XII) containing sequences of three consecutive **P** units.

$$3\blacklozenge\mathbf{P}\blacklozenge + 4\bullet\mathbf{R}^{B}\bullet \longrightarrow \bullet\mathbf{R}^{B}\text{–}\mathbf{P}\text{–}\mathbf{R}^{B}\text{–}\mathbf{P}\text{–}\mathbf{R}^{B}\text{–}\mathbf{P}\text{–}\mathbf{R}^{B}\bullet \quad (4(a))$$

(XI)

$$x(\mathrm{XI}) + y\bullet\mathbf{R}^{B}\bullet + (x+y)\blacklozenge\mathbf{R}^{A}\blacklozenge \longrightarrow$$

$$[(\text{–}\mathbf{R}^{A}\text{–}\mathbf{R}^{B}\text{–}\mathbf{P}\text{–}\mathbf{R}^{B}\text{–}\mathbf{P}\text{–}\mathbf{R}^{B}\text{–}\mathbf{P}\text{–}\mathbf{R}^{B}\text{–})_x\text{—}(\text{–}\mathbf{R}^{A}\text{–}\mathbf{R}^{B}\text{–})_y\text{—}]_n$$

(XII) (4(b))

As will be seen later, extensive variations can be made in the natures and relative proportions of the reactants, the M.W. of the oligomeric components, and the types of linking reactions used. Whilst the approach is less than ideal with respect to structural specificity, it is perhaps the most versatile of all methods available for the synthesis of block copolymers. Particularly important are the cases where $(\mathbf{P}\text{–}\mathbf{R}^{B})_x$ is of low T_g (and, if crystalline, of low T_m) and $(\mathbf{R}^{A}\text{–}\mathbf{R}^{B})_y$ is of high T_g, preferably crystalline and of high T_m, and hydrogen bonded. Polymers of this class are typically thermoplastic elastomers with the two types of blocks conforming to the 'soft' and 'hard' segment designations mentioned earlier.

1.2.4. Synthesis by Oligomer-linking Reactions

The simplest variety of oligomer-linking reactions (often termed 'chain extension' reactions) is that in which units of a single type of oligomer, which may be mono- or bifunctional, are connected into a larger molecular chain by means of a bifunctional reactant (eqns (5a)) and (5(b))). As already noted, it is questionable whether such products are rightly to be regarded as block copolymers.

$$2\mathbf{P}\blacklozenge + \bullet\mathbf{R}\bullet \longrightarrow \mathbf{P}\text{–}\mathbf{R}\text{–}\mathbf{P} \quad (5(a))$$

$$\blacklozenge\mathbf{P}\blacklozenge + \bullet\mathbf{R}\bullet \longrightarrow [\text{–}\mathbf{P}\text{–}\mathbf{R}\text{–}]_n \quad (5(b))$$

In a variation which certainly provides block copolymers, different types of similarly bifunctional oligomers in admixture are linked by a bifunctional co-reactant (eqn (6)). As with random block copolymer syntheses, this methodology leads to a statistical sequential distribution of the component oligomer units along the chain.

$$\blacklozenge\mathbf{P}^{A}\blacklozenge + \blacklozenge\mathbf{P}^{B}\blacklozenge + 2\bullet\mathbf{R}\bullet \longrightarrow [\text{–}\mathbf{P}^{A}\text{–}\mathbf{R}\text{–}/\text{–}\mathbf{P}^{B}\text{–}\mathbf{R}\text{–}]_n \quad (6)$$

The more certain approach leading to products in which the different component species must necessarily alternate along the chain is that involving the combination of two oligomers having mutually reactive end groups such that each species can react only with the other (eqn (7)). This procedure, which has been used particularly in recent work on polycarbonate–polysulphone and polysiloxane–polysulphone block copolymers, yields multiblock products with structures defined precisely by those of the starting reactants. As a practical matter, the formation of products of high M.W. free of contaminating oligomer residues imposes the need for (1) the prior synthesis of oligomers of high purity with accurately known M.W. and end group contents, (2) the means of bringing the oligomers to reaction with the required stoichiometry and in the absence of interfering impurities such as might be contained, for example, in insufficiently purified solvent media.

$$\blacklozenge P^{A} \blacklozenge + \bullet P^{B} \bullet \longrightarrow [-P^{A}-P^{B}-]_{n} \tag{7}$$

1.2.5. Post-reactions of Block Copolymers

Block copolymers of structures not attainable directly by polymerisation reactions can sometimes be prepared by the chemical modification of more readily accessible precursor block polymers. No general principles can be laid down, but examples include (1) the quaternisation of amine group-containing blocks to give polyionic products, (2) the hydrogenation of double bonds in block copolymers derived from dienes, yielding saturated sequences, (3) the removal of protecting groups in block copolymers to reveal sensitive groups that could not have survived during the construction of the main chain.

2. DEVELOPMENTS IN HETEROCHAIN BLOCK COPOLYMER SYNTHESIS

2.1. Blockiness in Randomisation and Step-copolymerisation Reaction Products

Copolymers of partial but imperfectly-defined block character are formed by various processes which involve interchange reactions between fissile groups in the main-chain structures of non-block precursors. Early examples included the partial inter-randomisation of pairs of (homo)-polyesters or (homo)polyamides at high temperatures in the melt.[50–55] Such reactions may occur by direct exchange between the relevant groups, or with the involvement of reactive chain end groups or of added catalysts. Analogous rearrangements leading to products in partial block character occur when strictly alternating copolyamides[56,57] or copolyesteramides[58,59] are melted, and when polyester–polyamide mixtures are heated in high-boiling solvents.[60] Comparable interchange ('shuffling') reactions can complicate the formation of polysiloxane copolymers, especially in the presence of strongly basic catalysts.[61–3] Interesting examples of the converse process are shown by the behaviour of the random copolyesters poly(*cis-co-trans*-cyclohexanedimethylene terephthalate) and poly(ethylene terephthalate-*co*-2-methylsuccinate) which, in critical ranges of temperature shortly below the melting points, undergo transfer of units from the lower-melting component into the crystal lattice of the higher-melting component resulting in conversion to partial block character.[64,65]

$$
\begin{array}{ccccc}
 & & \text{alternating copolymer} & & \\
 & & 2\updownarrow & & \\
\text{mixture of homopolymers} & \underset{}{\overset{1}{\rightleftharpoons}} & \text{random + block copolymer} & \overset{4}{\rightleftharpoons} & \text{random copolymer} \\
 & & 3\updownarrow & & \\
 & & \text{true block copolymer} & &
\end{array}
\qquad (8)
$$

Such interchange reactions resemble those leading to the equilibrium of M.W.'s and ring-chain equilibration in many step polymerisations, and are formally interrelated as shown in eqn (8). The position of equilibrium in each arm of the scheme is determined by the structures of the substances involved, the mechanistic and thermodynamic pathways available, and the conditions of reaction. The probability of some individual steps may seem indeed to be negligible though the de-randomisation reactions mentioned above should make for caution in considering what may be possible. The structure-dependence of the processes is illustrated by the relative rates of inter-randomisation of some polyesters with 2G.T (step **1**), which lie in the order 2G.4 > 2G.6 ≃ 2G.10 > poly(ethylene isophthalate) > poly(ethylene hexahydroterephthalate), whilst with respect to step **3** it has been claimed that block copolymers of 2G.T with sterically hindered aliphatic[66] and aromatic[67] polyesters, and of poly(*p*-xylylene hexadecamethylenedicarboxamide) with its *N*,*N*′-diethyl analogue,[68] are sufficiently stable that they do not undergo randomisation in the melt.

In general, however, there is an entropy-driven propensity for highly ordered block polymers containing fissile groups to become transformed in suitable conditions into random block copolymers and eventually into wholly random copolymers. Its importance lies in the potential for the loss of structural integrity when such materials are exposed during thermofabrication to temperatures higher than any encountered at earlier stages of their history.

The same considerations apply to blocky heterochain copolymers formed through the preferred reactivity of a particular reactant in ostensibly random step or condensation polymerisations. Typical of such reactions are the co-condensations of (a) *trans*-2,5-dimethylpiperazine with terephthaloyl chloride and ethylene bis(chloroformate), the first mentioned halide being more reactive,[69] (b) *cis*-2,6-dimethylpiperazine with terephthaloyl and sebacoyl chlorides, the aromatic dihalide being again the more reactive.[70] In each case, recognition of the reactivity differences permitted the control of experimental procedure by phased introduction of the reactants to yield products of well-developed (though not precisely defined) block character. The course of such reactions may also be influenced by the general parameters of operation as in the solution polycondensation of isophthaloyl chloride with mixed *m*- and *p*-phenylene diamines which yields highly blocked copolyamide when conducted in tetrahydrofuran but more random copolymers in polar media such as DMAc and HMPA.[71]

Such procedures are nevertheless highly empirical in character and until recently the formation of blocky structures could be inferred qualitatively only from observations of thermal transitions and crystal character, and quantitatively only from simple and often insensitive analyses. For wholly amorphous copolymers, only the latter method was available.

Compositions and sequence distributions in the products of random step copolymerisations (including random block copolymerisations) are important quantities which have been investigated theoretically with the aid of mathematical models.[72–7] However, the relevant kinetic parameters are rarely available to permit valid predictions to be made for the generality of reactions or for variance with practical conditions. The advent of NMR spectroscopy has opened new horizons in the experimental investigation of these problems. In one aspect, it is now often possible to study the competitive enrichment or depletion of a particular component during the course of a polymerisation; in another, the sequences of particular groups may be detected from the chemical shifts shown by specific components of those groups lying in different chemical environments. Thus, if two reactants X and Z of similar type are combined with a common co-reactant Y, the resulting polymer chains will contain the sequences X–Y–X, X–Y–Z (or Z–Y–X) and Z–Y–Z which, if distinguishable by NMR, can be determined and the results expressed in terms of a blockiness parameter B which is the ratio of the measured proportion of the linkages X–Y–Z and

Z–Y–X to that expected from Bernoullian statistics for a random copolymer. For a random copolymer $B = 1$, and for an alternating copolymer $B = 2$. Values of $B < 1$ indicate the presence of blocks, with decreasing values of B indicating increasing size of the blocks. In the limiting case, represented by a mixture of homopolymers, $B = 0$.[53,78] The use of this parameter thus permits the investigation of the influence of reaction conditions upon the blockiness of copolymers.

Table 1 shows some problems in heterochain copolymer composition and structure which have been studied by NMR. Other methods which have given useful information include the use of an isotopically-labelled monomer to investigate preferential incorporation during copolymerisation;[92] the application of isotopic dilution analysis to polymer degradation products to determine the proportion of isolated units of a particular component;[79] and the use of mass spectrometry to determine blockiness.[93]

2.2. Oligomer Synthesis and End Group Refunctionalisation Reactions

As outlined previously, many heterochain block copolymer syntheses depend upon the use of preformed oligomeric precursors furnished with suitable reactive end groups. The range of such substances available commercially is limited, consisting essentially of four groups: (a) aliphatic polyethers and copolyethers with hydroxyl (or, in some cases, amino) end groups, (b) hydroxyl-ended aliphatic polyesters and copolyesters, (c) hydroxyl-ended polybutadienes, (d) polysiloxanes. For research purposes other types of oligomers have been made by condensation reactions, using a deficit of one reactant to afford, for example, polyamides, polycarbonates and polysulphones with required end groups; examples of these will appear in later sections. Considerable use has also been made of the living polymerisations of styrene and dienes where, by suitable termination reactions, carbochain oligomers are obtained with end groups that can take part in subsequent condensation or other block-forming linking steps. Prominent amongst these are the 2-hydroxyethyl-ended oligomers formed by treatment of the oligomer carbanions with oxirane (ethylene oxide). Their preparations depend upon the fact that at relatively low temperatures the carbanion $\mathbf{P}^-$ is capable of opening the oxirane ring to give $\mathbf{P}—CH_2CH_2O^-$ which is insufficiently reactive in these conditions to initiate polymerisation of further oxirane molecules, though it can do so at higher temperatures.[94–97] Thus, by control of reaction temperature and nature of the counterion,[98,99] a single oxyethyl anion group can be attached at the reactive chain end(s) and converted to hydroxyethyl by treatment with a

TABLE 1

SOME NMR STUDIES OF HETEROCHAIN COPOLYMER COMPOSITION AND STRUCTURE

Substance or system investigated	*Method and feature studied*	*Reference*
Inter-randomisation of 2G.T and 2G.10 polyesters	^{1}H-NMR in $F_3C.COOH$ at 70 °C for determination of T-2G-T, T-2G-10 and 10-2G-10 sequences	53
Piperazine group-containing copolyamides and polyurethanes	^{13}C-NMR using CO group resonances to study composition, chemical stability and *cisoid-transoid* isomerism	57
De-randomisation of 2G.T–poly(ethylene 2-methylsuccinate) random copolymers	^{1}H-NMR in $F_3C.COOH$ at 37 °C using $O(CH_2)_2O$ protons for determination of sequence lengths	65
Condensation of isophthaloyl chloride with mixed *m*- and *p*-phenylene diamines and of *m*-phenylene diamine with mixed iso- and terephthaloyl chlorides	^{1}H-NMR in DMAc–LiCl using amide group proton resonances to determine blockiness	71
Copolymers from iso- or terephthaloyl chlorides with 2,2-dimethyl-1,3-propanediol and bisphenol A made in various conditions	^{1}H-NMR in $CDCl_3$ for determination of sequence lengths	77
Copolymers from isophthaloyl chloride and $O[\phi NH_2]_2 + SO_2[\phi NH_2]_2$ made in different solvent media	^{1}H-NMR in DMAc–LiCl at 80 °C for determination of sulphone → sulphone, sulphone → ether, ether → sulphone and ether → ether unions	78
Solution and interfacial condensation of $COCl_2$ + benzene-*m*-disulphonyl chloride with bisphenol A	^{13}C-NMR in $CDCl_3$ at 60 °C, using methyl resonances to determine carbonate → carbonate, sulphonate → carbonate and sulphonate → sulphonate linkages	79
Poly(*m*-phenylene isophthalamide)-*b*-poly(oxyethylene) and poly(*m*-phenylene isophthalamide)-*b*-poly(dimethylsiloxane)	^{1}H-NMR determination of aromatic proton:aliphatic ether and aromatic proton:siloxane methyl protons to confirm composition	80

Poly(BPAC-*b*-poly(dimethylsiloxane))	^{13}C- and ^{29}Si-NMR using $(acac)_3Cr$ shift reagent for complete analysis of composition and microstructure	81
Block polymers from $H_2N\phi-\phi NH_2$ with adipoyl and terephthaloyl chlorides	^{1}H-NMR in $ClSO_3H$ for determination of terephthaloyl group content	82
Solution copolymerisation of terephthaloyl chloride (T) with bisphenol A (B) and 4,4′-dihydroxy-1,1′-binaphthyl (N)	^{1}H-NMR in $C_2H_2Cl_4$ at 140 °C for determination of N–T–N, N–T–B and B–T–B sequences in relation to B:N ratio	83–5
Polyamide from $H_2N(CH_2)_2NH_2$ and $ClCO\phi CH_2SCH_2COCl$ made in various conditions	^{1}H-NMR in DMSO at 120–134 °C to determine head–head, head–tail, tail–tail combination sequences	86
Anionic copolymers from ε-caprolactam and ω-capryllactam with γ-butyrolactam	^{13}C-NMR in FSO_3H using CO group signals to determine composition and sequencing	87
Various aliphatic, alicyclic and aromatic copolyesters	^{13}C-NMR in $F_3C.COOH$ using CO and OCH_2 group signals to distinguish copolymers from mixed homopolymers and determine lengths of homogeneous blocks	88
Ternary polyesteramides with alternating sequences of aminobenzoyl, glycyl and hydroxy- or mercapto-acetyl residues	^{13}C-NMR in $F_3C.COOH$ to confirm sequence patterns	89
2G.T-*b*-poly(oxyethylene) random block copolymers	^{13}C-NMR in $CDCl_3$ for determination of average composition and degradation of POE units during polymerisation	90
2G.T-*co*-4G.T	^{13}C- and ^{1}H-NMR to determine blockiness and formation of oxydiethylene groups by side-reactions of polymerisation	91

source of protons. In practice, the procedure is affected by a troublesome increase in viscosity at the oxyanion stage and is limited by the restricted solubility of many organolithium initiators (particularly dilithio-initiators for bifunctional polymerisation) in hydrocarbon reaction media. Attempts at the amelioration of these difficulties include the use of soluble lithium-ended low M.W. polydienes as initiators for butadiene polymerisation,[100] of phenyloxirane as the source of oxyanions,[101] and of 6-tetra-hydropyranyloxy- or 6-[1-ethoxy(ethoxy)]hexyllithium as the initiator for butadiene polymerisation to give products reactable with oxirane at one end of the polydiene molecule, whilst carrying an acetal group at the other from which hydroxyl can finally be released by hydrolysis.[102]

The complex nature of these procedures, and the variable success obtained in other carbanion reactions (e.g. the sequence $\mathbf{P}^- \xrightarrow{CO_2} \mathbf{P}.CO_2^- \xrightarrow{H^+} \mathbf{P}.COOH$) has encouraged work towards alternative methods for the synthesis of useful reactive oligomers from readily accessible starting materials. Such approaches commencing with dienes include free radical polymerisations initiated by 4,4′-azo-bis(4-cyano-1-pentanol) or 4,4′-azo-bis(4-cyanopentanoic acid) giving polydiene oligomers terminated with OH and COOH groups respectively;[103–105] with living anionic polystyrene or polydienes, by direct oxygenation to α,ω-bis(hydroperoxy) oligomers which can be reduced to diols;[106,107] and with butadiene by polymerisation with H_2O_2–$HClO_4$ or by redox telomerisation.[108,109] Styrene has also been oligomerised by $AlCl_3$–$SOCl_2$ to α,ω-dichloro-polystyrenes which can be converted to α,ω-diamines and diols.[110] A novel development makes use of the termination-free polymerisation of iso-butene initiated by α,α'-dichloro-1,4-diisopropylbenzene + BCl_3 which gives rise to oligomers having 1-chloro-1,1-dimethylethyl end groups (eqn (9)). These products can be converted by way of dehydrochlorination and subsequent hydroboration to polyisobutenes having primary hydroxyl end groups (eqn (10)).[111]

$$Cl.Me_2C.\phi.CMe_2Cl + (2n+2)C_4H_8 \xrightarrow{BCl_3} Cl.CMe_2\text{—}CH_2[CMe_2CH_2]_n\text{—}CMe_2.\phi.CMe_2\text{—}[CH_2CMe_2]_n\text{—}CH_2.CMe_2.Cl \quad (9)$$

$$\mathbf{P}\text{—}CH_2.CMe_2.Cl \xrightarrow[\text{in THF}]{KO\textit{t}\text{-Bu}} \mathbf{P}\text{—}CH_2CMe{=}CH_2 \xrightarrow[\text{(2) NaOH/H}_2\text{O}_2]{\text{(1) 9-BBN* in THF}} \mathbf{P}\text{—}CH_2CHMe.CH_2OH \quad (10)$$

* 9-borabicyclo-3,3,1-nonane

Mention may also be made of the controlled ozonolytic or oxidative degradations of high M.W. isotactic poly(oxypropylene),[112] poly(vinyl chloride),[113] and poly(isobutene-*co*-2,3-dimethylbutadiene)[114] to give oligomers with functional end groups. However, no effective synthesis has yet been devised for α,ω-difunctional polymethylene oligomers which would be of considerable interest as precursors of block copolymers.

Much effort has also been given to extending the range of block-forming reactions of accessible oligomers by refunctionalisation of their end groups. Familiar examples include the transformations of hydroxyl-ended oligomers **P**—OH (only one end group is shown, though the materials are often bifunctional) with diisocyanates,[15,115–18] phosgene[95,97,117,119,120] or dicarbonyl chlorides[80,96,121–4] to give **P**—OCONH . R . NCO, **P**—OCOCl and **P**—OCO . R . COCl, respectively. Table 2 shows further refunctionalisation reactions and reaction sequences which have been explored in recent years; additional instances will be mentioned in later sections in connection with the synthesis of particular classes of block polymers.

2.3. Polyether and Polythioether Block Copolymers

The ring-opening polymerisation of oxirane initiated by terminally mono- or dicarbanionic carbochain oligomers provides a direct route to AB and BAB block copolymers (where A represents the carbochain entity and B is poly(oxyethylene)). Substances which have been prepared in this way include those with A units of polystyrene,[94,95,98,144] poly(ethyl methacrylate),[145] poly(2-vinylpyridine),[146] polydienes,[147,148] and preformed triblock isoprene–styrene copolymers.[149] A variation leading to three-armed $A(B)_2$ block copolymers has been achieved by attaching the 2-naphthylmethylene group at the ends of polystyryl or poly(4-t-butylstyryl) chains followed by metallation with potassium to generate dianions (XIII) having two reactive sites available for initiating the growth of POE chains.[150]

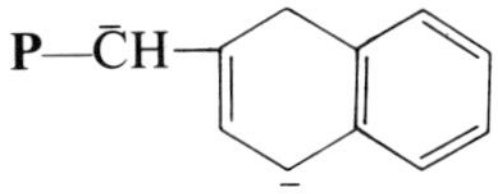

P—$\bar{C}$H—

(XIII)

The approach cannot ordinarily yield ABA copolymers because poly(oxyethylene)oxy anions are unable to activate most vinyl monomers for sequential polymerisation. However, in the presence of a strong base complexed with crown ether or cryptate ligands, POE chains with

TABLE 2

REFUNCTIONALISATION REACTIONS OF OLIGOMERS

	Refunctionalisation agent(s)	*Oligomer product*	*For reaction with*	*Reference*
Polymer carbanion (**P**$^-$)	1. CO_2, H^+ 2. Esterify 3. $LiAlH_4$	**P**-CH_2OH	OH, NCO	15
Polymer carbanion (**P**$^-$)	*cyclo*.[$SiMe_2O]_3$	**P**-$SiMe_2O^-$	Cyclopolysiloxanes (polymerisation)	125
Polymer carbanion (**P**$^-$)	$BrCH_2CH\!-\!CH_2$ (epoxide, O bridging CH–CH_2)	**P**-$CH_2CH\!-\!CH_2$ (epoxide, O bridging CH–CH_2)	NH_2	126
Polymer carbanion (**P**$^-$)	1. CO_2, H^+ 2. $SOCl_2$ 3. $H_2N(CH_2)_xNH_2$	**P**-$CONH(CH_2)_xNH_2$	Oxazolidin-2,5-diones (polymerisation)	127, 128
Polymer carbanion (**P**$^-$)	1. $MgBr_2$ 2. Br_2 3. $AgClO_4$	**P**$^+ClO_4^-$	Tetrahydrofuran (polymerisation)	129
Polymer carbanion (**P**$^-$)	1. CO_2, H^-, $SOCl_2$ 2. $AgClO_4$	**P**-$CO^+ClO_4^-$	Cationically polymerisable monomers	130
Polymer carbanion (**P**$^-$)	1. CH_3CN 2. $LiAlH_4$/THF	**P**-$CHMe.NH_2$		131

Polymer carbanion ($\mathbf{P}^-$)	1. CO_2, H^+ 2. $Bu_4N.OH$	$\mathbf{P}$-CO_2^- $^+NBu_4$	Pivalolactone (polymerisation)	132
Polymer carbanion ($\mathbf{P}^-$)	Cl.COOEt	$\mathbf{P}$-COOEt		133
$\mathbf{P}$-OH	1. $R(NCO)_2$ 2. Glycidol	$\mathbf{P}$-OCONH.R.NHCOO—$CH_2\underbrace{CH—CH_2}_{O}$	NH_2	134
$\mathbf{P}$-OH	OCN.$CH_2CH{=}CH_2$	$\mathbf{P}$.OCONH.$CH_2CH{=}CH_2$	Hydrosilanes	135, 136
$\mathbf{P}$-CH_2CH_2OH	MeϕSO_2Cl, pyridine	$\mathbf{P}$-$CH_2CH_2O.SO_2\phi Me$	1. 2-Oxazolines (polymerisation) 2. $PhCH_2NH_2$, giving $\mathbf{P}$-$NHCH_2Ph$	137, 138
$\mathbf{P}$-OH	1. $ClCO\phi NO_2$, pyridine 2. H_2/Pd–C	$\mathbf{P}$-$OCO\phi NH_2$		139
$\mathbf{P}$-OH	1. $OCN(CH_2)_6NCO$ 2. HCl/H_2O	$\mathbf{P}$-$OCONH(CH_2)_6NH_2$		131
$\mathbf{P}$-$(CH_2)_4Br$	1. $LiCR_2$– (N, Me_2, O) 2. H^+/H_2O	$\mathbf{P}$-$(CH_2)_4.CR_2COOH$	As Bu_4N^+ salt for polymerisation of pivalolactone	140
$\mathbf{P}$-OH	Succinic anhydride	$\mathbf{P}$-$OCO(CH_2)_2COOH$	As Bu_4N^+ salt for polymerisation of pivalolactone	141–3
$\mathbf{P}$-NH_2	1. $R(NCO)_2$ 2. Glycidol	$\mathbf{P}$-NHCONH.R.NHCOO$CH_2\underbrace{CH—CH_2}_{O}$	NH_2	157

$NHCH_2Ph$ or OH end groups become sufficiently nucleophilic to initiate the polymerisation of methyl and *t*-butyl methacrylates giving triblock polymers with central POE segments.[138,151]

Tetrahydrofuran cannot be polymerised by bases and the living anionic principle is therefore not available for the formation of block polymers with poly(oxytetramethylene) segments. Recent work, mostly involving cationic intermediates, has provided the following approaches to the synthesis of POTM-containing block copolymers:

(a) Polystyrenes with COCl end groups, when treated with silver salts such as $AgBF_4$, $AgSbF_6$ or $AgClO_4$ having anions of low nucleophilicity, are converted to oxocarbenium (acylium) species ($PSt{-}CO^+$) which can initiate the polymerisation of THF and other cation-sensitive monomers.[130]

(b) By the sequence of reactions shown in eqn (11), 2-hydroxyethyl-ended polystyrene is converted to (XIV) whose terminal dioxolenium group is used to initiate the polymerisation of THF to a PSt–POTM block copolymer. Analogous reactions lead to the dioxolenium-ended POTM (XV) which can initiate the polymerisation of 3,3-bis(chloromethyl)oxetane and of 7-oxa-[2.2.1]bicycloheptane yielding block copolyethers.[96]

$$PSt{-}(CH_2)_2OH \xrightarrow{\text{adipoyl chloride}} PSt{-}(CH_2)_2OCO(CH_2)_4COCl \xrightarrow{\text{2-bromoethanol}}$$

$$PSt{-}(CH_2)_2OCO(CH_2)_4COO(CH_2)_2Br \xrightarrow[\text{1-nitropropane}]{AgClO_4 \text{ in}}$$

$$\underset{(XIV)}{PSt{-}(CH_2)_2OCO(CH_2)_4{-}C\langle{}^{+}O{-}CH_2CH_2{-}O\rangle} \; ClO_4^- \tag{11}$$

$$\underset{(XV)}{POTM{-}OCO(CH_2)_4{-}C\langle{}^{+}O{-}CH_2CH_2{-}O\rangle} \; ClO_4^-$$

(c) The transfer-free polymerisation of THF initiated by $AgClO_4$ + 2-bromoethyl acetate or bis(2-bromoethyl) sebacate, or by $AgPF_6$ + benzyl bromide or *m*-xylylene dibromide gives POTM structures with cyclic oxonium groups at one or both ends of the molecule according to the functionality of the halide used. These active end groups react with carbanionic polystyrenes as shown in eqn (12) forming POTM block copolymers.[152-6]

$$\mathrm{PSt{-}CH_2\underset{\displaystyle Ph}{\underset{|}{C}}H^-Na^+} + X^- \;[\text{THF ring}]\overset{+}{O}\mathrm{{-}(CH_2)_4O{-}POTM} \longrightarrow$$

$$\mathrm{PSt{-}CH_2\underset{\displaystyle Ph}{\underset{|}{C}}H{-}(CH_2)_4O\,.\,POTM} + [\text{THF ring, O}] + \mathrm{Na}X \qquad (12)$$

(d) By reaction with $AgClO_4$, the terminal halogen atom of ω-bromopolystyrene, or preferably the *m*-xylylene homologue (XVI) which is not prone to hydrohalide elimination, is converted to a corresponding carbenium perchlorate, e.g. (XVII), which combines with a single THF molecule at low temperatures forming an oxonium complex (XVIII) that can initiate the polymerisation of THF at higher temperatures to form block copolymers (eqn (13)).[129,158]

$$\mathrm{PSt{-}CH_2\underset{\displaystyle Ph}{\underset{|}{C}}H.CH_2{-}}m{-}\phi\mathrm{{-}CH_2Br} \xrightarrow{AgClO_4} \quad \text{(XVI)}$$

$$\mathrm{PSt{-}CH_2\underset{\displaystyle Ph}{\underset{|}{C}}H.CH_2{-}}m{-}\phi\mathrm{{-}CH_2^+ClO_4^-} \xrightarrow{THF} \quad \text{(XVII)}$$

$$\mathrm{PSt{-}CH_2\underset{\displaystyle Ph}{\underset{|}{C}}H.CH_2{-}}m{-}\phi\mathrm{{-}CH_2{-}}\overset{+}{O}[\text{THF ring}]\ \mathrm{ClO_4^-} \quad \text{(XVIII)} \qquad (13)$$

(e) The polymerisation of THF initiated by trifluoromethane sulphonic anhydride gives the dicationic polymer (XIX) whose end groups can be displaced by 1,4,4-trimethylazetidine yielding (XX) which then serves as an initiator for polymerisation of the cyclic amine to polyether–polyamine triblock copolymers (XXI).[33]

$$[\text{THF ring}]\overset{+}{O}\mathrm{{-}[(CH_2)_4O]_a{-}(CH_2)_4{-}}\overset{+}{O}[\text{THF ring}]$$

$$\mathrm{2CF_3.SO_3^-}$$

(XIX)

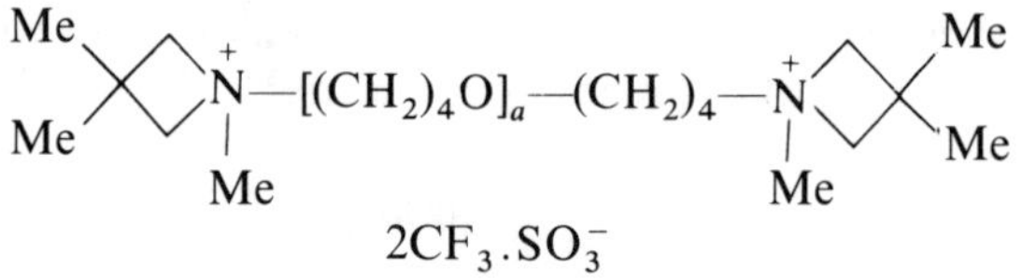

$$\mathrm{Me_2C(CH_2)_2\overset{+}{N}(Me)-[(CH_2)_4O]_a-(CH_2)_4-\overset{+}{N}(Me)(CH_2)_2CMe_2}\quad 2CF_3.SO_3^-$$

(XX)

$$\mathrm{Me_2C(CH_2)_2\overset{+}{N}(Me)-[-CH_2CMe_2CH_2N(Me)-]_b-[(CH_2)_4O]_a-(CH_2)_4-[-N(Me)-CH_2CMe_2CH_2-]_b-\overset{+}{N}(Me)(CH_2)_2CMe_2}\quad 2CF_3.SO_3^-$$

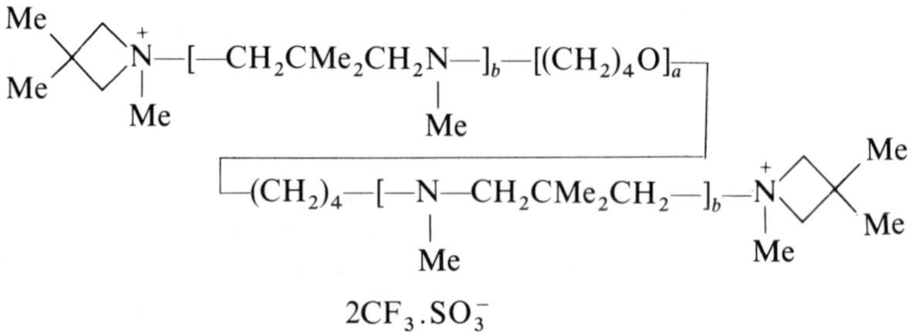

(XXI)

(f) α,ω-Dibromopolyethers (including POTM), formed by reaction of the α,ω-diols with thionyl bromide, are linked by quaternisation with tertiary amine-ended ionene oligomers to give polyionic block copolymers (XXII) with central polyether segments (eqn (14)).[34,159]

$$\begin{aligned}&2\mathrm{Cl(CH_2)_3\overset{+}{N}Me_2[(CH_2)_3\overset{+}{N}Me_2]_a-(CH_2)_3NMe_2}\ (X^-,\ X^-)\\&\quad+\mathrm{Br[(CH_2)_4O]_b-(CH_2)_4Br}\longrightarrow\\&\mathrm{Cl(CH_2)_3\overset{+}{N}Me_2[(CH_2)_3\overset{+}{N}Me_2]_a-(CH_2)_3\overset{+}{N}Me_2-[(CH_2)_4O]_b-}\ (X^-,\ X^-,\ Br^-)\\&\quad\mathrm{-(CH_2)_4-\overset{+}{N}Me_2(CH_2)_3[Me_2\overset{+}{N}(CH_2)_3]_a\overset{+}{N}Me_2(CH_2)_3Cl}\ (Br^-,\ X^-)\end{aligned}\tag{14}$$

(XXII)

Free radical polymerisation has been employed for syntheses of some POE block copolymers. In one approach, POE chains with peroxycarbamate[118] or *t*-butyldioxycarbonylbenzoate[160] end groups were used to initiate block polymerisations with styrene. Hydrophilic poly(2-hydroxyethyl methacrylate)POE block copolymers have been made by radical polymerisation of the methacrylate with 2-aminoethanethiol as chain transfer agent, giving amino-ended oligomers which were coupled to the polyether by reaction with a diisocyanate.[161] Free radical intermediates may also be involved in the primary stages of the complex reaction leading to the formation of block copoly(arylene ether)s when 2,6-diphenylphenol is co-oxidised with 2,6-dimethylphenol or 2-methyl-6-phenylphenol using oxygen and Cu(I)/amine catalysts.[162] In general, however, free radical reactions do not offer versatile and controllable means for the synthesis of structurally well-defined block copolymers.

Lastly, mention will be made of some methods for the formation of block copolymers containing poly(thioalkylene) sequences by the ring-opening polymerisations of cyclic sulphides.

In anionic systems, thiiranes and thietanes undergo ring-opening by different pathways, involving thiolate intermediates in the first case (eqn (15)) and carbanions in the second (eqn (16)).

$$\mathrm{EtLi} + \text{(methylthiirane)} \longrightarrow \mathrm{CH(Me){=}CH_2} + \mathrm{EtSLi} \xrightarrow{\text{methylthiirane}} \mathrm{EtS.CH_2CHMeSLi}, \text{ etc.} \quad (15)$$

$$\mathrm{EtLi} + \mathrm{Me}\text{–(thietane)} \longrightarrow \mathrm{EtS.CHMeCH_2CH_2Li}, \text{ etc.} \quad (16)$$

The blockwise polymerisation of either type of monomer can be induced by polyvinyl or polydienyl anions (A) to give AB or BAB copolymers, where B represents the poly(thioalkylene) sequences. In the thiirane case, the thiolate end group of the AB copolymer can readily be coupled with phosgene to give $A(B)_2A$ copolymers, whilst with thietanes the terminal carbanionic group can be used to initiate polymerisation of vinyl monomers to give ABA or ABC block polymers.[163,164] Block copolymers of methylthiirane with thiirane or with 2,2-dimethylthiirane have also been obtained by sequential polymerisations initiated in the first phase by 2-naphthylsulphonylmethylene anions.[165] Cationic polymerisation has found use in the conversion of 3,3-dimethylthietane to SH-ended oligomers which could be coupled with hydroxyl-ended POE chains by reaction with a diisocyanate,[166] but attempts to obtain block polymers of this thietane by direct initiation with cationic polystyrene were not successful.[167]

2.4. Polyester Block Copolymers

Apart from copolyesterurethanes, synthetic developments towards polymers containing polyester blocks have been dominated by two main

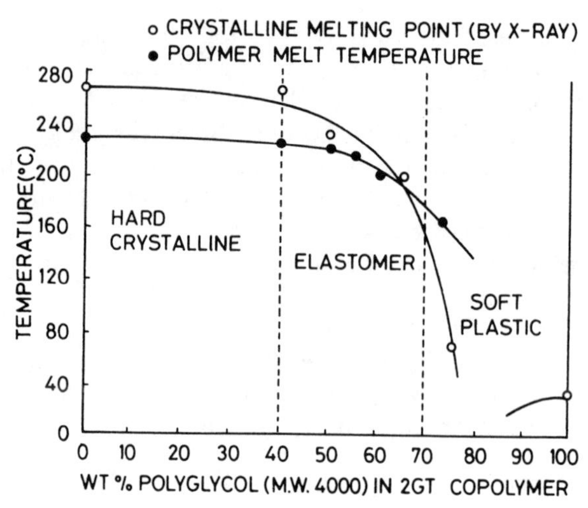

FIG. 3. Properties versus composition relationships for polyether–poly(ethylene terephthalate) random block copolymers. (With permission from Charch, W. H. and Shivers, J. C., *Textile Research Journal*, 1959, **29**, 538.)

themes, the first stemming from early work on block polyetheresters and the second from interests in the ring-opening polymerisation reactions of cyclic esters (lactones).

Polymers of the first group are typified by the poly(alkylene terephthalate)–poly[poly(oxyalkylene) terephthalate] family which are conveniently prepared by the copolyesterification of dimethyl terephthalate with mixtures of a short-chain diol and a poly(oxyalkylene)-α,ω-diol according to the general scheme of eqn (3).[7,8,90,168–75] Many different diols, polyether diols and aromatic dicarboxylate esters other than terephthalates have been employed in such investigations,[171,174–6] and a generalised view of the properties attainable is shown in Fig. 3.

Because of their hydrophilic nature the 2G.T-POE group of block copolymers are unsuitable for use as structural substances but this very property has indicated their application as electrostatic charge dispersants in synthetic fibres[177] and as potentially useful biodegradable surgical materials.[90]

Most attention has been devoted to the thermoplastic elastomeric group of random block polyetheresters, notably those based on 4G.T (or 4G.T-*co*-isophthalate) as the hard block components and POTM terephthalate as the soft segments which have emerged as important commercial products. These copolymers combine ready processability with excellent mechanical properties that are sustained over a wide temperature range (-55–$+150\,°C$), and with good resistance to many oils and chemicals. The optimum M.W. of the POTM component is approximately 1000, and variations in the hard segment content over the range 30–80 wt % gives copolymers of correspondingly increasing hardnesses and moduli, reflecting a progressive change from soft elastomeric to tough elastoplastic character. The scientific and technological properties of these materials are detailed in a number of publications[171,174,175,178–92] and are discussed fully in Chapter 7. Since their commercial début in the early 1970s, the annual US consumption has risen to 7000 tons.[193]

Elastomeric polymers with poly(arylene carbonate) hard blocks (XXIII) have been made according to eqn (17) by the co-phosgenation of bisphenols with soft segment oligomer α,ω-diols in the presence of an acceptor for the liberated hydrogen chloride.[194–6] Since these reactions take place at relatively low temperatures where structural reorganisation by ester–ester exchange does not occur, aliphatic polyester-diols can be employed to give poly(arylene carbonate)–aliphatic polyester block copolymers. Bisphenols which have been used in such preparations include bisphenol-A (4,4′-(1-methylethylidene)bisphenol), and the hindered compounds (XXIV) and (XXV) which give amorphous hard blocks. Other procedures giving polyester–polyester or polyester–polyether block polymers are (a) the condensation of hydroxyl-ended with chloroformate-ended oligomers,[197–201] (b) the coupling of hydroxyl-ended oligomers by means of dicarbonyl chlorides, diisocyanates or dichlorodimethylsilane.[202–205]

$$a\,\mathrm{HO.Ar.OH} + b\,\mathrm{HO{-}\mathbf{P}{-}OH} + (a+b)\,\mathrm{COCl_2} \longrightarrow [\!-\!(\!-\mathrm{O.Ar.O.CO}-\!)_a\!-\!(\!-\mathrm{O.\mathbf{P}.O.CO}-\!)_b\!-\!]_n \quad \text{(XXIII)} \qquad (17)$$

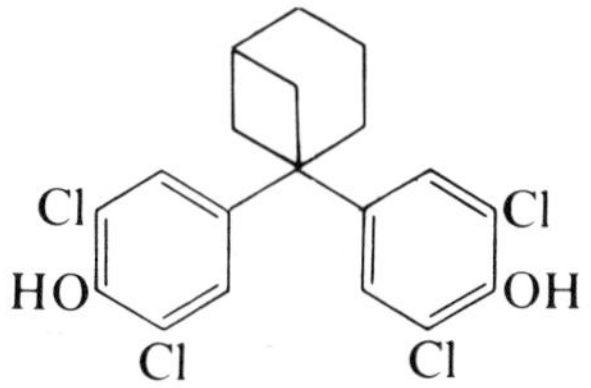

Cl, HO, Cl, Cl, OH, Cl

(XXIV)

R, HO, R, OH

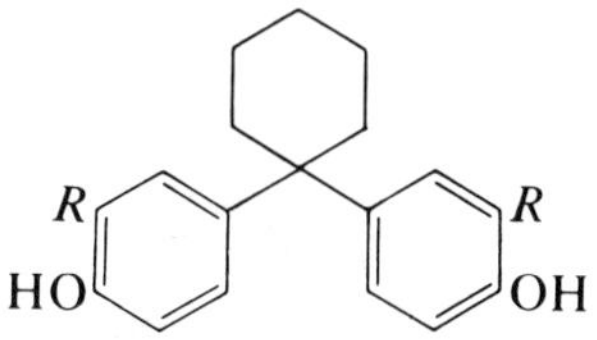

($R = R =$ H or Me)

(XXV)

The ionically-initiated ring-opening polymerisation reactions of lactones[206] should be adaptable in principle to the formation of block polymers. In practice such reactions are complicated by the unequal propensities of different lactones for polymerisation and by the occurrence of interchange side-reactions catalysed by the ionic species present. Binary copolyesters of partial block character are formed in some lactone–lactone copolymerisations carried to limited conversions,[92,207] but attempts to prepare block copolymers by the reactions of lactones with preformed polyvinyl, polydienyl or poly(oxyalkylene) anions have generally been unsatisfactory because of the considerable concurrent formation of higher cyclic polyester oligomers and of free homopolymers of the initiating species.[208-210] These difficulties, which may reflect the effects of proton abstraction from the active α-methylene position in the lactone or its derived polyester chain, can be avoided in the case of polyhydrocarbon anions by end-capping with oxirane before introduction of the lactone, and such modifications have permitted the formation of polystyrene-*b*-poly-butadiene-*b*-poly(ε-caprolactone) triblock polymers.[211-213]

A novel approach to the synthesis of block copolymers from lactones involves coordination catalysis by bimetallic oxoalkoxides which, in quite mild conditions, polymerise cyclic esters by insertion at the Al–O*R* bond of the catalytic complex (XXVI). The sequential introduction of different lactones leads to polyester–polyester block polymers as shown in eqn (18(a)), and the prior attachment of polystyryl unit (as alkoxide) to the aluminium atom opens an elegant route to polystyrene-*b*-poly(ε-caprolactone) block copolymers (eqn 18(b)).[214-17]

$$\sim\!\text{OZnO}\overset{|}{\text{Al}}\text{—OR} \xrightarrow{a\,\overline{\text{O(CH}_2)_5\text{CO}}} \sim\!\text{OZnO}\overset{|}{\text{Al}}\text{—[O(CH}_2)_5\text{CO]}_a\text{O}R$$

(XXVI)

$$\xrightarrow{b\,\overline{\text{O(CH}_2)_2\text{CO}}} \sim\!\text{OZnO}\overset{|}{\text{Al}}\text{—[O(CH}_2)_2\text{CO]}_b\text{—[O(CH}_2)_5\text{CO]}_a\text{—O}R \qquad (18(a))$$

$$\text{(XXVI)} + \text{HO(CH}_2)_2\text{—PSt} \longrightarrow \sim\!\text{OZnOAl—O(CH}_2)_2\text{PSt}$$

$$\xrightarrow{a\,\overline{\text{O(CH}_2)_5\text{CO}}} \sim\!\text{OZnO}\overset{|}{\text{Al}}\text{—[O(CH}_2)_5\text{CO]}_a\text{—O(CH}_2)_2\text{PSt} \qquad (18(b))$$

A further effective approach to lactone-based block copolymers makes use of the specific ability of α,α-dialkyl-β-propiolactones (3,3-dialkyloxetan-2-ones) to undergo living anionic polymerisations initiated by tetraalkylammonium carboxylate salts and propagated, through an alkyl-oxygen fission mechanism, by carboxylate anions. Much attention has been devoted to the case of pivalolactone (3,3-dimethyloxetan-2-one)

$$RCOO^{-}\,{}^{+}NAlk_4 + \underset{\text{(XXVII)}}{\begin{array}{l} \quad\;\; Me \\ Me-\!\!\!\!\!\quad | \!-CO \\ \qquad |\qquad | \\ \qquad \lfloor\!-\!-O \end{array}} \longrightarrow RCOOCH_2CMe_2COO^{-}\,{}^{+}NAlk_4$$

$$\xrightarrow{\text{monomer}} RCOO[CH_2CMe_2COO]_n^{-}\,{}^{+}NAlk_4 \qquad (19)$$

(a) $^{-}OOC\,.\,\mathbf{P}\,.\,COO^{-} + 2n(\text{XXVII}) \longrightarrow$

$$^{-}[OCOCMe_2CH_2]_n-\mathbf{P}-[CH_2CMe_2COO]_n^{-}$$

(b) $$PhCOO^{-} + m \begin{array}{l} \quad\;\; Me \\ Pr-\!\!\!\!\!\quad | \!-CO \\ \qquad |\qquad | \\ \qquad \lfloor\!-\!-O \end{array} \longrightarrow PhCOO[CH_2\underset{\displaystyle Pr}{\overset{\displaystyle Me}{C}}COO]_m^{-} \xrightarrow{n(\text{XXVII})}$$

$$PhCOO[CH_2\underset{\displaystyle Pr}{\overset{\displaystyle Me}{C}}COO]_m-[CH_2CMe_2COO]_n^{-}$$

(c) $$^{-}OOC(CH_2)_8COO^{-} + 2m \begin{array}{l} \quad\;\; Me \\ Bu-\!\!\!\!\!\quad | \!-CO \\ \qquad |\qquad | \\ \qquad \lfloor\!-\!-O \end{array} \longrightarrow$$

$$^{-}[OOC\underset{\displaystyle Bu}{\overset{\displaystyle Me}{C}}CH_2]_mOOC(CH_2)_8COO[CH_2\underset{\displaystyle Bu}{\overset{\displaystyle Me}{C}}COO]_m^{-} \xrightarrow{2n(\text{XXVII})}$$

$$^{-}[OOCCMe_2CH_2]_n-[OOC\underset{\displaystyle Bu}{\overset{\displaystyle Me}{C}}CH_2]_mOOC(CH_2)_8COO[CH_2\underset{\displaystyle Bu}{\overset{\displaystyle Me}{C}}COO]_m-[CH_2CMe_2COO]_n^{-} \qquad (20)$$

(XXVII) on account of the high-melting crystalline character of its polymer which, as a hard segment component in elastomeric block copolymers, provides a system of thermally reversible quasi-crosslinks. The general form of the reaction is shown in eqn (19), and eqn (20) (in which the ammonium counterions are omitted for simplicity) shows examples of its adaptation to the formation of various types of block copolymers.[218–21] The effectiveness of the initiating step (and hence the possibility of achieving sequential polymerisation) is affected by the proximity of the initiating carboxylate anion to other polar groups;[141] nevertheless, within certain limitations, the method provides a valuable source not only of the block copolymer types shown above but also of thermoplastic graft copolymers having poly(pivalolactone) branches attached to carbochain polymer backbones by reaction at pendant COOH groups.[218,222–5]

2.5. Polyamide, Polypeptide and Related Block Copolymers

It has long been known that the properties of partially *N*-alkylated linear polyamides vary with the pattern of distribution of the substituent groups, i.e. with the degree of randomness or blockiness of the unsubstituted and the substituted amide groups along the chain.[8,68,226–8] Indeed such polymers formed an early focus of interest in the development of non-hydrocarbon elastomers though they were soon overtaken in importance by the elastomeric block copolyurethanes which retain a similar dependence upon hydrogen bonding as the effective source of high-melting and cohesive character in their hard segments whilst being more readily and controllably prepared. Block copolymers containing polyamide or similar segments will here be considered as a class.

Adventitious routes to partially blocky copolyamides have been mentioned in an earlier section, and block copolymer syntheses by conventional random block copolymerisation and by oligomer combination reactions are summarised in Tables 3 and 4. It should be noted that the high melting points and restricted solubilities which are the source of useful properties in intermolecularly hydrogen bonded polyamides and analogous polymers are also a frequent source of practical difficulties in the preparation of their block copolymers.

The activated anionic polymerisation of cyclic amides (lactams) offers another recently employed approach to the synthesis of block polymers with polyamide segments. The mechanism of the reaction is complex (for an account of the chemistry see reference 206) but its overall result is shown in eqn (21) where the compound (XXVIII) (with R = alkyl, aryl or RNH) is an *N*-acyl or *N*-carbamoyl lactam which is used in catalytic quantities as

TABLE 3

POLYAMIDE AND SIMILAR BLOCK POLYMERS BY RANDOM BLOCK POLYMERISATION REACTIONS

Oligomeric reactant[a]	*Coreactant(s)*	*Product type*	*Reference*
$ClCO\text{-}m\text{-}\phi COO[POE]OCO\text{-}m\text{-}\phi COCl$	$m\text{-}\phi(NH_2)_2 + m\text{-}\phi(COCl)_2$	Polyether-*b*-polyamide	80
$H_2N[PSt]NH_2$	Nylon 6,6 salt	Polystyrene-*b*-polyamide	110
$ClCO(CH_2)_8CO[2G.T]O(CH_2)_2OCO(CH_2)_8COCl$	$H_2N(CH_2)_6NH_2 + ClCO(CH_2)_8COCl$	Polyester-*b*-polyamide	123
HO[POTM]OH	$MeOOC\phi CONH(CH_2)_6NHCO\phi COOMe + HO(CH_2)_6OH$	Polyether-*b*-polyesteramide	176
$ClCO[O(CH_2)_5CO]_nO(CH_2)_6O[CO(CH_2)_5O]_nCOCl$	$H_2N(CH_2)_6NH_2 + ClCO(CH_2)_4COCl$	Polyester-*b*-polyamide	200
HO[POE]OH	$Ar(COOH)_2 + HO(CH_2)_2NH_2$	Polyether-*b*-polyesteramide	230
HOOC[POE]COOH	Methylhexamethylene diamines + EtOOC . COOEt	Polyether-*b*-polyoxamide	231
$H_2N(CH_2)_3[POE](CH_2)_3NH_2$	$HOOC(CH_2)_xCOOH$ + aliphatic diamine or piperazine	Polyether-*b*-polyamide	232
$H\overset{i\text{-Bu}}{N}(CH_2)_6NH[CO(CH_2)_7CO\overset{i\text{-Bu}}{N}(CH_2)_6\overset{i\text{-Bu}}{N}—]_nH$	2,5-Dimethylpiperazine + $ClCO\phi COCl$	Polyamide-*b*-polyamide	232
$ClCOO(CH_2)_2OCO[\overset{Et}{N}(CH_2)_6\overset{Et}{N}COO(CH_2)_2O]_nCOCl$	Piperazine + $m\text{-}\phi(OCOCl)_2$	Polyurethane-*b*-polyurethane	233
HO[POE]OH	Nylon 6,10 salt	Polyether-*b*-polyamide	234
$H_3\overset{+}{N}(CH_2)_3[POE](CH_2)_3\overset{+}{N}H_3 . {}^-O_2C(CH_2)_4CO_2^-$	ε-Caprolactam	Polyether-*b*-polyamide	235
PSt^-, PBd^- or PSt-*b*-PBd^-	TDI + $m\text{-}\phi(NH_2)_2$ (directly, or by intermediate reaction with water)	Polyhydrocarbon-*b*-polyurea	236–9
HO[POTM]OH	Laurolactam + $HOOC(CH_2)_{10}COOH$	Polyether-*b*-polyamide	240

[a] Symbols such as HO[POE]OH indicate an α,ω-difunctional oligomer (see Section 1.1 for comment on the nomenclature and designation of such substances).

TABLE 4

BLOCK AND SEGMENTED POLYAMIDES AND RELATED POLYMERS BY OLIGOMER COMBINATION REACTIONS[a]

Oligomer A	*Oligomer B or other coreactant*	*Product type*	*Reference*
$H[NHCH_2\phi CH_2NHCO(CH_2)_{16}CO]_nNHCH_2\phi CH_2NH_2$	$HOOC(CH_2)_{16}CO[N(Et)CH_2\phi CH_2N(Et)CO(CH_2)_{16}CO]_nOH$	Polyamide-*b*-polyamide	68
$Cl[CO\text{-}m\text{-}\phi CONH\text{-}m\text{-}\phi NH]_nCO\text{-}m\text{-}\phi COCl$	HO[POE]OH	Polyether-*b*-polyamide	80
$ClCO(CH_2)_4COO[POE]OCO(CH_2)_4COCl$	$H_2N(CH_2)_6NH_2$, aqueous NaOH	Segmented poly(etheramide)	124
(epoxide)$-CH_2O[PSt]OCH_2-$(epoxide)	$H[NH(CH_2)_5CO]_2—NH(CH_2)_6NH—[CO(CH_2)_5NH]_2H$	Polystyrene-*b*-polyamide	126
$H_2N\phi COO[POE]OCO\phi NH_2$	Pyromellitic dianhydride	Segmented poly(etherimide)	139
$OCn\phi CH_2\phi NHCOO[POP]OCONH\phi CH_2\phi NCO$	Pyromellitic dianhydride	Segmented poly(etherurethane imide)	241
$OSN\phi O\phi CONH\phi O\phi NHCO\phi O\phi NSO$	Water + $m\text{-}\phi(COCl)_2$	Segmented polyamide	242
HN(piperazine)N—$[COO(CH_2)_4OCON$(piperazine)$N—]_nH$	ClCOO[POTM]OCOCl	Polyether-*b*-polyurethane	243
$HOOC(CH_2)_4CO[NH(CH_2)_6NHCO(CH_2)_4CO]_nOH$	$H[O(CH_2)_2O\,.\,CO\phi CO]_nO(CH_2)_2OH$	Polyester-*b*-polyamide	244
$H[NH\phi CH_2\phi NHCO$(phthalimide, CO/N/CO)$]_n\phi CH_2\phi NH_2$	ClCO[PBd]COCl	Polybutadiene-*b*-poly(amideimide)	245
$H[—N$(2,5-dimethylpiperazine, Me, Me)$NCO(CH_2)_4CO]_n—N$(2,5-dimethylpiperazine, Me, Me)NH	$Cl[CO\phi CON$(2,5-dimethylpiperazine, Me, Me)$N]_nCO\phi COCl$	Polyamide-*b*-polyamide	246
Nylon 11 oligomers with one or two COOH, or two ester end groups	[POE] and [POTM] with one or two OH end groups	Polyether-*b*-polyamide	247
Various homo- and copolyamides with two NH_2 end groups	$OCN(CH_2)_6NHCOO[POE]OOCNH(CH_2)_6NCO$	Urea-linked polyether-*b*-polyamide	248

[a] See footnote to Table 3 for comment on abbreviations.

the activator of polymerisation together with a catalytic amount of an initiating base (e.g. NaH) which generates anions from the lactam monomer, these being the effective intermediates of reaction.

$$R\text{CO—}\overline{\text{N—Alk—CO}} + n\text{H—}\overline{\text{N—Alk—CO}} \longrightarrow R\text{CO—[NH.Alk.CO]}_n\text{—}\overline{\text{N—Alk—CO}} \quad (21)$$

(XXVIII) (XXIX)

Evidently, if R is polymeric the product (XXIX) will be a block copolymer, of AB or ABA type according to the functionality of the activator. For the reaction to proceed satisfactorily the ring⇌chain equilibrium between lactam and polyamide must be favourable for polymerisation in the conditions employed, and there must be an absence of side reactions causing loss of propagative anionicity or transfer leading to the formation of free lactam homopolymer. Successful polymerisations have been achieved giving products with polyether,[249,250] polystyrene,[133,251-3] SBR,[250] polydiene[133,251] or polyisobutene[254] blocks, usually in combination with nylon 6 as the polyamide segment(s) or, in some cases, with nylon 4 or nylon 8.

Polybutadiene-*b*-nylon-6 copolymers prepared in this way are curable with dicumyl peroxide to hard resins of good flexural strength and impact resistance.[255]

The reaction cannot be extended to the formation of well-defined polyester–polyamide block copolymers. Thus, ε-caprolactone (2-oxepanone) (XXX) is an effective precursor (by combination with lactamate anion) of an acyllactam activator (XXXI) for the polymerisation of ε-caprolactam (hexahydro-2H-azepin-2-one) (eqn (22)), and this activator might be expected to initiate the polymerisation of further lactone molecules from its oxyanion end concurrently with that of lactam from the cyclic end, to give AB poly(caprolactone)–poly(caprolactam) block polymers. Under initiation by strong base the two monomers are indeed copolymerisable in all proportions, but the products—like those obtained using preformed poly(ε-caprolactone) as the activator—are random block copolymers resulting from interchange reactions promoted by the base.[256,257] Similar partial randomisations occur in copolymerisations with methylcaprolactones and with laurolactam (azacyclotridecan-2-one).[258]

$$\overline{\text{O(CH}_2)_5\text{CO}} + :\overline{\text{N(CH}_2)_5\text{CO}} \longrightarrow {}^{-}\text{O(CH}_2)_5\text{CO}\overline{\text{N(CH}_2)_5\text{CO}} \quad (22)$$

(XXX) (XXXI)

Synthetic polypeptides (carbon-substituted nylon 2 polymers) are substances of considerable importance to the biophysicist as models of protein structure and function. They are usually prepared by the amine-initiated ring opening polymerisation of 4-substituted oxazolidin-2,5-diones which are the *N*-carboxyanhydrides (NCA) of α-aminocarboxylic acids, a reaction which proceeds with the concurrent elimination of carbon dioxide. By the use of amine-ended oligomers as initiators, block polymers containing polypeptide sequences can be obtained as shown in eqn (23). Those with polystyryl or polydienyl segments are made using initiators of the types $\mathbf{P}.CONH.R.NH_2$ or $\mathbf{P}(CH_2)_2OCONH.R.NH_2$ made by refunctionalisation of carbochain oligomer anions as discussed earlier. If monofunctional they yield AB diblock polymers and, if bifunctional, ABA triblock polymers with a central carbochain unit.[127,128,259,260]

$$\mathbf{P}\text{—}NH_2 + n\ \text{(NCA; HN–CHR, O=C–O–C=O ring)} \longrightarrow \mathbf{P}\text{—}NH[CO.CHR.NH]_nH + nCO_2 \quad (23)$$

Since the polypeptide formed by the polymerisation of an NCA with a simple amine is itself amine-ended, sequential polymerisations can be conducted with different NCA monomers leading to AB diblock copolypeptides or, with a further cycle of operation using the original monomer, to ABA triblock copolymers. In the latter case, the outer A blocks may not be of equal lengths and, where such equality is required, a better procedure is to initiate the primary polymerisation bidirectionally with a diamine according to the scheme of eqn (24).[261–4]

$$H_2N.R.NH_2 + 2m\ \text{(NCA; HN–CHR}^1\text{)} \longrightarrow$$

$$H[NH.CHR^1.CO]_m\text{—}NH.R.NH\text{—}[CO.CHR^1.NH]_mH$$

$$\Big\downarrow 2n\ \text{(NCA; HN–CHR}^2\text{)} \quad (24)$$

$$H[NH.CHR^2.CO]_n\text{—}[NH.CHR^1.CO]_m\text{—}NH.R\text{—}NH\text{—}[CO.CHR^1.NH]_m\text{—}[CO.CHR^2.NH]_nH$$

By these various procedures block copolypeptides have been obtained with different combinations of crystalline, glassy or elastomeric, or hydrophilic and hydrophobic segments. In some cases, a sensitive group of the aminocarboxylic acid component has to be protected during the polymerisation step and then released by post-reaction; an example is the formation of polybutadiene-*b*-poly(L-lysine) shown in eqn (25).[128]

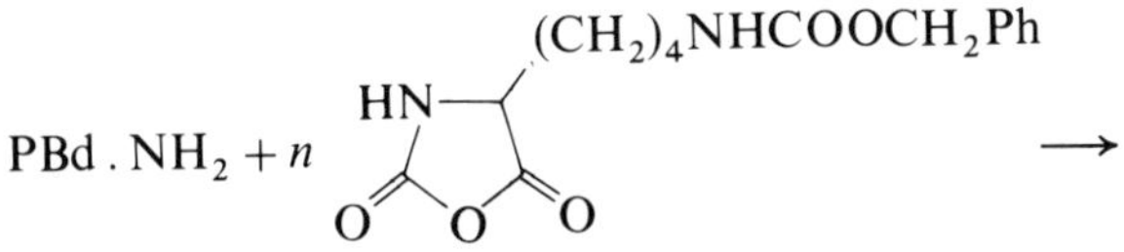

$$\mathrm{PBd.NH_2} + n\ \text{[NCA ring: HN–CH((CH}_2)_4\text{NHCOOCH}_2\text{Ph)–C(=O)–O–C(=O)]} \longrightarrow$$

$$\mathrm{PBd.NH[CO.CH((CH_2)_4NHCOOCH_2Ph).NH]_nH} \xrightarrow{\mathrm{HCl/HBr}} \mathrm{PBd.NH[CO.CH((CH_2)_4NH_2).NH]_nH} \quad (25)$$

Interesting aspects of such block polymer syntheses are found in the use of a glycopeptide having a terminal asparagine residue to initiate the polymerisation of γ-benzyl-L-glutamate NCA, yielding a polysaccharide–polypeptide block polymer,[265] and in the elegant demonstration that stereoselective blockiness is achievable in the first-named segments of poly(leucine)-*b*-poly(N^6-carbobenzoxy-DL-lysine) when DL-leucine NCA is polymerised with an optically active amine.[261]

Lastly, mention will be made of two further amide block copolymer syntheses not mentioned above. The first[266] employs living polymer anions from styrene, isoprene or methyl methacrylate to initiate the polymerisation of isocyanates to diblock polymers containing nylon 1 sequences. In this approach, selective polymerisation can be achieved at the unhindered isocyanate group of diisocyanates such as toluene-2,4-diisocyanate to give products with pendant NCO groups (XXXII) which are crosslinkable with diols to tough resins.

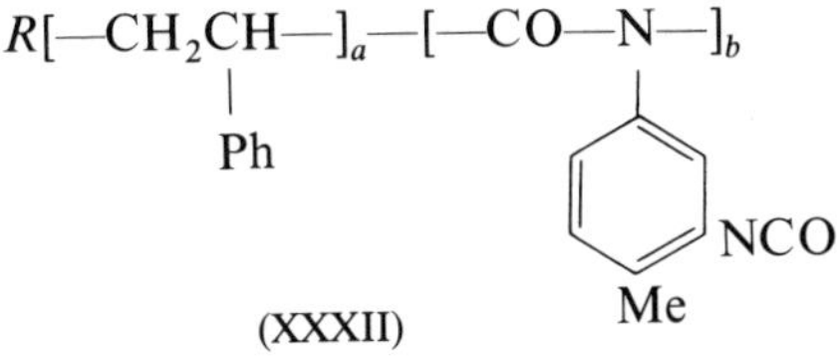

$$R[\!-\!\mathrm{CH_2CH(Ph)}\!-\!]_a\!-\![\!-\!\mathrm{CO}\!-\!\mathrm{N}(\mathrm{C_6H_3(Me)(NCO)})\!-\!]_b$$

(XXXII)

The second synthesis[267,268] commences with an acrylamido-ended prepolymer (XXXIII) which, by free radically-initiated polymerisation with styrene, gives bifurcated block products (XXXIV) of potential interest as novel non-thrombogenic biomedical materials.

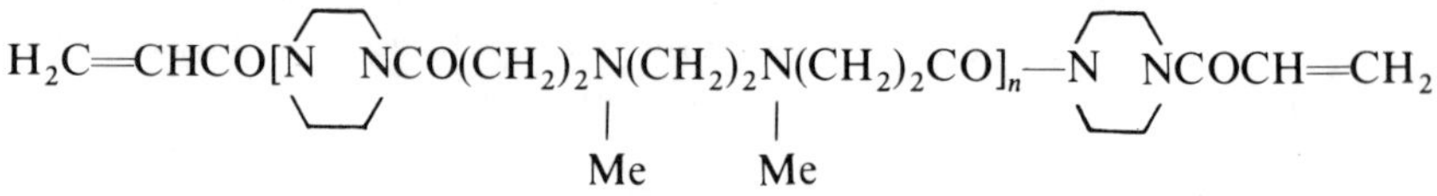

$$H_2C{=}CHCO[\overline{N\quad N}CO(CH_2)_2\underset{\substack{|\\ Me}}{N}(CH_2)_2\underset{\substack{|\\ Me}}{N}(CH_2)_2CO]_n{-}\overline{N\quad N}COCH{=}CH_2$$

(XXXIII)

$$\sim\!\mathbf{P}{-}\overline{N\quad N}COCH\begin{matrix} \diagup CH_2PSt \\ \diagdown PSt \end{matrix}$$

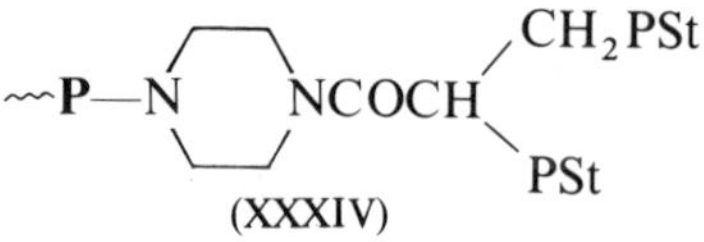

(XXXIV)

2.6. Polysiloxane Block Copolymers

Silicone polymers (polysiloxanes) form a group of materials of unusual properties which may include high thermal stabilities, low surface energies and low T_g values. Not surprisingly there has been long-standing interest in the incorporation of such components into block copolymer structures, and earlier work in this direction is reviewed in considerable detail in the two major monographs on block copolymers.[44,45] One particular group, the polysiloxane–poly(oxyalkylene) surfactants,[269–71] have already been mentioned as important aids in the manufacture of polyurethane foams, and much effort has been devoted to exploring the potentialities of other types of polysiloxane block polymers as elastomers and, more recently, as plastics materials. This section will review some methods of synthesis available in this field, noting first two factors which have to be taken into account in designing synthetic procedures and attainable molecular structures.

The first arises from the susceptibility of Si–O–Si linkages to undergo scission and bond reformation by the action of strong bases (eqns 26(a) and 26(b)). These processes can lead to a randomisation ('shuffling' or equilibration) of siloxane units between polysiloxane chains and hence to a

$$KOH + {-}\overset{|}{\underset{|}{Si}}{-}O{-}\overset{|}{\underset{|}{Si}}{-} \rightleftharpoons {-}\overset{|}{\underset{|}{Si}}{-}O^-K^+ + HO{-}\overset{|}{\underset{|}{Si}}{-} \qquad (26(a))$$

$${-}\overset{|}{\underset{|}{Si}}{-}O^-K^+ + {-}\overset{|}{\underset{|}{\mathbf{Si}}}{-}O{-}\overset{|}{\underset{|}{\mathbf{Si}}}{-} \rightleftharpoons {-}\overset{|}{\underset{|}{Si}}{-}O{-}\overset{|}{\underset{|}{\mathbf{Si}}}{-} + K^{+\,-}O{-}\overset{|}{\underset{|}{\mathbf{Si}}}{-} \qquad (26(b))$$

broadened distribution of M.W.'s and often to the liberation of polysiloxane homopolymer, either during synthesis or on subsequent contact with bases. Care has therefore to be taken in the selection of basic initiators or catalysts for use in polymerisation reactions in this series.

The second factor relates to the mode of attachment of polysiloxanyl units to the other components of structure in block copolymers; Si–O–C connections are relatively readily hydrolysed and, in the interests of product stability, are better replaced where possible by more stable Si–C bonds.[272–5]

The principle of random block polymerisation has been used for polysiloxane block polymer synthesis as in eqn (27) where the block connection is made by condensation between two types of silanol groups. This reaction, which is catalysed by weak bases such as hexylamine carboxylate salts or tetramethylguanidine to avoid Si–O bond randomisation, leads to multiblock polymers (XXXV) having *p*-phenylene groups in the main chain.[61]

$$a\,\mathrm{HOSiMe_2\phi SiMe_2OH} + \mathrm{HO[SiMe_2O]}_b\mathrm{H} \longrightarrow \{\!-\![\mathrm{SiMe_2\phi SiMe_2O}]_a\!-\![\mathrm{SiMe_2O}]_b\!-\!\}_n \quad \text{(XXXV)} \tag{27}$$

In other approaches, the polysiloxane (e.g. poly(dimethyl siloxane), poly(oxydimethylsilylene)) is furnished with functional organic end groups which are employed in subsequent conventional copolycondensation reactions. Thus polycarbonate–PDMS block polymers, which have been studied in considerable detail,[22,81,276–8] are obtained by end-capping α,ω-dichloro-PDMS with bisphenol A to give reactive oligomers (XXXVI) which are converted to block copolymers (XXXVII) by co-phosgenation with further quantities of the bisphenol (eqn (28)).

$$a\,\mathrm{HO\phi CMe_2\phi[PDMS]O\phi CMe_2\phi OH}\ \text{(XXXVI)} + b\,\mathrm{HO\phi CMe_2\phi OH} + (a+b)\,\mathrm{COCl_2} \longrightarrow \{\!-\![\!-\!\mathrm{O\phi CMe_2\phi O[PDMS]O\phi CMe_2\phi O.CO}\!-\!]_a\!-\![\mathrm{O\phi CMe_2\phi O.CO}\!-\!]_b\!-\!\}_n \ \text{(XXXVII)} \tag{28}$$

The copolymers are clear tough materials with properties, depending on the $a{:}b$ ratio, ranging from flexible elastomers to tough plastics. Use of the hindered bisphenol (XXXVIII) in place of bisphenol A gives rise to

TABLE 5

POLYSILOXANE BLOCK POLYMERS BY OLIGOMER COMBINATION REACTIONS

Polysiloxane oligomer	*Additional oligomers and other reactant(s)*	*Product type*	*Reference*
ClCO-*m*-ϕCOO[PDMS]OOC-*m*-ϕCOCl	H[HN*m*-ϕNHCO-*m*-ϕCO]$_n$NH-*m*-ϕNH$_2$	PDMS-*b*-polyamide	80
$^-O_4Cl\ ^+SiMe_2$[PDMS]$SiMe_2^+ClO_4^-$	$^-$PSt$^-$	PDMS-*b*-polystyrene	156
Cl[PDMS]$SiMe_2Cl$	$HOSiMe_2$. PSt . $SiMe_2OH$/pyridine	PDMS-*b*-polystyrene	282
Cl[PDMS]$SiMe_2Cl$	$HO(CH_2)_6O[CO(CH_2)_8COO(CH_2)_6O]_nH$/pyridine	PDMS-*b*-polyester	283, 284
Cl[PDMS]$SiMe_2Cl$	Hydroxyl-ended poly(ethylene adipate-*co*-maleate)/*N*-methylmorpholine	PDMS-*b*-unsaturated copolyester	285
HO[PDMS]$SiMe_2OH$	$HO[CH_2CH_2]_nOH$/tin(II) octoate	PDMS-*b*-polyethylene	286

transparent thermoplasts of exceptionally high T_g and possessing an impressive level of fire-safety performance.[279–81]

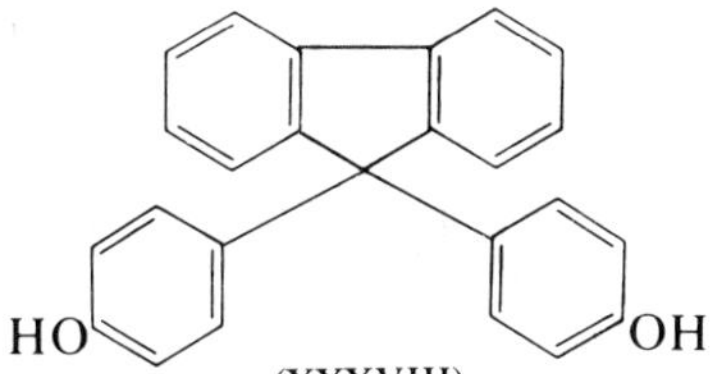

(XXXVIII)

In an analogous synthesis α,ω-di(*m*-aminophenyl)PDMS oligomers have been condensed with *m*-phenylene diamine and isophthaloyl chloride to give poly(*m*-phenylene isophthalamide)–PDMS block copolymers.[80]

The next class of reactions to be considered is that yielding polysiloxane block or segmented polymers by oligomer combination. Some simple examples are given in Table 5. These have been supplemented by two types of reactions that are specific to organosilicon chemistry.

The first comprises the reaction of addition which occurs between hydridosilane and vinyl groups, usually catalysed by chloroplatinic acid, forming Si–C bonds (eqn (29)).

$$-\overset{|}{\underset{|}{Si}}-H + H_2C{=}CH- \xrightarrow{H_2PtCl_6} -\overset{|}{\underset{|}{Si}}-CH_2CH_2- \qquad (29)$$

The reaction can be adapted to block polymer formation in various ways, e.g. by forming polystyrenes with $-SiMe_2CH{=}CH_2$ or $-CH_2SiMe_2CH{=}CH_2$ end groups and combining these with mono- or dihydrido-ended PDMS oligomers, or inversely by reacting α,ω-di(hydridodimethylsilyl)polystyrenes with PDMS oligomers having allyloxy end groups.[275,282,287] Another variant uses the hydrosilylation reaction to attach epoxide end groups to the ends of a PDMS chain, giving oligomers (XXXIX) which can form block copolymers by ring-opening polyaddition to carboxyl-ended polyesters (eqns (30(a)) and (30(b))).[274]

$$H[PDMS]SiMe_2H + 2\,H_2C{=}CH{-}CH_2OCH_2\underset{\diagdown O \diagup}{CH{-}CH_2} \xrightarrow{H_2PtCl_6}$$

$$\underset{\diagdown O \diagup}{H_2C{-}CH}CH_2O(CH_2)_3{-}[PDMS]{-}SiMe_2(CH_2)_3OCH_2\underset{\diagdown O \diagup}{CH{-}CH_2} \qquad (30(a))$$

(XXXIX)

$$(\text{XXXIX}) + \text{HOOC[polyester]COOH} \xrightarrow[\text{HO}^- \text{ catalyst}]{R_3\text{N or}}$$

$$\{\text{—OC[polyester]—COOCH}_2\underset{\displaystyle \text{OH}}{\underset{|}{\text{C}}}\text{HCH}_2\text{O(CH}_2)_3\text{—[PDMS]—SiMe}_2\text{(CH}_2)_3\text{OCH}_2\underset{\displaystyle \text{OH}}{\underset{|}{\text{C}}}\text{HCH}_2\text{O—}\}_n \quad (30(b))$$

The second specific reaction is that of condensation between polysiloxane oligomers with dialkylamino end groups and hydroxyl-ended co-reactant oligomers; dialkylamine is eliminated as Si–O bonds are formed. The OH groups of the co-reactants may be phenolic or alcoholic, the former being the more reactive. The approach has been used to obtain multiblock polymers of PDMS with a variety of other species;[284,288] an example leading to a polyester-*b*-polysiloxane product is shown in eqn (31), and one using silanol end groups in eqn (32).[289]

$$\text{HO—}\langle\text{C}_4\text{Me}_4\rangle\text{—}\left[\text{O.CO.O}\langle\text{C}_4\text{Me}_4\rangle\text{O—}\right]_a\text{H}$$

$$+ \text{Me}_2\text{N[PDMS]SiMe}_2\text{NMe}_2 \xrightarrow[\text{or THF}]{\text{reaction in PhCl}}$$

$$2\,\text{Me}_2\text{NH} + \left\{\text{—O—}\langle\text{C}_4\text{Me}_4\rangle\text{—}\left[\text{O.CO.O}\langle\text{C}_4\text{Me}_4\rangle\text{O—}\right]_a\text{—[PDMS]—}\right\}_n \quad (31)$$

Ultrahigh M.W. segmented poly(carboranesiloxane) polymers have also been obtained by the coupling reaction shown in eqn (33).[290]

As with other classes of polymers considered above, ring opening polymerisation has also been employed as the basis of another general approach to the synthesis of siloxane block copolymers—in this case by the anionic polymerisation of cyclopolysiloxanes in ways which yield either polysiloxane-*b*-polysiloxane products or those containing polysiloxane and other structural classes of blocks.

$$\mathrm{HO[PDMS]SiMe_2OH + Et_2N{-}\overset{\overset{\large Me}{|}}{\underset{\underset{\large (CH_2)_2Py}{|}}{Si}}{-}NEt_2 \longrightarrow}$$

$$\left\{ \mathrm{-O[PDMS]SiMe_2O{-}\overset{\overset{\large Me}{|}}{\underset{\underset{\large (CH_2)_2Py}{|}}{Si}}{-}} \right\}_n \tag{32}$$

Py = pyridyl

$$\mathrm{HOSiMe_2 . CB_{10}H_{10}C . SiMe_2[O\overset{\overset{\large R}{|}}{\underset{\underset{\large Me}{|}}{Si}}OSiMe_2 . CB_{10}H_{10}C . SiMe_2]_aOH}$$

$$+ \left[\left(\text{pyrrolidine} \right)\mathrm{N . CO\underset{\underset{\large Ph}{|}}{N}{-}} \right]_2 \mathrm{SiMePh} \longrightarrow \tag{33}$$

$$\mathrm{[{-}O\overset{\overset{\large R}{|}}{\underset{\underset{\large Me}{|}}{Si}}OSiMe_2 . CB_{10}H_{10}C . SiMe_2{-}]_n + 2 \left(\text{pyrrolidine} \right)NCONHPh}$$

The proneness of siloxane bonds to undergo cleavage in strongly basic conditions imposes the need for a specific choice of polymerisable cyclopolysiloxane and of the initiator counterion if well-defined products are to result. The most readily available monomer, octamethylcyclotetrasiloxane, is particularly susceptible to bond randomisation, leading to the problems already mentioned. However, these can be avoided by use of the cyclic trimer hexamethylcyclotrisiloxane (XL). Through its greater reactivity, this substance is open to polymerisation by weaker bases such as lithium silanolates or the lithium salts of hydrocarbon polymer anions which do not promote the equilibration reaction.

$$\underset{\text{(XL)}}{\mathrm{\llcorner[SiMe_2O]_3\lrcorner}} \qquad \underset{\text{(XLI)}}{\mathrm{[PDMS]OSiMe_2(CH_2)_{10}CO\overline{N(CH_2)_5C}O}}$$

By initiating the polymerisation of (XL) with $MePh_2SiOLi$ or $LiOSiPh_2OLi$ and reacting the resultant $LiOSiMe_2$-ended polymer with hexaphenylcyclotrisiloxane, polymers of $[SiMe_2O]_a$-*b*-$[SiPh_2O]_b$ structure have been formed.[62] Correspondingly, initiation with PSt^-Li^+, $Li^{+-}PSt^-Li^+$ or analogous polydienyl species gives rise to PSt–PDMS, PDMS–PSt–PDMS and polydiene–PDMS block copolymers[63,125,291–4] Silanolate end groups are not themselves sufficiently nucleophilic to initiate the polymerisation of vinyl monomers, and the approach cannot therefore be extended to the direct synthesis of ABA copolymers with PDMS units as the central segments. To obtain such materials, recourse must be made to the coupling of silanol- or silanolate-ended diblock polymers by reagents such as Me_2SiCl_2 or $Ph_2Si(OAc)_2$;[295,296] multiblock polysiloxane copolymers have been made by the similar reactions of BAB copolymers having outer siloxane blocks.[125]

Lastly, reference may be made to the use of the *N*-acyllactam-ended oligomer (XLI), itself prepared with the aid of several of the reactions already discussed, as an activator for the anionic polymerisation of ε-caprolactam, yielding PDMS–nylon 6 block copolymers.[297]

2.7. Polysulphone Block Copolymers

The strong dipoles of sulphone $\leftarrow SO_2 \rightarrow$ groups incorporated as recurring features of structure in polymer chains exercise a powerful elevating effect on intermolecular cohesion, comparable with that conferred by the CONH groups in polyamides. Of the many known types of polymers containing sulphone groups, particular interest attaches to the poly(arylene sulphones) and poly(aryleneether sulphones) which were discovered in the 1960s and are now well established as high-performance engineering thermoplasts.[206] Investigation in this field has recently been extended to block copolymers, particularly of poly[oxy-1,4-phenylenesulphonyl-1,4-phenyleneoxy-1,4-phenylene(1-methylethylidene)-1,4-phenylene] (XLII) which, for ease of presentation, will be the only polysulphone considered in this section. The polymer is prepared by the polyetherification of bisphenol A with 4,4′-dichlorodiphenyl sulphone in the presence of a base. By control of the reactant stoichiometry, its oligomers can be obtained either with phenolic or with *p*-chlorophenylsulphonyl groups at both ends of the molecule; both forms have been employed in the synthesis of block copolymers.

$$[-O\phi SO_2\phi O\phi CMe_2\phi-]_n$$

(XLII)

The formation of polysulphone-*b*-PDMS block copolymers has thus been accomplished by use of the silylamine-OH reaction as shown in eqn (34). The products are transparent materials with properties ranging from elastomeric to rigid thermoplastic as the polysulphone content increases.[24,288,298]

$$\mathrm{HO\phi CMe_2\phi[polysulphone]OH + Me_2N[PDMS]SiMe_2NMe_2 \longrightarrow}$$
$$\mathrm{2\,Me_2NH + \{—O\phi CMe_2\phi[polysulphone]O—}b\mathrm{—[PDMS]SiMe_2—\}_n} \quad (34)$$

Polysulphone–BPAC block copolymers have been prepared by three types of reactions which lead to products having different degrees of structural order: (a) by the random block copolymerisation of hydroxyl-ended polysulphone oligomers with bisphenol A and phosgene,[299,300] (b) by the co-phosgenation of mixtures of hydroxyl-ended polysulphone and BPAC oligomers,[301] (c) by the condensation of hydroxyl-ended polysulphone with α,ω-chloroformate-ended BPAC oligomers.[300,302] From the nature of both constituent units, these block copolymers are tough thermoplasts, those with polysulphone blocks of M.W. $< 16\,000$ being homogeneous single-phase materials. Polysulphone block copolymers with polystyrene or with 4G.T sequences have been prepared by similar reactions. Polysulphone–poly(carboranesiloxane) block polymers have also been obtained by a sequence of reactions commencing with hydroxyl-ended polysulphone oligomers (eqn (35)); these materials possess high resistance towards thermal oxidation.[303]

$$\mathrm{HO\phi CMe_2\phi[polysulphone]OH} + 2\left[\;\bigcirc\!\mathrm{NCON}(\mathrm{Ph})—\;\right]_2 \mathrm{Si}R_2 \longrightarrow$$
$$\bigcirc\!\mathrm{NCON(Ph)Si}R_2\mathrm{O\phi CMe_2\phi[polysulphone]OSi}R_2\mathrm{N(Ph)CON}\!\bigcirc + 2\,\bigcirc\!\mathrm{NCONH}R$$
$$\Big\downarrow\ \mathrm{HOSiMe_2.CB_{10}H_{10}C.SiMe_2[OSi}R_2\mathrm{OSiMe_2.CB_{10}H_{10}C.SiMe_2]_{\mathit{a}}OH}$$
$$\{—\mathrm{[carboranesiloxane]OSi}R_2\mathrm{O\phi CMe_2\phi[polysulphone]OSi}R_2\mathrm{O}—\}_n + 2\,\bigcirc\!\mathrm{NCONHPh} \quad (35)$$

Polysulphone–nylon 6 block copolymers have been obtained by utilising the interesting ability of *p*-chlorophenylsulphonyl-ended polysulphone oligomers to serve as activators for the anionic polymerisation of ε-caprolactam. The activation depends on the formation of *N*-(*p*-phenylenesulphonyl-*p*-phenylene)-ε-caprolactam end groups, and these can also be formed by means of an alternative and simpler procedure involving the addition of sodium hydride to molten polysulphone–lactam mixtures when ether groups of the polysulphone chain are cleaved by the lactamate anions generated *in situ* (eqn (36)); the reaction has a formal resemblance to the

$$\sim\phi SO_2\phi O\phi CMe_2\sim + :\overline{N(CH_2)_5CO} \longrightarrow \sim\phi SO_2\phi\overline{N(CH_2)_5CO} + {}^{-}O\phi CMe_2\sim \quad (36)$$

poly(caprolactone)–lactam copolymerisation mentioned earlier. The polysulphone–nylon 6 copolymers are two-phase materials which are compatible with both of the related homopolymers and display some advantages of mechanical properties over those of polysulphone resins.[304,305]

2.8. Other Block Copolymers

Some further approaches to the synthesis of heterochain block copolymers, not included in earlier sections, are summarised below:

(a) Under initiation by ionic or ionogenic substances such as protic acids, oxazolinium salts, alkyl iodides or alkyl *p*-toluenesulphonates, 2-oxazolines undergo living polymerisation to products with oxazolinium end groups (XLIII) (eqn (37)). These can

$$X^+Y^- + (a+1)\,\text{N}\overset{\frown}{\underset{R^1}{\smile}}\text{O} \longrightarrow X[-\underset{}{\overset{COR^1}{N}}(CH_2)_2-]_a-\text{N}\overset{\frown}{\underset{R^1}{\overset{+}{\smile}}}\text{O}\,Y^- \quad (37)$$

(XLIII)

initiate the polymerisation of a second oxazoline or of an oxazine monomer in a sequential manner to give block copolymers composed of different types of poly(*N*-acylalkyleneimine) segments as shown in eqn (38).[306,307] Polymers of the structure (XLIV) with R^1 = undecyl and R^2 = ethyl represent combinations of hydrophobic and hydrophilic blocks and are surfactants.

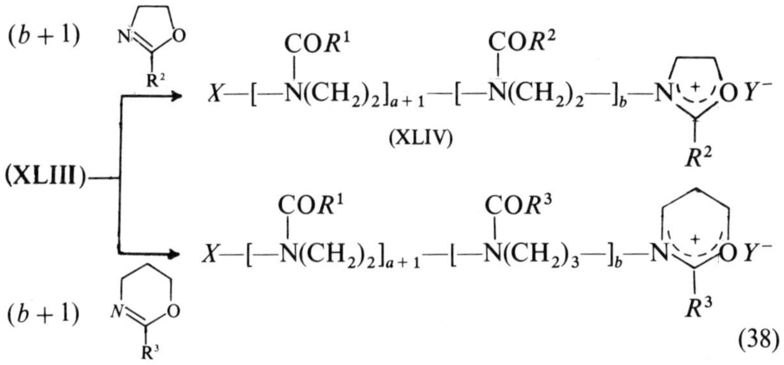

$$(b+1)\ \text{2-}R^2\text{-2-oxazoline} \xrightarrow{} X\text{—}[\text{—}\underset{\displaystyle COR^1}{N}(CH_2)_2]_{a+1}\text{—}[\text{—}\underset{\displaystyle COR^2}{N}(CH_2)_2\text{—}]_b\text{—}N^{+}\text{(oxazolinium, } R^2\text{)}\ Y^- \quad \text{(XLIV)}$$

$$\text{(XLIII)}$$

$$(b+1)\ \text{2-}R^3\text{-5,6-dihydro-4H-1,3-oxazine} \xrightarrow{} X\text{—}[\text{—}\underset{\displaystyle COR^1}{N}(CH_2)_2]_{a+1}\text{—}[\text{—}\underset{\displaystyle COR^3}{N}(CH_2)_3\text{—}]_b\text{—}N^{+}\text{(oxazinium, } R^3\text{)}\ Y^- \quad (38)$$

An alternative synthesis of block copolymers based on 2-oxazolines employs oligomeric initiator cations or procations, as in the polymerisation of 2-methyl-2-oxazoline with iodo-ended poly(phenyloxirane),[308] and in that of 2-oxazoline with the ditosylate of α,ω-di(2-hydroxyethyl)-1,2-polybutadiene; the last-mentioned reaction gives rise to poly(*N*-formylethyleneimine)-*b*-polybutadiene-*b*-poly(*N*-formylethyleneimine) which can be converted, by removal of its formyl groups, to the block copolyamine (XLV).[137]

$$HO[CH_2CH_2NH]_a\text{—}[\text{—}CH_2\text{—}\underset{\displaystyle H_2C{=}CH}{CH}\text{—}]_b\text{—}[NHCH_2CH_2]_a\,OH$$

(XLV)

(b) The decompositions of the bis(diazonium chloride) (XLVI) and of bis(*N*-nitroso-*N*-acetylamines) (XLVII) in the presence of dienes have been studied in attempts to obtain flexible electronically-conducting polymers which might be miscible with (and hence dispersible in) non-conducting organic polymers.[309,310] The ensuing reactions of radical formation and recombination give rise to block polymers of imperfect structure containing the structural units shown in (XLVIII).

$$Cl^-\ {}^+N_2\phi\text{—}\phi N_2^+\ Cl^- \qquad \text{(XLVI)}$$

$$CH_3CO\underset{\displaystyle NO}{N}\text{—}\phi X\phi\text{—}\underset{\displaystyle NO}{N}COCH_3 \qquad \text{(XLVII)}$$

$$[\text{—}\phi X\phi\text{—}]_a/[\text{—}\phi X\phi N{=}N\text{—}]_b/\text{polydiene}$$

(XLVIII)

(c) Polymers containing cellulose or amylose blocks have been prepared by the controlled hydrolysis of the polysaccharide triacetates to hydroxyl-ended triacetate oligomers of controlled chain lengths. By coupling with diisocyanate end-capped poly-(oxypropylene)- or polybutadiene diols these were converted to products of the type exemplified in (XLIX) and thence, by methanolysis of the acetate groups, to polysaccharide–polyether or polysaccharide–polybutadiene multiblock copolymers some of which have interest as biodegradable film-forming materials.[35,38,40]

$$\left\{ \left[-O-C_6H_7O(CH_2OAc)(AcO)_2- \right]_a OCONH.R.NHCOO[POP]OOCNH.R.NHCO- \right\}_n$$

(XLIX)

3. SOME STRUCTURE–PROPERTY RELATIONSHIPS IN HETEROCHAIN BLOCK COPOLYMERS

3.1. General Considerations

A prominent factor distinguishing block from random copolymers is the frequent, though not invariable, separation of the materials of their constituent blocks into microphases or domains, resulting in a supramolecular texture that is specific to block copolymers and which has an important influence upon their properties and uses.

Most comparative studies of block copolymer properties have been concerned with the effects of structural variations within particular classes such as the all-carbochain block polymers, block polyetheresters, or block copolyurethanes. This section will attempt a wider comparison of the characteristics of some individual types of blocks combined in copolymers, with particular regard to (a) the mutual compatibility or otherwise of blocks of similar and of different chemical types, (b) the characteristics of individual types of blocks in varied molecular environments, (c) the effects of block combination on the crystallisation behaviour of crystallisable components, (d) the minimum segment length that permits individual block components to display properties characteristic of their corresponding homopolymers. All these features are of practical interest for the molecular design of new types of block copolymers. For the most part they cannot be predicted reliably from fundamental or *a priori* considerations, and perforce they will here be considered only qualitatively and empirically on the basis of available experimental evidence. In the space available it is not possible to consider in detail the properties and morphology of the polymers, but a few idealised morphological models will be included as indicators of current thinking on the varieties of textures believed to exist in block copolymers.

It will be helpful to begin with a résumé of the essential characteristics of the main types of block polymers, commencing with the now classical materials composed of polystyrene (A) and polydiene (B) units, and then proceeding to their simple structural analogues and to the more complex multiblock copolymers.

Briefly, whilst PSt–PBd AB diblock and BAB triblock copolymers may display phase separation of their components, the resulting texture does not provide the stress-resisting network structure that is possible in their ABA counterparts and the materials are subject to viscous flow under applied stress, becoming usefully elastomeric only when chemically crosslinked. In contrast, provided that certain criteria of composition and block M.W. are satisfied, phase separation occurs in the ABA copolymers to give a texture of the type shown schematically in Fig. 4, with well-defined domains of glassy PSt units set in and connected molecularly with a continuum of rubbery PBd units, each terminus of the latter being anchored in rigid-phase agglomerate. At ordinary temperatures, applied mechanical stresses are taken up by the PBd matrix, the deformation of which is limited by the physical network imposed by the PSt domains. At temperatures exceeding T_g of the PSt aggregates these become fluid and ductile, thereby providing the characteristic thermally-reversible elastomeric nature of the materials. In practice, the phase separation which is essential for useful elastomeric properties requires A block M.W.'s of at least 6000–7000, and typical PSt–PBd–PSt elastomers have A block M.W.'s of 10 000–25 000, B block M.W.'s of 50 000–100 000, and PSt contents of 20–35 wt %. The real geometrical forms of the domains vary markedly with the volume fractions of the A units and with the history of the sample as discussed in Chapters 1 and 4, but their general dimensions are broadly within the range 100–300 Å. In summary, therefore, the basic architectural model leading to useful thermoplastic elastomer properties in these materials comprises a hard segment-*b*-soft segment-*b*-hard segment triblock structure of high total M.W. (approximately 100 000), composed of non-polar and amorphous constituents, and of polyphasic morphology.

We shall now examine the extent to which this model can be modified by structural variation with the retention of technically useful properties.

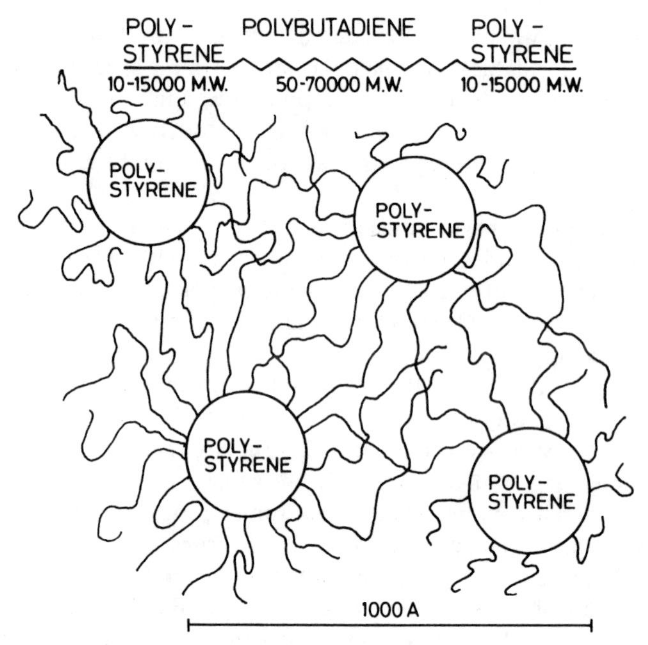

FIG. 4. Generalised phase separated morphological model for polystyrene-*b*-polybutadiene-*b*-polystyrene thermoplastic elastomeric triblock copolymers. (From Morton, M., *J. Polym. Sci., Polym. Symp.*, 1977, **60**, 1; with permission, John Wiley & Sons, Inc.)

Replacement of PBd as the B block with polyisoprene or hydrogenated high-vinyl 1,2-polybutadiene, or of PSt as the A block with poly(α-methylstyrene), poly(*p*-t-butylstyrene) or poly(vinylcyclohexane) has no adverse effect and may indeed improve some practical properties,[311,312] but PSt–PDMS–PSt block polymers and their analogues with PαMeSt outer blocks, whilst being elastomeric and polyphasic, have considerably reduced tensile strengths as compared with PSt–PBd–PSt copolymers;[295,296] interestingly, $[\text{PSt–PDMS}]_n$ multiblock copolymers have mechanical properties resembling those of the triblock series with similar compositions.[291] The polymer PαMeSt-*b*-polyisobutene-*b*-PαMeSt with central and outer block M.W.'s of 60 000 and 4000, respectively, is also elastomeric, though weak, with tensile behaviour closely similar to that of a comparably constituted PαMeSt-*b*-PBd-*b*-PαMeSt; it is not clear, however, whether phase separation occurs in this instance.[313]

The outer blocks of the ABA model may be composed of high-melting crystalline polyester as in poly(pivalolactone)-*b*-PIp-*b*-poly(pivalolactone), giving elastomers of very high strengths,[132] and this theme has been extended to further ABA structures where A is poly(pivalolactone) and B is poly(α-methyl-α-propyl-β-propiolactone) or the α-methyl-α-butyl analogue.[220] In these cases, the centre blocks are either amorphous or of low crystallinity and the products are elastomeric but with tensile strengths lower than for PSt–PBd–PSt or poly(pivalolactone)–PIp–poly(pivalolactone) copolymers. In contrast, the polyether-containing block polymers POE-*b*-PIp-*b*-POE and PSt–POTM–PSt appear to have no elastomeric properties,[148,152] and copolymers with central PBd or PIp segments and outer poly(thioalkylene) blocks—crystalline and polyphasic in the case of poly(thiotrimethylene)—have insignificant properties as elastomers.[164,314] Evidently the possession of a high M.W. hard block–soft block–hard block phase-separated model is not of itself a sufficient condition for the display of useful elastomeric character.

The alternative major class of structures leading to such properties comprises the $[AB]_n$ multiblock copolymers, as in the polyetheresters and polyester- or polyether-urethanes, which are generally (though not invariably) composed of soft blocks with M.W.'s in the order of 1000–3000 and hard blocks having a distribution of DP values but with average M.W.'s (depending on the repeat unit mass and the proportion of hard block component) typically within the limits of about 1000–5000. The average sizes of both types of components are thus decisively lower than in the copolymers discussed above, as are the total copolymer M.W.'s which may be as low as about 20 000, ranging perhaps up to 60 000–80 000 in some copolyurethanes.

The phase-separated state (and the resulting morphology) of block copolymers of the PSt–PBd–PSt type is a consequence of the very small entropy of mixing and the positive interaction energies of the sub-species. The relevant thermodynamic theories predict that the tendency to phase separation will diminish as the block sizes and total copolymer M.W.'s decrease and as the number of blocks per molecule increase.[44] Moreover, the theories are predicated on the assumption of non-compatibility of the components—an assumption which is imperfect even for all-hydrocarbon copolymers and still more uncertain for materials composed of blocks having polar, hydrogen bonding, or ionic character. An instructive demonstration of the effects of such factors on the phase state is given in eqn (39) which shows the phase states reported as existing in a closely related series of carbochain block copolymers.[26,27]

In the light of these indications, the probability of extensive and well-defined species-from-species phase separation in the typically polar $[AB]_n$

$$[-CH_2-\underset{\underset{N}{|}}{CH}-]_m-b-[-CH_2-\overset{\overset{Me}{|}}{\underset{\underset{COOSiMe_3}{|}}{C}}-]_n \quad \text{domain structure}$$

$$\xrightarrow{HO^-} [-CH_2-\underset{\underset{N}{|}}{CH}-]_m-b-[-CH_2-\overset{\overset{Me}{|}}{\underset{\underset{COOH}{|}}{C}}-]_n \quad \text{homogeneous}$$

$$\xrightarrow{CH_3I} [-CH_2-\underset{\underset{N^+-CH_3 \;\; I^-}{|}}{CH}-]_m-b-[-CH_2-\overset{\overset{Me}{|}}{\underset{\underset{COOH}{|}}{C}}-]_n \quad \text{domain structure}$$

$$\xrightarrow[(-HI)]{NEt_3} [-CH_2-\underset{\underset{\overset{+}{N}-CH_3}{|}}{CH}-]_m-b-[-CH_2-\overset{\overset{Me}{|}}{\underset{\underset{COO^-}{|}}{C}}-]_n \quad \text{homogeneous}$$

$$\underset{\text{base} \; (-HCl)}{\overset{+HCl}{\rightleftharpoons}} [-CH_2-\underset{\underset{\overset{+}{N}-CH_3 \;\; Cl^-}{|}}{CH}-]_m-b-[-CH_2-\overset{\overset{Me}{|}}{\underset{\underset{COOH}{|}}{C}}-]_n \quad \text{domain structure}$$

(39)

block copolymers may be expected to be less than in non-polar materials, though such separation may occur when the block M.W.'s are high and where there is a substantial difference between the solubility parameters of the constituents. Of greater importance, however, are the effects of crystallisation when this occurs in one or more of the varieties of combined blocks. In this case, the morphology and physical characteristics of the block copolymers are dominated by the pattern of organisation assumed by the crystallising component(s) which may have to accommodate more or less discrete inclusions of non-crystalline material within the spherulites or between lamellae. By definition, crystallisation in a block copolymer implies the separation of the crystallising species as a distinct phase, such separation being driven by the energy gain from lattice formation, and most commonly the restriction to movement of molecular segments in the crystallising component by its connection to the component results in a lower rate and extent of crystallisation than occurs with the same species as a homopolymer. The gross structure of a partly crystalline block copolymer therefore consists of one or more crystalline phases of the crystallising component(s) together with a composite amorphous phase made up of the non-crystallising component(s) together with that fraction of the crystallisable component that has not attained crystalline order. Whether the amorphous phase is itself homogeneous or heterogeneous depends on the nature of the components and the block M.W.'s, but for most polar block copolymers the heterogeneity—when present—leads to an irregular-shaped amorphous phase morphology with nodular or threadlike aggregates having diffuse boundaries and widths commonly in the range 30–100 Å. Although these are often termed domains, the term has a less precise geometrical significance than for the phase-separated wholly-amorphous block copolymers discussed above.

The existence of crystalline phases is readily perceived by the appearance of spherulites or by the recognition of thermal transitions or WAXD characteristics typical of the corresponding homopolymeric species. The presence of the crystalline entity usually interferes with attempts to investigate the texture of the amorphous phase by direct methods; its nature is therefore generally inferred by indirect methods such as the study of relaxations by dynamic mechanical or thermal analysis, from the differential behaviour of the components of structure under applied mechanical stresses, or by SAXS (see Chapter 6).

It should be noted that the molecular organisation of a crystallising block may be affected by the neighbouring blocks or even by single connecting groups between blocks, as for example in block or chain-extended poly(oxyethylenes) which may thus be directed from extended-chain to folded molecular form.[120,315–18] Additionally, some crystalline polymers are subject to polymorphism, yielding different patterns of lattice organisation and/or configuration in response to thermal, mechanical or other influences. Substances occurring as components of block copolymers, which show polymorphism as homopolymers, include nylon 6 and the polypeptides, as well as POE,[319] 2G.6,[320] 4G.6,[321] poly(pivalolactone)[132,322] and 4G.T.[323,324]

There are thus numerous respects in which the supramolecular texture of heterochain $[AB]_n$ block copolymers and its relationship to their properties differs profoundly from that in carbochain ABA block copolymers.

3.2. Polymers Containing Polyether Blocks

Polymers composed of POE (A) and PSt (B) blocks exemplify block copolymers having one component crystallisable and the other inherently amorphous with no extensive degree of mutual polar interaction. Since $T_{g_B} > T_{m_A}$, when these copolymers are cooled from the melt the PSt component will form a glassy matrix in which crystallisation of the POE component must then occur; correspondingly at temperatures between T_{m_A} and T_{g_B} a liquefied POE phase may exist in the presence of glassy segments.

Investigations of such copolymers have covered AB,[325–8] ABA[101,329–31] and $[AB]_n$[101,331,332] types. Crystallisation of the POE segments does not occur at block M.W.'s of less than about 1000 and is developed fully only at M.W.'s approaching 6000 though to lesser extents than in POE homopolymer and decreasing both in rate and extent with increase in PSt content and with the number of blocks per molecule. The morphology of the crystallised copolymers is of a sandwich type (Fig. 5) with layers of PSt lying between regularly spaced lamellae of POE composed of folded molecules whose fold numbers are determined by the compositions and POE block M.W.'s. Whilst a high degree of phase separation obviously exists there is nevertheless evidence for some measure of plasticisation of the PSt component, indicating some diffusion of the polyether segments into the hydrocarbon polymer phase. POE–polydiene block polymers have been less intensively studied, but those with PBd[147] and PIp[148] components have a general similarity to the materials just discussed. POTM–PSt block copolymers have also been studied.[152,153,333] In POTM-rich copolymers of high M.W. extensive phase separation occurs leading to a crystalline spherulitic morphology made up of polyether fibrils with an amorphous PSt coating. In these materials the thermal transitions and degree of crystallinity of the POTM component are substantially the same as in homo-POTM, but in copolymers of high PSt content the rates of crystallisation are reduced in the order homo-POTM > POTM-*b*-PSt-*b*-POTM > POTM-*b*-PSt.

With this background, attention will now be given to polymers containing polyether blocks combined with more polar co-units.

Diblock copolymers of poly(ethyl methacrylate) (M.W. 8000) with POE blocks of various M.W.'s have been studied.[334] Provided that the polyether

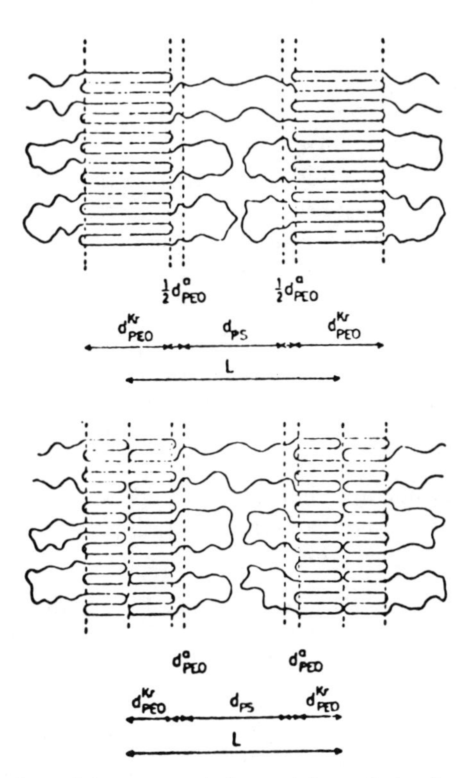

FIG. 5. Structural models proposed for poly(oxyethylene)–polystyrene block copolymers showing layers of folded-chain POE lamellae alternating with layers of disordered PSt segments. (From Neidlinger, H. H., Höcker, H. and Zachmann, H. G., *Polym. Preprints*, 1977, **18**(1), 610; with permission, American Chemical Society.)

units were substantially larger than the polymethacrylate, the rates and extents of POE crystallisation were high, but both characteristics were greatly reduced at a POE M.W. of 5500 as was the perfection of spherulitic structure.

Turning next to polyether–polyester block copolymers, indications of the separation of POE as a discrete crystalline phase have been found in alternating $[POE–BPAC]_n$ copolymers at polyether block M.W.'s of as low as 620 and a content of 17 wt %.[199] For random block POE-4G.T copolymers, polyether block crystallisation occurs at 47 wt % polyether content for POE M.W. 1000, 37 wt % for POE M.W. 2800, and 30 wt % for POE M.W. 6120.[335] With 6G.T co-units the corresponding appearance content for POE M.W. 2800 is at 18 wt %.[170] Although the degrees of polyether crystallinity are lower than for corresponding homo-POE values, it does seem that significant POE crystallisation is attained in these polyetheresters at lower polyether block M.W. than in POE–polyvinyl block copolymers. Moreover, in certain ranges of composition the polyether and polyester segments coexist as discrete crystalline phases, but over the whole range of compositions the amorphous phase shows only a single T_g which increases monotonically with polyester block content, indicating its mixed-component character.

As a general observation, the polyether blocks in corresponding POTM–polyester random block copolymers have notably lower propensities for crystallisation. Superficially this might seem to be correlated with the lower T_m of POTM which, even for the homopolymer, lies not much above ambient temperatures. However, the following observations emphasise the considerable influence of copolymer composition, block M.W., nature of the co-unit, and the orientational state on the occurrence or otherwise of a distinct POTM crystalline phase:

(a) In combination with 4G.T or 6G.T as hard segments, POTM blocks of M.W. 1000 do not crystallise at any composition.[180,184,185,336,337]
(b) In combination with 4G.T the appearance content for crystallisation of POTM M.W. 2000 is 30 wt %, but with 6G.T the value is 60 wt %. Increase of POTM block M.W. to 3000 permits its crystallisation at 50 wt % content in combination with 6G.T and at 10 wt % with 4G.T.[172,336]
(c) Whilst POTM blocks of M.W. 2000 do not crystallise spontaneously at 50 wt % combination with 6G.T, reversible strain-induced crystallisation occurs at elongations of from 100–300 %. The polyester component remains unaffected at these strains but undergoes irreversible orientation at higher strains (600 %), thus indicating the differential behaviour of the sub-species.[336,337]

With these copolymers, as with those of POE, the rates and extents of crystallisation of both components are lowered relative to those of the homopolymers, and the amorphous phase is again homogeneous and of mixed composition. An idealised representation of the texture of POTM-6G.T copolymers is given in Fig. 6.

Widening the range of polyester units combined with POE or POTM (Table 6) reveals further aspects of the structure–crystallisability relationships in polyetheresters.[169,335] With the exception of the first-listed

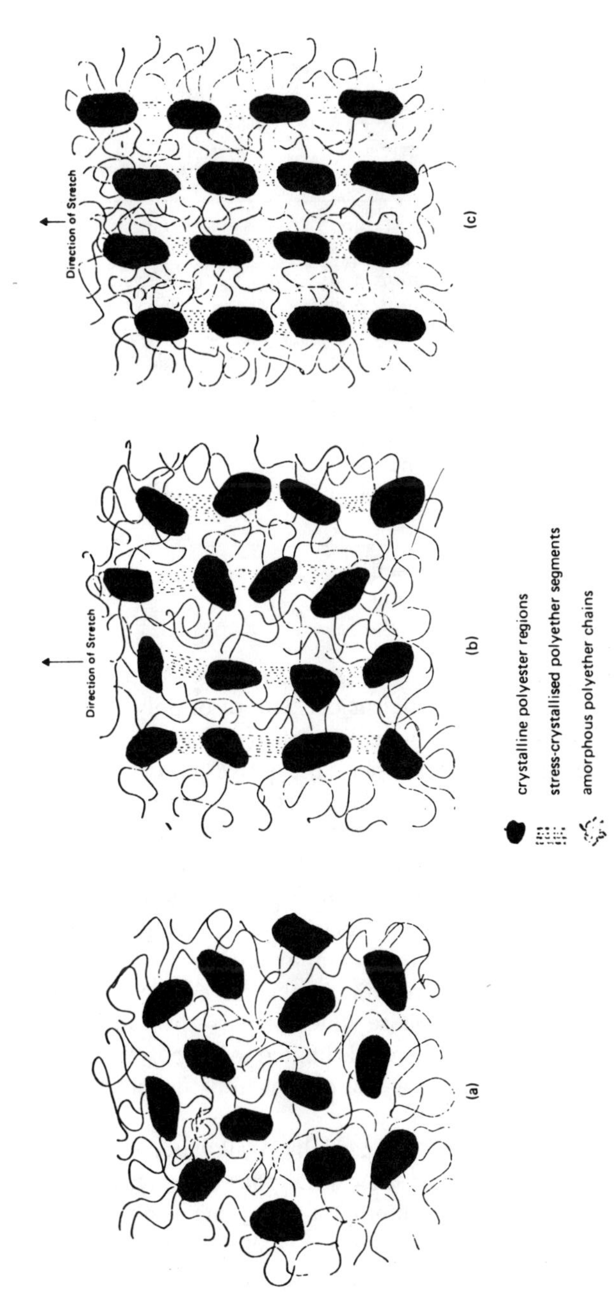

FIG. 6. Schematical textural models for poly(oxytetramethylene)terephthalate–poly(hexamethylene terephthalate) random block copolymers (a) unstretched, (b) at about 200% extension, (c) at high extensions. (With permission, From Ghaffar, A., Peters, R. H. and Goodman, I., *Br. Polym. J.*, 1978, **10**, 127.)

TABLE 6

ROOM TEMPERATURE PHASE STATES OF THE COMPONENTS IN 50 wt % RANDOM BLOCK COPOLYMERS OF POE (M.W. = 2800) OR POTM (M.W. = 2000) WITH VARIOUS POLYESTERS

Copolymer	*Polyester block*	*Polyether block*
POE–2G.T	Crystalline	Crystalline
POE–4G.T	Crystalline	Amorphous
POE–1,2P.T	Amorphous	Crystalline
POE–1,2P.6	Amorphous	Crystalline
POTM–4G.T	Crystalline	Amorphous
POTM–5G.T	Crystalline	Amorphous
POTM–6G.T	Crystalline	Amorphous
POTM–6G.6	Crystalline	Amorphous
POTM–$Me_3$6G.T[a]	Amorphous	Crystalline
POTM–1,2P.T	Amorphous	Crystalline
POTM–202G.6[b]	Amorphous	Crystalline
POTM–4G.I[c]	Crystalline	Amorphous

[a] Mixed 2,2,4-/2,4,4-trimethyl isomers.
[b] Poly(oxy-2,2′-diethylene adipate).
[c] Poly(tetramethylene isophthalate).

copolymer (which can be explained on the basis of the slow crystallisation of 2G.T from the melt) there is a clear-cut pattern of polyether crystallisation where the polyester component is inherently amorphous, and of amorphous polyether character where the polyester can lay down a crystalline matrix which presumably impedes the ordered aggregation of the terminally-bonded polyether segments. A further feature of polyether-esters is the not uncommon observation of dense opacity in the melts. This phenomenon, which indicates some form of phase separation that is not yet well understood, is found particularly with the higher M.W.'s of POTM,[169] with POP as the ether component,[185,203] and with poly(cyclohexane-1,4-dimethylene terephthalate) as the ester component.[175]

Information concerning the phase states of polyethers in block combination with the more polar polymers is limited to the polyetherurethanes which are considered later, and a few observations on block polyether amides.

In POTM–nylon-12[240] and POE-partially *N*-alkylated nylon[338] block polymers there are indications of substantial phase mixing, and for ABA and BAB copolymers of POE with poly(*m*-phenylene isophthalamide) no polyether crystallinity occurs at POE block M.W.'s below 6000 whilst that of the polyamide segment(s) is induced only after swelling with formic acid.[80,339,340] These indications of component miscibility may reflect the effects of hydrogen bonding between amide and ether groups.

3.3. Polymers Containing Poly(dimethylsiloxane) Blocks

From the low solubility parameter of PDMS as compared with many other polymers, its block copolymers may be expected to display a marked tendency to phase separation as indeed occurs with the amorphous hydrocarbon co-units PSt[341,342] and PαMeSt.[296,343] Depending on composition and the processing history of these materials, the PDMS component may form either the continuous or the dispersed phase. Thus, films of diblock PDMS–PSt cast from bromobenzene as selective solvent for the PSt phase are hard and brittle, but the same copolymers cast from cyclohexane (a selective solvent for the PDMS component) are soft and rubbery.

Block copolymers of PDMS with other crystallising polysiloxane co-units show marked phase separation, determined by the crystallisation of the higher melting component, as is evidenced by poly(diphenylsiloxane)-*b*-PDMS-*b*-poly(diphenylsiloxane)[344] and the PDMS–poly(tetramethyl-*p*-silphenylenesiloxane) random block copolymers (XXXV).[345–50] In the latter series, the PDMS blocks had a fixed M.W. of 1480 and comprised from 10 to 70 wt % of the copolymers which had total M.W.'s in excess of 10^5; the average segment lengths (ASLs) of the hard segments ranged correspondingly from 64 to 3. The structures of these copolymers formally resemble those of the 4G.T-POTM polyetheresters, being derived hypothetically by insertion of —$SiMe_2\phi SiMe_2O$— units into the PDMS chain. Their general characteristics are also similar, ranging from stiff elastoplastic to elastomeric in properties with increase in PDMS content. Spherulitic crystallisation of the hard block component occurred over the whole range of compositions examined (though with diminishing perfection as its proportion decreased), and lamellar single crystals were obtainable across this range. For PDMS contents ≤ 20 wt % two amorphous relaxations were found, indicating separation of the components, but for contents ≥ 50 wt % the amorphous phase was homogeneous. Generally similar behaviour has been found in polyester–polysiloxane $[6G.6\text{-}PDMS]_n$ multiblock copolymers having polyester block M.W.'s of 3160 and PDMS block M.W.'s from 1400 to 10 600;[283,284] in this series phase separation appeared to be virtually complete in the solid state and was inferred also to occur in the melts which were opaque.

Block copolymers of PDMS with BPAC (XXXVII) and with polysulphone (eqn (34)) co-units have been mentioned above as forming families of

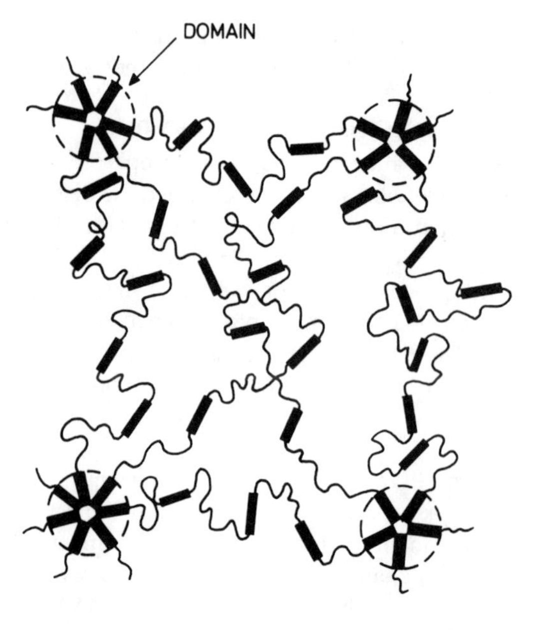

FIG. 7. Schematic representation of morphology in bisphenol A polycarbonate–poly(dimethylsiloxane) block copolymers showing BPAC hard block domains dispersed in a mixed-component matrix. (From LeGrand, D. G., *Polym. Letters*, 1969, **7**, 584; with permission, John Wiley & Sons, Inc.)

materials ranging in properties from elastomeric to tough rigid thermoplastics depending on the proportion of combined aromatic segments. The incorporation of as little as 15 wt % of PDMS into BPAC chains lowers the temperature of the ductile–brittle transition from -10 to -50 °C and thus improves considerably the toughness of the materials at sub-ambient temperatures. In these copolymers, as in those of PDMS with the polycarbonate of (XXXVIII), the relaxation characteristics correspond with those of phase-separated materials constituted as shown in Fig. 7 with domain aggregates of the polycarbonate component set in a mixed-component amorphous matrix.[22,276,279,351] It is interesting that such phase separation occurs, with PDMS block M.W.'s of 1400, at BPAC average segment lengths of as little as about 6, and that the PDMS component does not crystallise at low temperatures even at block M.W.'s of 2960. Alternating multiblock polysulphone–PDMS copolymers have rather similar characteristics; for a polysulphone block M.W. of 4700 only single-phase texture is found at a PDMS block M.W. of 350, but phase separation is apparent when the PDMS block M.W. is increased to 1700 and occurs to a high degree at a M.W. of 5100.[24,298]

It is surprising in the light of these findings that only single-phase texture occurs in poly(*m*-phenylene isophthalamide)-*b*-PDMS-*b*-poly(*m*-phenylene isophthalamide) for PDMS block M.W.'s < 9000 or contents up to 13 wt %. Beyond these limits, spheroids of PDMS are found in a featureless polyamide matrix.[80,339]

3.4. Polymers Containing Polyester Blocks

Aromatic polyester–poly(oxyalkylene) block copolymers have already been discussed with respect to the phase state of the polyether components. For those with poly(alkylene terephthalate) hard segments, crystallisation of the polyester units occurs at ASLs of as low as 2–3. The mixed-component character of the amorphous phase in these materials naturally implies that not all of the polyester component in fact crystallises and recent DSC studies with 4G.T-POTM copolymers suggest that the segments of polyester actually crystallising are of rather uniform sequence length, those of shorter or greater lengths being the fraction that is rejected into the amorphous phase.[187,352]

High M.W. BPAC homopolymer is ordinarily amorphous (though crystallisable by treatment with certain solvents). It is therefore of interest that the polycarbonate segments in block copolymers of BPAC with POE or POTM crystallise at ASLs of about 6 and that the tendency for crystallisation increases as the BPAC block M.W.'s increase. BPAC units also crystallise in block copolymers with aliphatic polyesters. With 2G.6 co-units, both components crystallise at a 2G.6 block M.W. of 6000 though not at 2000.[197,199] With poly(ε-caprolactone) of M.W. 15 000, the BPAC component is amorphous and the polylactone crystalline for BPAC block M.W.'s up to 1540 but the situation is reversed for a BPAC M.W. of 2420.[200] PBAC–PDMS block polymers, although phase-separated, show no BPAC crystallinity,[351] and BPAC–polysulphone copolymers are monophasic at M.W.'s for each block of 5000 but biphasic at block M.W.'s of 16 000.[300,301] The behaviour of BPAC units in block copolymers is thus acutely sensitive to block size and the nature of the adjoined units.

Continuing with polyester–polyester structures, in multiblock 2G.T-*b*-2G.6 copolymers made by linking of oligomers of block M.W.'s of 2890 and 2970, respectively, with a diisocyanate, only the aliphatic component crystallised at 2G.T contents less than 15 mol % but both components did so at higher 2G.T contents.[51] Another example of the positive effect of block polymerisation on crystallisation is found in 6G.10-*b*-poly(2-methyl-2-ethyl-1,3-propylene sebacate) where the second component is structurally isomeric with the first but inherently amorphous. Here the 6G.10 units crystallise from the melt in essentially pure form, and the fraction of 6G.10 actually crystallising increases as its proportion in the copolymers decreases.[332,353,354] ABA block polymers having outer segments of poly(pivalolactone) and the central segment of poly(α-methyl-α-propyl-β-propiolactone) or poly(α-methyl-α-butyl-β-propiolactone) also show marked separation of both species as crystalline phases even with A blocks as small as M.W. 5000 (DP = 50) in the presence of B blocks with M.W.'s up to 284 000. In the first-mentioned case, the occurrence of two T_m and two T_g values implies species separation in both the crystalline and the amorphous phases, and the poly(pivalolactone) blocks exist in different polymorphic forms depending on the thermal and mechanical history.[220,221]

Lastly, attention will be given to copolymers containing polyester and polyamide blocks. In these substances competition for hydrogen bonding with the amide NH groups can occur between the carbonyl groups of CONH and COO, and any substantial extent of cross-species (i.e. amide to ester) bonding might be expected to impede phase separation or crystallisation of one or both segment types.

A first contra-indication is found in the 2G.T–nylon-6,10 random multiblock polyesteramides which have been prepared by polycondensations with individual $[2G.T]_a$ oligomers. For the cases of $a = 1$ or 2 only the polyamide component crystallised, but with $a = 3$ and at fairly high 2G.T contents both components did so.[123]

Two series of aliphatic polyesteramides have been examined, with closely similar geometries of the co-units so that co-crystallisation might occur, or at least that crystallisation should not be impeded by grossly dissimilar unit shapes. The first series,[355] prepared from synthetic COOEt-ended nylon 6,6 oligomers by condensation with HO-ended 6G.6 in conditions where randomisation was unlikely to occur, was of structure (L) with $a = 1$ or 2 and b corresponding to polyester block M.W.'s of 1390 or 1900. These

$$\{\!-OC(CH_2)_4CO[NH(CH_2)_6NHCO(CH_2)_4CO]_a\!-\!O(CH_2)_6O[CO(CH_2)_4COO(CH_2)_6O\!-]_b\!-\}_n$$

(L)

polymers can thus be regarded either as poly(hexamethylene adipate)s modified by replacement of —O— with —NH— at regular positions along the chain or as regularly alternating block polymers of 6G.6 with small nylon 6,6 units.

In contrast to 6G.6 homopolymer, which is a highly crystalline relatively stiff solid of $T_m = 62\,°C$, the ordered polyesteramides were all translucent and rubbery. Two T_m values were detected by DSC in the products with $a = 1$, and a single T_m (in the range 160–165 °C) for $a = 2$; WAXD indicated the coexistence in all the materials of crystalline phases with Bragg spacings close to those of nylon 6,6 and 6G.6 homopolymers. These substances stand on the borderline between chain-extended polymers and true block copolymers. The reduction of crystallinity of the main (polyester) component is not unexpected in either case, but the manifestation of some characteristics of nylon 6,6 at block lengths as low as 1·5 (for $a = 1$) and 2·5 (for $a = 2$) is striking.

The second series[256,257] comprises random block copolymers of $[\text{—NH(CH}_2)_5\text{CO—}]_a$ and $[\text{—O(CH}_2)_5\text{CO—}]_b$ units prepared by the ε-caprolactam-ε-caprolactone copolymerisation reaction, mentioned earlier; the components have the same structural relationship as the series just discussed. The T_g versus composition dependence indicated a mixed amorphous phase over the whole range of $a{:}b$ ratios, whilst T_m and WAXD data showed only poly(ε-caprolactone) crystallinity for nylon 6 contents less than 15 wt %, coexisting α-nylon 6 and polyester crystallinities over 15–60 wt %, and only polyamide crystallinity at higher nylon 6 contents. The onset of nylon 6 crystallinity at about 20 wt % coincided with a marked discontinuity in the pattern of thermal and mechanical properties and with brittleness that was not found above or below this composition. These effects may be due to one or both of two causes: (a) spatial incompatibility between the α-nylon 6 monoclinic and the polycaprolactone orthorhombic unit cells in a composition range where the ASL of the polyamide units (about 5) is not yet sufficient to establish a coherent network and that of the polyester units is diminishing from the level required for structural cohesion, (b) the hydrogen bonding state, which was found by infrared to be substantially between NH and ester CO groups at low amide contents, again precluding the establishment of a true polyamide network.

3.5. Polymers Containing Polyamide Blocks

Few detailed investigations have been made on the textures of block polymers containing conventional polyamide segments. Nylon 6 copolymers with PBd,[133,239,251] PSt[251] or polyisobutene[254] are phase-separated as might be expected from the nature of their components. The same is true for nylon-6-polysulphone[304,305] and for nylon-6,10-PSt graft copolymers which, for a given composition, have the polyamide or the hydrocarbon component as the continuous phase depending on whether moulding is performed below or above the melting point of the nylon component.[356] The non-crystallising nature of poly(*m*-phenylene isophthalamide) in block polymers with POE or PDMS, and some characteristics of block polyesteramides have been mentioned earlier.

The most interesting developments in this field have been concerned with substances containing polypeptide blocks having the general structure $[\text{—NH.CHR.CO—}]_n$. These substances are of significance as models for the structure of proteins and of biological membranes. A particular feature of polypeptides is their tendency to adopt in the solid state a specific backbone chain conformation which may be either the intramolecularly hydrogen-bonded α-helical or the intermolecularly hydrogen-bonded β-chain form.[357,358] The normally stable forms of (homo)polypeptides of interest in the present context are summarised in Table 7. Interconversions between the α and β forms, or between α-helical and random coil forms occur when some polypeptides are taken into solution in particular solvents, or subjected to chemical or mechanical treatments; it is therefore of interest to ascertain how the conformations are affected by combination into block structures.

AB copolymers of poly(γ-benzyl-L-glutamate) with PBd or PSt,[128,259,359] and the corresponding ABA copolymers with the central block of PBd[260,360] or poly(butadiene-*co*-acrylonitrile)[361] all retain the α-helical form of the polypeptide entity as do the analogous diblock copolymers of poly(carbobenzoxy-L-lysine). In the dry state, or when

TABLE 7
NORMAL SOLID-STATE CONFORMATIONS OF HOMOPOLYPEPTIDES $[\text{—NH.CHR.CO—}]_n$

Polypeptide	R	*Conformation*
Poly(L-alanine)	Me	α (β on stretching)
Poly(L-valine)	$CHMe_2$	β
Poly(L-leucine)	CH_2CHMe_2	α
Poly(L-glutamic acid)	$(CH_2)_2COOH$	α
Poly(γ-benzyl-L-glutamate)	$(CH_2)_2COOCH_2Ph$	α
Poly(L-lysine)	$(CH_2)_4NH_2$	α[a]
Poly(carbobenzoxy-L-lysine)	$(CH_2)_4NHCOOCH_2Ph$	α

[a] Salts readily transformed to β-chain form in the presence of moisture.

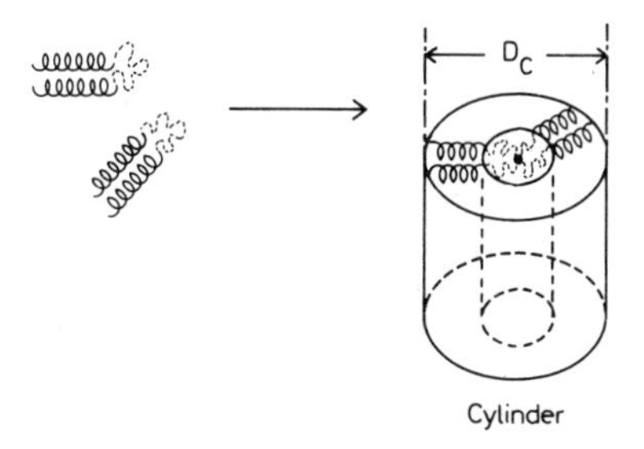

FIG. 8. Model proposed for the aggregation of poly(γ-benzyl-L-glutamate)–polybutadiene–poly(γ-benzyl-L-glutamate) triblock copolymers into structures comprising α-helical polypeptide hollow cylinders with disordered polydiene cores. (From Nakajima, A., Kugo, K. and Hayashi, T., *Macromolecules*, 1979, **12**, 845; with permission, American Chemical Society.)

swollen with selective solvents for the carbochain component, the diblock copolymers display phase-separated lamellar structures comprising layers of the amorphous segment alternating with layers of polypeptide block material, the chains of the latter being hexagonally packed and with 0–5 folds in the layer thickness depending on the block M.W.'s A similar lamellar structure has been identified in the brittle films of poly(γ-benzyl-L-glutamate)-*b*-poly(butadiene-*co*-acrylonitrile)-*b*-poly(γ-benzyl-L-glutamate) cast from dioxane or xylene; films cast from chloroform were tough and apparently of mixed-phase type. On the other hand, poly(γ-benzyl-L-glutamate)-*b*-PBd-*b*-poly(γ-benzyl-L-glutamate) has cylindrical morphology with the rubbery component forming a core within an α-helical polypeptide matrix; the structural model envisaged for the aggregation of this material is shown in Fig. 8.

In each of the above copolymers the conformational state of the polypeptide blocks is unperturbed from that in the corresponding homopolymer. Some deviation from this condition occurs in diblock copolymers of poly(L-lysine) with PSt or PBd.[127,359] Here the polypeptide components occur in layered structures alternating in composite lamellae with layers of the carbochain components, but whereas homopoly(L-lysine) is crystalline and α-helical in the dry state and transformable to the β-form by contact with water, the polypeptide units in the block polymers are disorganised and made up of a mixture of α-helical, β-chain and randomly coiled forms which is unaffected by contact with water.

Studies have also been made of some block copolypeptides. Substances composed of poly(γ-benzyl-L-glutamate) (A) and poly(L-leucine) (B) blocks linked in ABA, BAB, AB–BA and BA–AB modes show a clear separation of the components, both being crystalline and in α-helical form, and with whichever species forms the outer blocks constituting the matrix with discrete inclusions of the other component. Poly(γ-benzyl-L-glutamate) (A)–poly(L-valine) (C) block copolymers of ACA and CA–AC types are also phase-separated, each component retaining its characteristic homo-polypeptide chain conformation unaffected by the presence of the other.[263,264,361,362] The incompatibility between the different polypeptide blocks in these various copolymers resembles that in the triblock poly-pivalolactone block copolymers discussed earlier.

3.6. Polyurethane Block Copolymers

Of the large body of scientific and technological investigation that has been devoted in recent years to the commercially important group of linear block copolyurethanes (see Chapters 6, 7 and 8), attention will be given only to those questions of phase state that have been reviewed above for other classes of block copolymers.

As with the polyetheresters, the multiblock copolyurethanes of greatest technical importance are elastomers composed of hard and soft blocks whose ratios and relative molecular segmental mobilities determine the moduli, hardnesses and general characteristics of the materials. Unlike the polyetheresters, some measure of sequence distribution is possible during synthesis and a measure of molecular interaction between hard and soft blocks is frequently to be expected as a consequence of inter-species hydrogen bonding. The soft blocks of copolyurethanes of greatest technical importance are composed of aliphatic polyester or polyether material, combined into the overall structure by reaction of terminal hydroxyl groups with a diisocyanate. The hard blocks are of two types, formed by reaction of the diisocyanate with either a short-chain diol or with a diamine. The products of the first type (LII) have urethane connecting groups throughout the molecular chain and are familiar as thermoplastic elastomers; those the second type (LII) have polyurea hard blocks with

$$\{\text{—[—OCNH}\,.\,R\,.\,\text{NHCOO}\,.\,R'\,.\,\text{O—]}_a\text{—[—OCNH}\,.\,R\,.\,\text{NHCOO—soft block—}O\text{—]}_b\text{—}\}_n$$

(LI)

$$\{\text{—[—OCNH}\,.\,R\,.\,\text{NHCONH}\,.\,R'\,.\,\text{NH—]}_a\text{—[—OCNH}\,.\,R\,.\,\text{NHCOO—soft block—}O\text{—]}_b\text{—}\}_n$$

(LII)

urethane-linked soft blocks and usually form the basis of elastomeric fibres. The two classes will be distinguished as polyester- or polyetherurethanes and as polyester- or polyetherurethaneureas, respectively.

The diisocyanates most commonly used are diphenylmethane-4,4′-diisocyanate (1,1′-methylene bis[4-isocyanatobenzene]) (MDI), and the toluene diisocyanates (diisocyanatomethylbenzenes) (TDI) as separate 2,4- or 2,6- isomers or as technical mixtures of the two. Table 8 lists the precursors of some polyurethanes and polyurethaneureas which have been the subject of particularly detailed scientific investigations. Numerous variations of structures, block M.W.'s, proportions and sequence distributions have been explored in these materials and the findings cannot be surveyed in detail; nevertheless the following broad generalisations emerge:

(a) Block polyurethanes and polyurethaneureas are usually multiphase materials, owing their properties to a dispersion of the hard block components within an elastomeric matrix. Where hard block domains have been recognised by electron microscopy or studied by SAXS, they are irregularly shaped and with diffuse boundaries; their dimensions are typically within the range 20–150 Å, the smaller values occurring with polyester and the larger values with polyether soft blocks. Some limited penetration of soft block component into the hard block domains may occur, and there is usually a greater degree of component mixing in the continuous phase.

(b) The extent of phase separation increases with soft block M.W. and is sensitive to the thermal and mechanical history of the materials. The components of structure often respond differentially to applied tensile strains even where no discrete soft block phase is present and the elastomeric continuum is apparently of mixed component character.

(c) Phase mixing is greater in polyesterurethanes than in polyetherurethanes because urethane NH–ester carbonyl hydrogen bonds are stronger than urethane NH–ether oxygen bonds.

(d) Phase separation is greater (i) in polyetherurethaneureas than in polyetherurethanes, (ii) in polyurethanes based on 2,6-TDI than in those from 2,4-TDI, (iii) in MDI-based polyesterurethanes than in those based on TDI, (iv) in polyurethanes based on aliphatic or alicyclic diisocyanates than in those from aromatic diisocyanates.

(e) The thermal transitions due to the soft blocks in polyester- or

TABLE 8

PRECURSORS OF SOME RECENTLY-STUDIED POLYURETHANE AND POLYURETHANEUREA BLOCK COPOLYMERS

Hard block precursors		Soft block precursors (as α, ω-diols)	References
Diisocyanate	Chain extender		
(a) *Polyetherurethanes:*			
MDI	$HO(CH_2)_4OH$	POTM	363–7, 372, 377, 380, 409
2,4- or 2,6-TDI	$HO(CH_2)_4OH$	POTM	371, 375, 379, 382, 383
MDI	$HO(CH_2)_4OH$	(-) (*S*)-POP	374
MDI	$HO(CH_2)_4OH$	POE-*b*-POP-*b*-POE	376, 407, 409
(b) *Polyesterurethanes:*			
MDI	$HO(CH_2)_4OH$	4G.6	363–7, 372, 373, 384
MDI	$HO(CH_2)_4OH$	Poly(ε-caprolactone)	368, 377, 378
TDI(80:20 2,4-/2,6-mixture)	$HO(CH_2)_4OH$	Poly(ε-caprolactone)	369
TDI(80:20 2,4-/2,6-mixture)	$HO(CH_2)_4OH$	2G.6	
$OCN(CH_2)_6NCO$	$HO(CH_2)_4OH$	2G.6	370
OCN–(cyclohexane ring)–CH_2–(cyclohexane ring)–NCO	$HO(CH_2)_4OH$	2G.6	
(ring structure: Me, Me, NCO, Me, CH_2NCO)	$HO(CH_2)_4OH$	2G.6	
2,4- or 2,6-TDI	$HO(CH_2)_4OH$	4G.6	371

(c) *Polyetherurethaneureas:*			
MDI	$H_2N(CH_2)_2NH_2$	Isotactic POP	112
MDI	$H_2N(CH_2)_2NH_2$	POTM	385
MDI	N_2H_4, $H_2N(CH_2)_2NH_2$, $H_2NCHMeCH_2NH_2$ or $H_2N\phi CH_2\phi NH_2$	POTM	386, 387
MDI	$H_2N(CH_2)_6NH_2$, $H_2N\text{-}C_6H_{10}\text{-}CH_2\text{-}C_6H_{10}\text{-}NH_2$ or $MeN[(CH_2)_3NH_2]_2$	POP	388
2,4-TDI	$H_2N(CH_2)_2NH_2$	POTM	389–91
$OCN\text{-}C_6H_{10}\text{-}CH_2\text{-}C_6H_{10}\text{-}NCO$	$H_2N(CH_2)_2NH_2$ or $H_2N\phi NH_2$	POE	392
$OCN\text{-}C_6H_{10}\text{-}CH_2\text{-}C_6H_{10}\text{-}NCO$	Various aromatic diamines	POE/POP mixtures	392
MDI	$H_2N(CH_2)_3NH_2$	POTM	408
(d) *Polyesterurethaneureas:*			
MDI	$H_2N(CH_2)_xNH_2$, $x = 0$, 2–9	2G.6	116, 410
2,4-TDI	$H_2N(CH_2)_2NH_2$	4G.6	390
(e) *Other or unspecified:*			
MDI	$HO(CH_2)_4OH$	PBd or PIp	105
MDI	$HO(CH_2)_4OH$	'Mixed polyester'	381
MDI	N_2H_4 or $H_2N(CH_2)_2NH_2$	'Mixed polyester' or 'polyether'	393–6

TABLE 9

PHASE STATES IN SOME BLOCK COPOLYURETHANES

Hard block precursors		*Soft block*	*Crystallinity of copolymer containing 50 wt % soft block (M.W. = 2000)*		*Wt % soft block (M.W. = 20000) conferring crystallinity if not present at 50 wt %*	*Notes*
Diisocyanate	*Diol*		*Soft block*	*Hard block*		
(a) *Polyesterurethanes:*						
MDI	4G	2G.6	+	+		(a) Hard block crystalline over range 20–85 wt % polyester; (b) Copolymer with 50 wt % 2G.6 M.W. = 1000 showed orientation but no crystallinity up to 300 % strain
MDI	6G	2G.6	–	+		No 2G.6 crystallinity or orientation up to 300 % strain (MDI/6G and MDI/10G)
MDI	10G	2G.6	–	+		
MDI	$Me_2C[\phi OCH_2CHMeOH]_2$	2G.6	–	Inherently amorphous	≥ 70	
MDI	$Me_3$6.G	2G.6	–	Inherently amorphous		Two T_g values
MDI	4G	PCL	–	+	≥ 58	
HMDI	4G	2G.6	–	+	≳ 60	
HMDI	4G	4G.6	–	+	≳ 60	
HMDI	4G	6G.6	+	+		
HMDI	4G	6G.6	–	+	≳ 60	
HMDI	6G	PCL	+ (Extremely slow crystallisation)	+	≥ 60 for significant crystallisation	Strain-induced crystallisation for PCL M.W. ≥ 1250 at wt % ≥ 50
$Me_3$6 . HMDI	$Me_3$6.G	2G.6	+	Inherently amorphous		Soft block crystallisation from 25 wt % 2G.6
(b) *Polyetherurethanes:*						
MDI	6G	POTM	–	+		
HMDI	6G	POTM	–	+		Strain-induced soft block crystallisation for POTM M.W. ≥ 1250

$Me_3$6 = mixed 2,2,4-/2,4,4-trimethylhexamethylene isomers, PCL = poly(ε-caprolactone); 6G.C = poly(hexamethylene carbonate); HMDI = 1,6-diisocyanatohexane. Other abbreviations as in the general text.

polyetherurethaneureas are relatively insensitive to hard block content whereas there is a marked dependence on this parameter in polyester- or polyetherurethane block copolymers.

The crystallisation behaviour of the components in polyurethanes is naturally of interest. The tendency of linear aliphatic polyester blocks to undergo slow spontaneous crystallisation at ambient temperatures has been known since the early days[4,5,397] and is responsible for the deleterious phenomenon of cold hardening; the tendency increases with polyester block M.W. and, with different polyesters, appears to be greater for those of higher T_m. Crystallisation of the soft and/or the hard block components may also be induced by strain (work hardening) and, if resulting in permanent orientation, may cause an undesirable permanent set.

Species specificity is apparent with regard to all these effects. Even in the simplest types of segmented polyesterurethanes—as in poly(ethylene adipate)s chain-extended with diisocyanates, where no true hard blocks are present—the polyester T_m values are lowered and the T_g values raised to different extents with different isocyanate reactants.[398] More extensive evidence of the variability of crystallisation with block polyurethane structure is shown in Table 9.[399] As a generalisation, the rates and extents of crystallisation in both the hard and soft block components are reduced from those of the corresponding homopolymers (where these are crystalline), and the crystallisability of the soft block components is influenced acutely—and often unpredictably—by the nature of the hard block material.

Spontaneous crystallisation of soft block polyesters or polyethers in copolyurethanes rarely occurs at block M.W.'s of less than 2000 though shorter blocks may crystallise, in some cases, in strained materials.

$$\{\!-\!O[(CH_2)_4O]_a\!-\!CON\langle\rangle N\!-\![\!-\!COO(CH_2)_4OCON\langle\rangle N\!-\!]_b\!-\!CO\!-\!\}_n$$

(LIII)

Corresponding information for hard block components is available with precision only from studies with the *N*-substituted polyetherurethanes (LIII) in which poly(tetramethylene piperazine-1,4-diyl dicarbamate) units of unique sizes of defined size distributions alternate with POTM units along the chains.[243,403–6]

From the standpoint of the present discussion the most interesting conclusions concerning these compounds are as follows:

(a) polymers having unique values of b from 1 to 4 each show distinctive characteristics of hard block crystallinity by DSC (though microscopically visible spherulitic structure exists only for $b \geq 2$);
(b) in blends of polymers with different values of b, discrete crystallisation of the different hard blocks occurs in some cases and discrete crystallisation plus co-crystallisation in others, and
(c) whilst, for average a values of 14 or 24, crystallisation of the POTM blocks does not occur in polymers with integral values of b such crystallisation does occur in materials having a broad distribution of b values.

It may, of course, be considered that since the polymers (LIII) are not hydrogen-bonded, and hence not susceptible to the urethane–urethane or urethane–soft-block interactions characteristic of normal copolyurethanes, their properties are not directly pertinent to the situation in the latter. However, the findings have a resemblance to those for the ordered polyesteramides (L), and scrutiny of the experimental data from various investigations[371,378,382,399,407] on random block polyester- and polyetherurethanes covering a wide range of structures leads to the conclusion that the polyurethane hard blocks derived from MDI or 2,6-TDI with 1,4-butanediol display crystalline characteristics from threshold contents in the range 23–42 wt %, corresponding to hard block ASLs of 1·7–5·4. Even though these values are statistical and the criteria used by different investigators for recognising crystallinity are diverse and not of common sensitivity, the conclusion is clear that hard block crystallisation can occur in copolyurethanes at quite low average block sizes, of a similar order to those found in polyetherester and polyesteramide block copolymers.

Fundamental understanding of the properties of copolyurethanes, as for other block copolymers, must ultimately involve recourse to the details of molecular geometry, configuration and packing. For the commoner soft block components such as 2G.6,[411] poly(ε-caprolactone)[412] and POTM[413,414] such information has been available for some time, as has also been the case for linear aliphatic polyurethane hard block materials.[415] However, for the technically important MDI-based hard block components such perception has been based until recently only upon hypothetical but hitherto untested molecular models[384,392–6,407,416] which must now be modified in the light of important new findings on the crystal

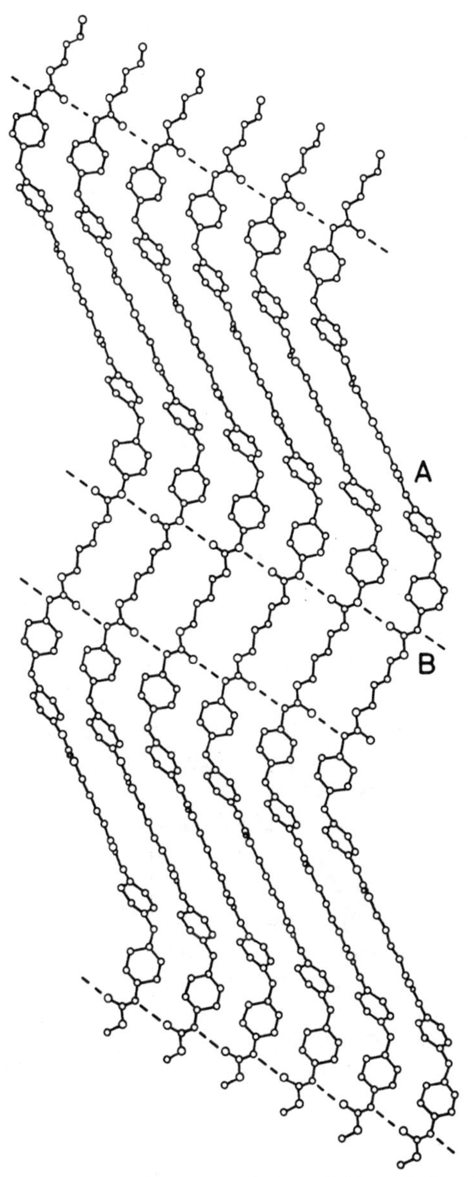

FIG. 9. Chain packing in poly[tetramethylene-1,1′-methylene-bis(4-phenylcarbamate)], the hard block component of MDI/4G-based block copolyurethanes. (From Blackwell, J. and Gardner, K. H., *Polymer*, 1979, **20**, 16; with permission IPC Business Press Ltd ©.)

structures of the model bisurethane (LIV) and the MDI-1,4-butanediol hard block component which can be represented as (LV).[417–20]

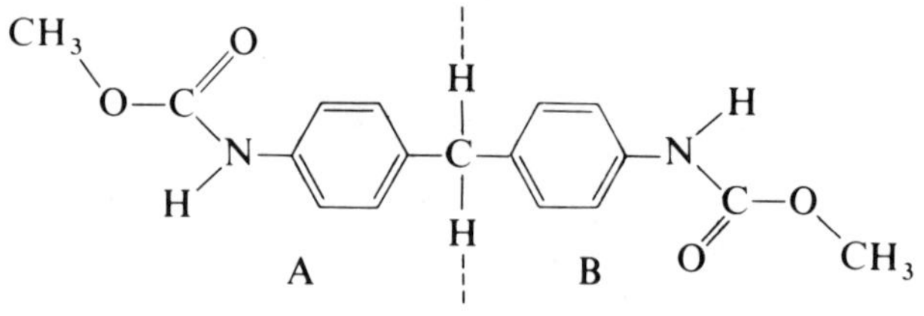

(LIV)

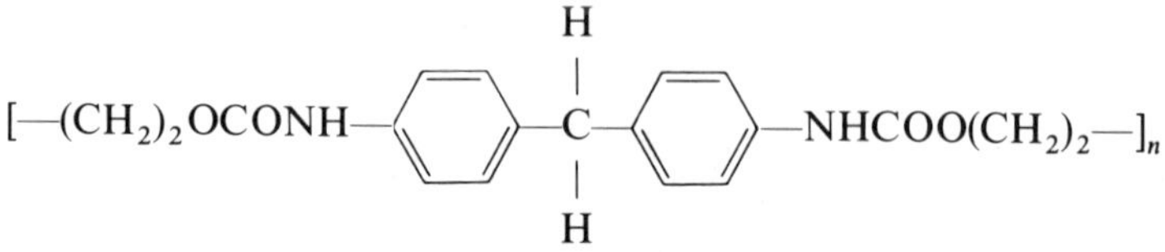

$[-(CH_2)_2OCONH-$ … $-NHCOO(CH_2)_2-]_n$

(LV)

Briefly, whilst (LIV) can exist in different crystalline modifications, that most relevant to the state of (LV) in block copolyurethanes has dissimilar halves A and B. The *p*-phenylene ring in A is twisted at 75·1° to the central C–CH_2–C plane with its attached planar urethane group at 36·1° to the ring; the ring in B is twisted at 34·7° to the C–CH_2–C plane (and virtually perpendicular to the ring in A) whilst its urethane group is inclined at 10·5° to the ring. The hydrogen bonds formed by each NHCOO group are of N—H··· O═C(—O) and not of N—H··· O—C(═O) type and are of different strengths; those of group B form a network with the B halves of other molecules in one plane, and the urethane groups of A form a second network in a perpendicular plane.

The transposition of these findings to (LV) leads to a configurational assembly of hard block molecular segments associated through B type hydrogen bonds as shown in Fig. 9, and this should be understood to be connected by A type bonds with another molecular assembly, forming a three-dimensional multilayer structure. The extension and evolving implications of these findings will be awaited with interest.

4. CONCLUDING REMARKS

Of the many types of heterochain block copolymers surveyed in this chapter, the preponderant effort of research and technological investigation in recent years has been devoted to the elastomers and elastoplastics composed of hard and soft molecular segments. These materials owe their mechanical properties to the separation of the constituent species into more or less distinct phases, and their thermal properties to the polar interactions and, frequently, crystalline character of one or more of the components.

Not all heterochain block copolymers are crystalline or carbochain block copolymers amorphous, but it is particularly through the combined effects of crystallinity and polar character—where both of these are present—that heterochain block polymers differ most significantly from the generality of carbochain block copolymers and merit recognition as a distinct class of materials. The distinction is most apparent in the magnitudes of block sizes permitting phase separation which takes place, even in wholly amorphous heterochain block copolymers, at markedly lower segment lengths than in the carbochain variety. Where phase separation takes the form of crystallisation this may occur at hard block ASLs of five repeating units or less, and in soft blocks at values of the order of 10–20. Whether crystallisation occurs in a constituent of a heterochain block copolymer that is ordinarily crystallisable as a homopolymer is another matter. As has been shown above, this characteristic is profoundly (and often unpredictably) affected by the nature of the other combined species; but where crystallisation occurs it dominates the morphology of the materials, especially that acquired by cooling from the melt which is the circumstance of thermoplastics fabrication. Such morphology is a modified form of that found normally in semi-crystalline polymers and has little in common with the typical domain morphology of amorphous carbochain block materials.

Impressive progress has been made in the past decade in elucidating details of molecular and morphological structures in heterochain block copolymers, and correlating these with bulk properties. Nevertheless, even within such industrially well-established groups as the block polyetheresters and block copolyurethanes important problems remain to be solved such as (1) the determination of actual block sizes and sequence distributions in random block copolymers and their relationship to polymerisation procedures, (2) the details of morphological assembly in copolymers having more than one crystalline phase, (3) the detailed molecular and textural character of the amorphous phase(s) coexisting with crystalline phase(s). Studies in the last-mentioned direction would be assisted considerably if heavy-atom complexing reagents were found that would bond differentially to saturated polyether or polyester chains to enhance density contrast for electron-microscopical examination. A paradox not much discussed in the literature of block copolyurethanes and kindred polymers is the coexistence of inter-species hydrogen-bonded segments with discretely separated crystalline phases of the individual segments. The interpretation may lie chemically in the equilibrium energetics of the various competing interactions, and texturally at the diffuse crystalline–amorphous boundary.

Although thermoplastic elastomers represent the largest volume usage of heterochain block polymers, many diverse applications exist for other suitably constituted members of the group, particularly those of hydrophilic or surfactant character which have interest as detergents, antistatic and soil-releasing agents for textiles, emulsifiers and blending agents, cellular foam growth-controlling agents, and impact-resistance modifiers for plastics materials. Each of these areas contain scope for improvement and development.

In a period of uncertain economic confidence, ideas for novel directions of heterochain block copolymer utilisation must necessarily be speculative, but these may well include biomedical applications (which have already been considered by some investigators), novel membrane systems for industrial separation processes, and the possible introduction into device-technology of the still embryonic class of block copolymers containing anisotropic self-orienting structural components.[421–4] There is no limit in principle to the range of characteristics that may be incorporated into polymeric substances by block combination, but its realisation depends on the continued evolution of skills in synthesis which, as shown earlier in the chapter, remains an area of unflagging vigour.

REFERENCES

1. HARPER, D. A., *Inst. Rubb. Ind. Trans.*, 1948, **24**, 181.
2. WHITE, H. G., *J. Oil Col. Chem. Assoc.*, 1949, **32**, 461.
3. HARPER, D. A., SMITH, W. F. and WHITE, H. G., *Proc. 2nd Rubb. Technol. Conf.*, 1948, p. 61.
4. BAYER, O., MÜLLER, E., PETERSEN, S., PIEPENBRINK, H. F. and WINDEMUTH, E., *Angew. Chem.*, 1950, **62**, 57.
5. MÜLLER, E., BAYER, O., PETERSEN, S., PIEPENBRINK, H. F., SCHMIDT, F. and WEINBRENNER, E., *Angew. Chem.*, 1952, **64**, 523.
6. BAYER, O. and MÜLLER, E., *Angew. Chem.*, 1960, **72**, 934.
7. COLEMAN, D., *J. Polym. Sci.*, 1954, **14**, 15.
8. CHARCH, W. H. and SHIVERS, J. C., *Text. Res. J.*, 1959, **29**, 536.
9. VAUGHN, T. H., SUTER, H. R., LUNDSTED, L. G. and KRAMER, M. G., *J. Am. Oil Chem. Soc.*, 1951, **28**, 294.

10. Vaughn, T. H., Jackson, D. R. and Lundsted, L. G., *J. Am. Oil Chem. Soc.*, 1952, **29**, 240.
11. Lundsted, L. G. and Schmolka, I. R., in: *Block and Graft Copolymerization*, vol. 2, chapts. 1 and 2, ed. R. J. Ceresa, Wiley, London, 1976.
12. Szwarc, M., *Proc. Roy. Soc., Ser. A.*, 1964, **279**, 260.
13. Szwarc, M., *Carbanions, Living Polymers and Electron Transfer Processes*, Wiley, New York, 1968.
14. Fetters, L. J., *J. Polym. Sci., C.*, 1969, **26**, 1.
15. Hayashi, K. and Marvel, C. S., *J. Polym. Sci., A* (General Papers), 1964, **2**, 2571.
16. Bailey, D. L. and O'Connor, F. M. (Union Carbide Co.), US Patent 2 834 748, 1958.
17. Bailey, D. L. and O'Connor, F. M. (Union Carbide Co.), British Patent 802 688, 1958.
18. Haluska, L. A. (Dow Corning Corp.), US Patent 2 846 458, 1958.
19. Bailey, D. L. and O'Connor, F. M. (Union Carbide Co.), British Patent 880 022, 1961.
20. Hostettler, F. M. (Union Carbide Co.), British Patent 892 136, 1962.
21. du Pont de Nemours & Co., British Patent 1 121 866, 1968.
22. Kambour, R. P., *Polym. Prepr., Am. Chem. Soc. Div. Polym. Chem.*, 1969, **10**(2), 885; *J. Polym. Sci., B* (*Polym. Lett.*), 1969, **7**, 573.
23. Farbenfabriken Bayer AG, French Patent 1 590 390, 1970.
24. Noshay, A., Matzner, M. and Merriam, C. N., *J. Polym. Sci., A-1* (*Polym. Chem.*), 1971, **9**, 3147.
25. Matzner, M., Noshay, A. and McGrath, J. E., *Polym. Prepr., Am. Chem. Soc. Div. Polym. Chem.*, 1973, **14**(1), 68.
26. Kamachi, M., Kurihara, M. and Stille, J. K., *Macromolecules*, 1972, **5**, 161.
27. Kurihara, M., Kamachi, M. and Stille, J. K., *J. Polym. Sci., Polym. Chem. Ed.*, 1973, **11**, 587.
28. Lorenz, O. and Rose, G., *Angew. Makromol. Chem.*, 1975, **45**, 85.
29. Marie, P., Herrenschmidt, Y.-Lê and Gallot, Y., *Makromol. Chem.*, 1976, **177**, 2773.
30. Fielding-Russell, G. S. and Pillai, P. S., *Polymer*, 1977, **18**, 859.
31. Varoqui, R., Tran, Q. and Pefferkorn, E., *Macromolecules*, 1979, **12**, 831.
32. Selb, J. and Gallot, Y., *Makromol. Chem.*, 1980, **181**, 809, 2605.
33. Goethals, E. J., Schacht, E. H., Bogaert, Y. E., Ali, S. I. and Tezuko, Y., *Polym. J.*, 1980, **12**, 571.
34. Kawaguchi, M., Oohira, M., Tajima, M. and Takahashi, A., *Polym. J.*, 1980, **12**, 849.
35. Kim, S., Stannett, V. T. and Gilbert, R. D., *J. Macromol. Sci.-Chem.*, 1976, **A10**, 671.
36. Douy, A. and Gallot, B., *Polym. Eng. Sci.*, 1977, **17**, 523.
37. Gallot, B. R. M., *Adv. Polym. Sci.*, 1978, **29**, 85.
38. Amick, R., Gilbert, R. D. and Stannett, V., *Polymer*, 1980, **21**, 648.
39. Douy, A., Gervais, M. and Gallot, B., *Makromol. Chem.*, 1980, **181**, 1199.
40. Lynn, M. M., Stannett, V. T. and Gilbert, R. D., *J. Polym. Sci., Polym. Chem. Ed.*, 1980, **18**, 1967.

41. Kelly, J., MacKenzie, W. M., Sherrington, D. C. and Reiss, G., *Polymer*, 1979, **20**, 1048.
42. Arai, K. and Ogiwara, Y., *J. Polym. Sci., Polym. Chem. Ed.*, 1979, **17**, 404.
43. Arai, K. and Ogiwara, Y., *J. Polym. Sci., Polym. Chem. Ed.*, 1980, **18**, 1643.
44. Allport, D. C. and Janes, W. H. (Eds), *Block Copolymers*, Applied Science Publishers, London, 1973.
45. Noshay, A. and McGrath, J. E., *Block Copolymers: Overview and Critical Survey*, Academic Press, New York, 1977.
46. IUPAC, Tentative nomenclature of regular single-strand organic polymers, *J. Polym. Sci., Polym. Lett. Ed.*, 1973, **11**, 389.
47. IUPAC, *Pure & Appl. Chem.*, 1974, **40**, 479.
48. *Chem. Abstr.*, 8th Collective Index, Vols 66–75 (1967–71).
49. *Chem. Abstr.*, 9th Collective Index, Vols 76–85 (1972–76) (see, in particular, the policy statements in paragraphs 222 and 277 of the Index Guide Appendices).
50. Kresse, P., *Faserforsch. u. Textiltech.* 1960, **11**, 353.
51. Iwakura, I., Taneda, Y. and Uchida, S., *J. Appl. Polym. Sci.*, 1961, **5**, 108.
52. Kiyotsukuri, T., Kashiwabara, H. and Imamura, R., *Kogyo Kagaku Zasshi*, 1966, **69**, 1812.
53. Yamadera, R. and Murano, M., *J. Polym. Sci., A-1* (*Polym. Chem.*), 1967, **5**, 2259.
54. Beste, L. F. and Houtz, R. C., *J. Polym. Sci.*, 1952, **8**, 395.
55. Ayers, C., *J. Appl. Chem.*, 1954, **4**, 444.
56. Kagiya, T., Izu, M., Matsuda, T. and Fukui, K., *J. Polym. Sci., A-1* (*Polym. Chem.*), 1967, **5**, 15.
57. Kricheldorf, H. R. and Rieth, K. H., *J. Polym. Sci., Polym. Lett. Ed.*, 1978, **16**, 379.
58. della Fortuna, G., Oberrauch, E., Salvatori, T., Sorta, E. and Bruzzone, M., *Polymer*, 1977, **18**, 269.
59. de Chirico, A., *Eur. Polym. J.*, 1978, **14**, 329.
60. Jasse, B., *Compt. rend. acad. sci.* (*Paris*), 1969, **268**, 319.
61. Merker, R. L., Scott, M. J. and Haberland, G. G., *J. Polym. Sci., A* (General Papers), 1964, **2**, 31.
62. Bostick, E. E., *Polym. Prepr., Am. Chem. Soc. Div. Polym. Chem.*, 1969, **10**, 877.
63. Bajaj, P., Varshney, S. K. and Misra, A., *J. Polym. Sci., Polym. Chem. Ed.*, 1980, **18**, 295.
64. Lenz, R. W. and Go, S., *J. Polym. Sci., Polym. Chem. Ed.*, 1973, **11**, 2927.
65. Lenz, R. W. and Go, S., *J. Polym. Sci., Polym. Chem. Ed.*, 1974, **12**, 1.
66. Eastman Kodak Co., British Patent 982 575, 1965.
67. Quisenberry, R. K. (du Pont de Nemours & Co.), US Patent 3 265 762, 1966.
68. Saotome, K. and Komoto, H., *J. Polym. Sci., A-1* (*Polym. Chem.*), 1967, **5**, 107.
69. Lyman, D. S. and Jung, S. L., *J. Polym. Sci.*, 1959, **40**, 407.
70. Morgan, P. and Kwolek, S. L., *J. Polym. Sci., A* (General Papers), 1964, **2**, 181.
71. Ogata, N., Sanui, K. and Kamiyama, S., *J. Polym. Sci., Polym. Chem. Ed.*, 1978, **16**, 1991.

72. PEEBLES, L. H., *Macromolecules*, 1974, **7**, 872.
73. KOTLIAR, A. M., *J. Polym. Sci., Polym. Chem. Ed.*, 1975, **13**, 973.
74. PEEBLES, L. H., *Macromolecules*, 1976, **9**, 58.
75. SNOW, A. W., *Macromolecules*, 1977, **10**, 1371.
76. SORTA, E. and MELIS, A., *Polymer*, 1978, **19**, 1153.
77. MACKEY, J. H., PATTISON, V. A. and PAWLAK, J. A., *J. Polym. Sci., Polym. Chem. Ed.*, 1978, **16**, 2849.
78. CURNUCK, P. A. and JONES, M. E. B., *Br. Polym. J.*, 1973, **5**, 21.
79. KODAIRA, Y. and HARWOOD, H. J., *Polym. Prepr., Am. Chem. Soc. Div. Polym. Chem.*, 1973, **14**(1), 323.
80. ZDRAHALA, R. J., FIRER, E. M. and FELLERS, J. F., *J. Polym. Sci., Polym. Chem. Ed.*, 1977, **15**, 689.
81. WILLIAMS, E. A., CARGIOLI, J. D. and HOBBS, S. Y., *Macromolecules*, 1977, **10**, 782.
82. BOLLINGER, J.-C. and AUBINEAU, C., *J. Macromol. Sci.-Chem.*, 1977, **A11**, 1177.
83. JEDLÍNSKI, Z., ŞEK, D. and DZIEWIĘCKA, B., *Eur. Polym. J.*, 1977, **13**, 871.
84. ŞEK, D., *Eur. Polym. J.*, 1977, **13**, 967.
85. DANIELEWICZ, M. and SĘK, D., *Eur. Polym. J.*, 1979, **15**, 639.
86. PINO, P., LORENZI, G. P., SUTER, U. W., CASARTELLI, P. G., STEINMANN, A., BONNER, F. G. and QUIROGA, J. A., *Macromolecules*, 1978, **11**, 624.
87. KRICHELDORF, H. R. and HULL, W. E., *J. Polym. Sci., Polym. Chem. Ed.*, 1978, **16**, 2253.
88. KRICHELDORF, H. R., *Makromol. Chem.*, 1978, **179**, 2133.
89. KRICHELDORF, H. R. and KASCHIG, J., *Eur. Polym. J.*, 1978, **14**, 923.
90. GILDING, D. K. and REED, A. M., *Polymer*, 1979, **20**, 1389.
91. NEWMARK, R. A., *J. Polym. Sci., Polym. Chem. Ed.*, 1980, **18**, 559.
92. DUBOSC, J. P. and PRAT, M., *Bull. soc. chim. France*, 1967, 4357.
93. LEE, A. K. and SEDGWICK, R. D., *J. Polym. Sci., Polym. Chem. Ed.*, 1978, **16**, 685.
94. RICHARDS, D. H. and SZWARC, M., *Trans. Faraday Soc.*, 1959, **55**, 1644.
95. FINAZ, G., REMPP, P. and PARROD, J., *Bull. soc. chim. France*, 1962, 262.
96. YAMASHITA, Y., in: *Polymerization Reactions and New Polymers*, ed. N. A. J. Platzer, Am. Chem. Soc., Adv. Chem. Ser., No. 129, 1973, p. 248.
97. PINAZZI, P., ESNAULT, J. and PLEURDEAU, A., *Makromol. Chem.*, 1976, **177**, 663.
98. O'MALLEY, J. J., CRYSTAL, R. G. and ERHARDT, P. F., *Polym. Prepr., Am. Chem. Soc. Div. Polym. Chem.*, 1969, **10**, 796.
99. CAMBERLIN, Y. and PASCAULT, J. P., *Makromol. Chem.*, 1979, **180**, 397.
100. REED, S. F., *J. Polym. Sci., A-1 (Polym. Chem.)*, 1972, **10**, 1187.
101. SHIMURA, Y. and LIN, W., *J. Polym. Sci., A-1 (Polym. Chem.)*, 1970, **8**, 2171.
102. SCHULZ, D. N., HALASA, A. F. and OBERSTER, H. E., *J. Polym. Sci., Polym. Chem. Ed.*, 1974, **12**, 153.
103. REED, S. F., *J. Polym. Sci., A-1 (Polym. Chem.)*, 1971, **9**, 2029.
104. REED, S. F., *J. Polym. Sci., A-1 (Polym. Chem.)*, 1971, **9**, 2147.
105. DEQUATRE, C., CAMBERLIN, Y., PILLOT, C. and PASCAULT, J. P., *Angew. Makromol. Chem.*, 1978/79, **72**, 11.
106. BROSSAS, J. and CLOUET, G., *Makromol. Chem.*, 1974, **175**, 3067.

107. CATALA, J.-M., RIESS, G. and BROSSAS, J., *Makromol. Chem.*, 1977, **178**, 1249.
108. BOUCHAL, K., ŽŮRKOVA, E., KÁLAL, J., SUFČÁK, M. and SEYČEK, O., *Angew. Makromol. Chem.*, 1980, **86**, 33.
109. ANTHOINE, J. C. and VERNET, J. L., *Eur. Polym. J.*, 1980, **16**, 519.
110. KATAYAMA, S., SERITA, H. and TAKAHASHI, Y., *J. Polym. Sci., Polym. Chem. Ed.*, 1977, **15**, 2109.
111. IVÁN, B., KENNEDY, J. P. and CHANG, V. S. C., *J. Polym. Sci., Polym. Chem. Ed.*, 1980, **18**, 3177.
112. SHIBATANI, K., LYMAN, D. J., SHIEH, D. F. and KNUTSON, K., *J. Polym. Sci., Polym. Chem. Ed.*, 1977, **15**, 1655.
113. MICHEL, A., CASTANEDA, E. and GUYOT, A., *Eur. Polym. J.*, 1979, **15**, 935.
114. GUIZARD, C. and CHERADAME, H., *Eur. Polym. J.*, 1981, **17**, 121.
115. FRAZER, A. H. and SHIVERS, J. C. (du Pont de Nemours & Co.), US Patent 2 928 803, 1960.
116. HEIKENS, D., MEIJERS, A. and VON RETH, P. H., *Polymer*, 1968, **9**, 15.
117. HUET, J. M. and MARECHAL, E., *Eur. Polym. J.*, 1974, **10**, 757.
118. ORHAN, E. H., YILGÖR, I. and BAYSAL, B. M., *Polymer*, 1977, **18**, 286.
119. MAGNUSSON, A. B., *J. Appl. Polym. Sci.*, 1967, **11**, 2175.
120. ASHMAN, P. C. and BOOTH, C., *Polymer*, 1976, **17**, 105.
121. FRANKENBURG, P. E. and FRAZER, A. H. (du Pont de Nemours & Co.), US Patent 2 957 852, 1960.
122. SCHAEFGEN, J. R. and SHIVERS, J. C. (du Pont de Nemours & Co.), US Patent 3 044 987, 1962.
123. KIYOTSUKURI, T. and SHIMOMURA, Y., *Kobunshi Kagaku*, 1971, **28**(314), 516.
124. CASTALDO, L., MAGLIO, G. and PALUMBO, R., *J. Polym. Sci., Polym. Lett. Ed.*, 1978, **16**, 643.
125. DEAN, J. W., *J. Polym. Sci.*, (*Polym. Lett.*), 1970, **8**, 677.
126. SHIMURA, Y. and IKEDA, N., *J. Polym. Sci., Polym. Chem. Ed.*, 1973, **11**, 1271.
127. BILLOT, J.-P., DOUY, A. and GALLOT, B., *Makromol. Chem.* 1976, **177**, 1889.
128. BILLOT, J.-P., DOUY, A. and GALLOT, B., *Makromol. Chem.* 1977, **178**, 1641.
129. BURGESS, F. J., CUNLIFFE, A. V., MACCALLUM, J. R. and RICHARDS, D. H., *Polymer*, 1977, **18**, 719.
130. HALLENSLEBEN, M. L., *Makromol. Chem.*, 1977, **178**, 2125.
131. BROZE, G., LEFÈBVRE, P. M., JÉRÔME, R. and TEYSSIÉ, PH., *Makromol. Chem.*, 1977,**178**, 3171.
132. FOSS, R. P., JACOBSON, H. W., CRIPPS, H. N. and SHARKEY, W. H., *Macromolecules*, 1979, **12**, 1210.
133. PETIT, D., JEROME, R. and TEYSSIÉ, PH., *J. Polym. Sci., Polym. Chem. Ed.*, 1979, **17**, 2903.
134. PATTISON, D. B. (du Pont de Nemours & Co.), US Patent 2 830 038, 1958.
135. VAUGHN, H. A., JR., US Patent 3 419 634, 1968.
136. PANDE, K. C. and KALLENBACH, S. E. (Powers Chemco Inc.), US Patent 3 776 889, 1973.
137. SAEGUSA, T. and IKEDA, H., *Macromolecules*, 1973, **6**, 805, 808.
138. SUZUKI, T., MURAKAMI, Y. and TAKEGAMI, Y., *J. Polym. Sci., Polym. Lett. Ed.*, 1979, **17**, 241.
139. DE VISSER, A. C., GREGONIS, D. E. and DRIESSEN, A. A., *Makromol. Chem.*, 1978, **179**, 1855.

140. Broze, G., Jérôme, R. and Teyssié, Ph., *Makromol. Chem.*, 1978, **179**, 1383.
141. Broze, G., Lefèbvre, P. M., Jérôme, R. and Teyssié, P., *Macromolecules*, 1979, **12**, 1047.
142. Kern, W., Munk, R., Sabel, A. and Schmidt, K. H., *Makromol. Chem.*, 1955/56, **17**, 201.
143. Conix, A., *Makromol. Chem.*, 1958, **26**, 226.
144. Wesslén, B. and Månsson, P., *J. Polym. Sci., Polym. Chem. Ed.*, 1975, **13**, 2545.
145. Seow, P. K., Gallot, Y. and Skoulios, A., *Makromol. Chem.*, 1975, **176**, 3153.
146. Lingelser, J.-P.,Marie, P. and Gallot, Y., *Compt. rend. acad. sci.* (*Paris*), *Ser. C.*, 1976, **282**, 579.
147. Gervais, M. and Gallot, B., *Makromol. Chem.*, 1977, **178**, 1577.
148. Hirata, E., Ijitsu, T., Soen, T., Hashimoto, T. and Kawai, T., *Polym. Prepr., Am. Chem. Soc. Div. Polym. Chem.*, 1974, **15**, 177.
149. Koetsier, D. W., Bantjes, A., Feijen, T. and Lyman, D. J., *J. Polym. Sci., Polym. Chem. Ed.*, 1978, **16**, 511.
150. Gia-H.B., Jérôme, R. and Teyssié, Ph., *J. Polym. Sci., Polym. Chem. Ed.*, 1980, **18**, 3483.
151. Suzuki, T., Murakami, Y. and Takegami, Y., *Polym. J.*, 1980, **12**, 183.
152. Takahashi, A. and Yamashita, Y., *Polym. Prepr., Am. Chem. Soc. Div. Polym. Chem.*, 1974, **15**, 184.
153. Yamashita, Y., *J. Macromol. Sci.-Chem.*, 1979, **A13**, 401.
154. Richards, D. H., Kingston, S. B. and Souel, T., *Polymer*, 1978, **19**, 68.
155. Richards, D. H., Kingston, S. B. and Souel, T., *Polymer*, 1978, **19**, 806.
156. Kucera, M., Bozek, F. and Majerová, K., *Polymer*, 1979, **20**, 1013.
157. Tesoro, G. T., Sello, S. B. and Wooster, R. F., German OLS 1 956 153, 1970.
158. Burgess, F. J., Cunliffe, A. V., Dawkins, J. V. and Richards, D. H., *Polymer*, 1977, **18**, 733.
159. Takahashi, A., Kawaguchi, M., Kato, T., Kuno, M. and Matsumoto, S., *J. Macromol. Sci.-Phys.*, 1980, **B17**, 747.
160. Ladousse, A., Maillard, B., Villenave, J-J. and Filliatre, C., *Makromol. Chem.*, 1980, **181**, 903.
161. Ikemi, M., Odagiri, N. and Shinohara, I., *Polym. J.*, 1980, **12**, 777.
162. Cooper, G. D., Bennett, J. G. and Factor, A., in: *Polymerization Kinetics and Technology*, ed. N. A. J. Platzer, Am. Chem. Soc., Adv. Chem. Ser., No. 128, 1973, p. 230.
163. Morton, M., Kammereck, R. F. and Fetters, L. J., *Br. Polym. J.*, 1971, **3**, 120.
164. Morton, M. and Mikesell, S. L., *J. Macromol. Sci.-Chem.*, 1973, **A7**, 1391.
165. Roggero, A., Mazzei, A., Bruzzone, M. and Cernia, E., in: *Copolymers, Polyblends and Composites*, ed. N. A. J. Platzer, Am. Chem. Soc., Adv. Chem. Ser., No. 142, 1975, p. 330.
166. Rahman, R. and Avny, Y., *J. Macromol. Sci.-Chem.*, 1979, **A13**, 971.
167. Bossaer, P. K., Goethals, E. J., Hackett, P. J. and Pepper, D. C., *Eur. Polym. J.*, 1977, **13**, 489.
168. Leibnitz, E. and Reinisch, G., *Faserforsch. u. Textiltech.*, 1970, **21**, 426.

169. GHAFFAR, A., GOODMAN, I. and HALL, I. H., *Br. Polym. J.*, 1973, **5**, 315.
170. GOODMAN, I., PETERS, R. H. and SCHENK, V. T. J., *Br. Polym. J.*, 1975, **7**, 329.
171. HOESCHELE, G. K. and WITSIEPE, W. K., *Angew. Makromol. Chem.*, 1973, **29/30**, 267.
172. BOUSSIAS, C. M., PETERS, R. H. and STILL, R. H., *J. Appl. Polym. Sci.*, 1980, **25**, 855.
173. VARMA, D. S., MAHESWARI, A., GUPTA, V. and VARMA, I. K., *Angew. Makromol. Chem.*, 1980, **90**, 23.
174. WITSIEPE, W. K., '*Polymerization Reactions and New Polymers*, ed. N. A. J. Platzer, Am. Chem. Soc., Adv. Chem. Ser., No. 129, 1973, p. 39.
175. WOLFE, J. R., *Multiphase Polymers*, eds. S. L. Cooper and G. M. Estes, Am. Chem. Soc., Adv. Chem. Ser., No. 176, 1979, p. 129.
176. SORTA, E. and DELLA FORTUNA, G., *Polymer*, 1980, **21**, 728.
177. LEEMING, P. A., OLDHAM, J. and SAGAR, H. (ICI Ltd.), British Patent 1 176 648, 1970.
178. HOESCHELE, G. K., *Polym. Eng. Sci.*, 1974, **14**, 544.
179. du Pont de Nemours & Co., *du Pont Hytrel Polyester Elastomers*, Technical Brochure A-99054, printed in USA.
180. BUCK, W. H., CELLA, R. J., GLADDING, E. K. and WOLFE, J. R., *J. Polym. Sci., Polym. Symp.*, 1974, **48**, 47.
181. SHEN, M., MEHRA, U., NIINOMI, M., KOBERSTEIN, J. T. and COOPER, S. L., *J. Appl. Phys.*, 1974, **45**, 4182.
182. SEYMOUR, R. W., OVERTON, J. R. and CORLEY, L. S., *Macromolecules*, 1975, **8**, 331.
183. LILAONITKUL, A., WEST, J. C. and COOPER, S. L., *J. Macromol. Sci.-Phys.*, 1976, **B12**, 563.
184. LILAONITKUL, A. and COOPER, S. L., *Rubb. Chem. Technol.*, 1977, **50**, 1.
185. WOLFE, J. R., *Rubb. Chem. Technol.*, 1977, **50**, 688.
186. HOESCHELE, G. K., *Angew. Makromol. Chem.*, 1977, **58/59**, 299.
187. WEGNER, G., FUJII, T., MEYER, W. and LIESER, G., *Angew. Makromol. Chem.*, 1978, **74**, 295.
188. NORTH, A. M., PETHRICK, R. A. and WILSON, A. D., *Polymer*, 1978, **19**, 923.
189. LILAONITKUL, A. and COOPER, S. L., *Macromolecules*, 1979, **12**, 1146.
190. MASUKO, T., *Makromol. Chem.*, 1979, **180**, 2183.
191. BOUSSIAS, C. M., PETERS, R. H. and STILL, R. H., *J. Appl. Polym. Sci.*, 1980, **25**, 869.
192. HÄSSLIN, H.-W. and DRÖSCHER, M., *Makromol. Chem.*, 1980, **181**, 2357.
193. *Modern Plastics International*, January 1981, p. 39.
194. GOLDBERG, E. P., *J. Polym. Sci., C*, **4**, 1964, 707.
195. PERRY, K. P., JACKSON, W. J. and CALDWELL, J. R., *J. Appl. Polym. Sci.*, 1965, **9**, 3451.
196. READER, A. M. and RULISON, R. N., *J. Polym. Sci., A-1* (*Polym. Chem.*), 1967, **5**, 927.
197. MERRILL, S. H., *J. Polym. Sci.*, 1961, **55**, 343.
198. RICHES, K. and HAWARD, R. N., *Polymer*, 1968, **9**, 103.
199. MERRILL, S. H. and PETRIE, S. E., *J. Polym. Sci., A* (General Papers), 1965, **3**, 2189.
200. HUET, J. M. and MARECHAL, E., *Eur. Polym. J.*, 1974, **10**, 771.

201. FRADET, A. and MARECHAL, E., *Eur. Polym. J.*, 1978, **14**, 749.
202. FRADET, A. and MARECHAL, E., *Eur. Polym. J.*, 1978, **14**, 755.
203. WATTS, M. P. C. and WHITE, E. F. T., in: *Multiphase Polymers*, eds. S. L. Cooper and G. M. Estes, Am. Chem. Soc., Adv. Chem. Ser., No. 176, 1979, p. 153.
204. BOSNYAK, C. P., PARSONS, I. W., HAY, J. N. and HAWARD, R. N., *Polymer*, 1980, **21**, 1448.
205. O'MALLEY, J. J., *J. Polym. Sci., Polym. Lett. Ed.*, 1974, **12**, 381.
206. GOODMAN, I., in: *Developments in Polymerisation—2*, chapt. 4, ed. R. N. Haward, Applied Science Publishers, London, 1979.
207. GILDING, D. K. and REED, A. M., *Polymer*, 1979, **20**, 1459.
208. TABUCHI, T., NOBUTOKI, K. and SUMITOMO, H., *Kogyo Kagaku Zasshi*, 1968, **71**, 1926.
209. PERRET, R. and SKOULIOS, A., *Makromol. Chem.*, 1972, **156**, 143.
210. GOODMAN, I., JOHNSON, A. F. and YORK, D., unpublished results.
211. MUELLER, F. X. and HSIEH, H. L. (Phillips Petroleum Co.), US Patent 3 585 257, 1971.
212. HSIEH, H. L., *J. Appl. Polym. Sci.*, 1978, **22**, 1119.
213. CLARK, E. and CHILDERS, C. W., *J. Appl. Polym. Sci.*, 1978, **22**, 1081.
214. TEYSSIÉ, P., BIOUL, J. P., HAMITOU, A., HEUSCHEN, J., HOCKS, L., JÉRÔME, R. and OUHADI, T., *Polym. Prepr., Am. Chem. Soc. Div. Polym. Chem.*, 1977, **18**(1), 65.
215. TEYSSIÉ, P., BIOUL, J. P., HAMITOU, A., HEUSCHEN, J., HOCKS, L., JÉRÔME, R. and OUHADI, T., in: *Ring-Opening Polymerization*, eds T. Saegusa and E. Goethals, Am. Chem. Soc., Symposium, Ser. 59, 1977, p. 165.
216. HAMITOU, A., JÉRÔME, R. and TEYSSIÉ, P., *J. Polym. Sci., Polym. Chem. Ed.*, 1977, **15**, 1035.
217. HERMAN, J-J., JÉRÔME, R., TEYSSIÉ, P., GERVAIS, M. and GALLOT, B., *Makromol. Chem.*, 1978, **179**, 1111.
218. FOSS, R. P., JACOBSON, H. W., CRIPPS, H. N. and SHARKEY, W. H., *Macromolecules*, 1976, **9**, 373.
219. KING, C. (du Pont de Nemours & Co.), US Patent 3 418 393, 1968.
220. LENZ, R. W., DROR, M., JORGENSEN, R. and MARCHESSAULT, R. H., *Polym. Eng. Sci.*, 1978, **18**, 937.
221. ALLEGREZZA, A. E., LENZ, R. W., CORNIBERT, J. and MARCHESSAULT, R. H., *J. Polym. Sci., Polym. Chem. Ed.*, 1978, **16**, 2617.
222. SUNDET, S. A., THAMM, R. C., MEYER, J. M., BUCK, W. H., CAYWOOD, S. W., SUBRAMANIAN, P. M. and ANDERSON, B. C., *Macromolecules*, 1976, **9**, 371.
223. CAYWOOD, S. W., *Rubb. Chem. Technol.*, 1977, **50**, 127.
224. BUCK, W. H., *Rubb. Chem. Technol.*, 1977, **50**, 109.
225. HARRIS, J. F. and SHARKEY, W. H., *Macromolecules*, 1977, **10**, 503.
226. BIGGS, B. S., FROSCH, C. J. and ERICKSON, R. H., *Ind. Eng. Chem.*, 1946, **38**, 1016.
227. WITTBECKER, E. L., HOUTZ, R. C. and WATKINS, W. W., *Ind. Eng. Chem.*, 1948, **40**, 875.
228. CHAUVEL, B., *Ann. Chim.*, 1961, **6**, 893.
229. WRIGHT, P. and CUMMING, A. P. C., *Solid Polyurethane Elastomers*, Maclaren, London, 1969.

230. Holmen, R. E. (Minnesota Mining & Manufacturing Co.), US Patent 2 692 253, 1954.
231. Coleman, D. (ICI Ltd.), British Patent 793 451, 1958.
232. Shivers, J. C. (du Pont de Nemours & Co.), US Patent 3 044 989, 1962.
233. Steuber, W. (du Pont de Nemours & Co.), US Patent 3 044 990, 1962.
234. Garforth, J. D. (ICI Ltd.), British Patent 1 270 097, 1972.
235. Toray Industries Inc., British Patent 1 276 214, 1972.
236. Ambrose, R. J. and Hergenrother, W. L., *Polym. Prepr., Am. Chem. Soc. Div. Polym. Chem.*, 1974, **15**, 185.
237. Ambrose, R. J. and Hergenrother, W. L., *J. Appl. Polym. Sci.*, 1975, **19**, 1931.
238. Ambrose, R. J. and Hergenrother, W. L., *J. Polym. Sci., Polym. Lett. Ed.*, 1976, **14**, 603.
239. Ambrose, R. J. and Hergenrother, W. L., *J. Polym. Sci., Polym. Symp.*, 1977, **60**, 15.
240. Mumzu, S., Burzin, K., Feldman, R. and Feinauer, R., *Angew. Makromol. Chem.*, 1978/79, **74**, 49.
241. de Visser, A. C., Driessen, A. A. and Wolke, J. G. C., *Makromol. Chem., Rapid Commun.*, 1980, **1**, 177.
242. Lorenz, G. and Nischk, G. E., *Makromol. Chem.*, 1969, **130**, 55.
243. Harrell, L. L., *Macromolecules*, 1969, **2**, 607.
244. Ateya, K., *Angew. Makromol. Chem.*, 1969, **9**, 56.
245. Jablonski, R. J., Witzel, J. M. and Kruh, D., *J. Polym. Sci., B (Polym. Lett.)*, 1970, **8**, 191.
246. Bollinger, J.-C. and Aubineau, C., *J. Macromol. Sci.-Chem.*, 1977, **A11**, 1159.
247. Deleens, G., Foy, P. and Maréchal, E., *Eur. Polym. J.*, 1977, **13**, 337, 343, 353.
248. Masař, B., Čefelín, P. and Šebenda, J., *J. Polym. Sci., Polym. Chem. Ed.*, 1979, **17**, 2317.
249. Yamashita, Y., Matsui, H. and Ito, K., *J. Polym. Sci., Polym. Chem. Ed.*, 1972, **10**, 3577.
250. Allen, W. T. and Eaves, D. E., *Angew. Makromol. Chem.*, 1977, **58/59**, 321.
251. Hergenrother, W. L. and Ambrose, R. J., *J. Polym. Sci., Polym. Chem. Ed.*, 1974, **12**, 2613.
252. Stehlíček, J. and Šebenda, J., *Eur. Polym. J.*, 1977, **13**, 949, 955.
253. Nitadori, Y., Franta, E. and Rempp, P., *Makromol. Chem.*, 1978, **179**, 927.
254. Wondraczek, R. H. and Kennedy, J. P., *Polym. Bull.*, 1980, **2**, 675.
255. Hergenrother, W. L. and Ambrose, R. J., *J. Appl. Polym. Sci.*, 1975, **19**, 3225.
256. Goodman, I. and Hurworth, N. R. (ICI Ltd.), British Patent 1 099 184, 1968.
257. Goodman, I. and Vachon, R. N., *Abstr. XXIVth International Symposium on Macromolecules*, Jerusalem, 1975, p. 211.
258. Goodman, I., Kettle, S. J. and Valavanidis, A. P., unpublished results.
259. Perly, B., Douy, A. and Gallot, B., *Makromol. Chem.*, 1976, **177**, 2569.
260. Nakijima, A., Hayashi, T., Kugo, K. and Shinoda, K., *Macromolecules*, 1979, **12**, 840.

261. ELIAS, H. G., BUEHRER, H.-G. and SEMEN, J., *Appl. Polym. Symposium*, 1975, **26**, 269.
262. JONES, N. B. and JONES, M. N., *Progr. Colloid & Polym. Sci.*, 1979, **66**, 403.
263. URALIL, F., HAYASHI, T., ANDERSON, J. M. and HILTNER, A., *Polym. Eng. Sci.*, 1977, **17**, 515.
264. HAYASHI, T., WALTON, A. G. and ANDERSON, J. M., *Macromolecules*, 1977, **10**, 346.
265. DOUY, A. and GALLOT, B., *Makromol. Chem.*, 1977, **178**, 1595.
266. GODFREY, R. A. and MILLER, G. W., *J. Polym. Sci., A-1* (*Polym. Chem.*), 1969, **7**, 2387.
267. FERRUTI, P., MARTUSCELLI, E., NICOLAIS, L., PALMA, M. and RIVA, F., *Polymer*, 1977, **18**, 387.
268. FERRUTI, P., ARNOLDI, D., MARCHISIO, M. A., MARTUSCELLI, E., PALMA, M., RIVA, F. and PROVENZALE, L., *J. Polym. Sci., Polym. Chem. Ed.*, 1977, **15**, 2151.
269. Dow Corning Corp., British Patents 955 916, 1964; 983 850, 1965; 1 149 744, 1969.
270. BAILEY, D. L. and PATER, A. S. (Union Carbide Co.), British Patent 1 015 611, 1966.
271. BAILEY, D. L. (Union Carbide Co.), British Patent 1 034 781, 1966.
272. STEFFEN, K-D., *Angew. Makromol. Chem.*, 1972, **24**, 21.
273. NOSHAY, A. and MATZNER, M., *Angew. Makromol. Chem.*, 1974, **37**, 215.
274. MADEC, P-J. and MARECHAL, E., *J. Polym. Sci., Polym. Chem. Ed.*, 1978, **16**, 3165.
275. CHAUMONT, P., HERZ, J. and REMPP, P., *Eur. Polym. J.*, 1979, **15**, 537.
276. VAUGHN, H. A., *J. Polym. Sci., B* (*Polym. Lett.*), 1969, **7**, 569.
277. MERRITT, W. D. (General Electric Co.), US Patent 3 832 419, 1974.
278. MERRITT, W. D. and VESTERGAARD, J. H. (General Electric Co.), US Patent 3 821 325, 1974.
279. KAMBOUR, R. P. and NIZNIK, G. E., General Electric Co., *Technical Information Report 74CRD055*, 1974.
280. KOURTIDES, D. A. and PARKER, J. A., *Polym. Eng. Sci.*, 1978, **18**, 855.
281. KOURTIDES, D. A., GILWEE, W. J. and PARKER, J. A., *Polym. Eng. Sci.*, 1979, **19**, 24.
282. GREBER, G. and BALCIUNAS, A., *Makromol. Chem.*, 1964, **79**, 149.
283. O'MALLEY, J. J. and STAUFFER, W. J., *Polym. Eng. Sci.*, 1977, **17**, 510.
284. O'MALLEY, J. J., PACANSKY, T. J. and STAUFFER, W. J., *Macromolecules*, 1977, **10**, 1197.
285. MADEC, P-J. and MARECHAL, E., *J. Polym. Sci., Polym. Chem. Ed.*, 1978, **16**, 3157.
286. BUSFIELD, W. K. and COWIE, J. M. G., *Polym. Bull.*, 1980, **2**, 619.
287. CHAUMONT, P., BEINERT, G., HERZ, J. and REMPP, P., *Eur. Polym. J.*, 1979, **15**, 459.
288. NOSHAY, A., MATZNER, M. and WILLIAMS, T. C., *Ind. Eng. Chem., Prod. Res. Develop.*, 1973, **12**, 268.
289. MAZUREK, M., NORTH, A. M. and PETHRICK, R. A., *Polymer*, 1980, **21**, 369.

290. Stewart, D. D., Peters, E. N., Beard, C. D., Dunks, G. B., Hedaya, E., Kwiatkowski, G. T., Moffitt, R. B. and Bohan, J. J., *Macromolecules*, 1979, **12**, 373.
291. Saam, J. C., Ward, A. H. and Fearon, F. W. G., *J. Inst. Rubb. Ind.*, 1973, **7**(2), 69.
292. Saam, J. C., Ward, A. H. and Fearon, F. W. G., in: *Polymerization Reactions and New Polymers*, ed. N. A. J. Platzer, Am. Chem. Soc., Adv. Chem. Ser., No. 129, 1973, p. 239.
293. Marsiat, A. and Gallot, Y., *Makromol. Chem.*, 1975, **176**, 1641.
294. Varshney, S. K. and Khanna, D. N., *J. Appl. Polym. Sci.*, 1980, **25**, 2501.
295. Morton, M., Kesten, Y. and Fetters, L. J., *Polym. Prepr., Am. Chem. Soc. Div. Polym. Chem.*, 1974, **15**, 175.
296. Morton, M., Kesten, Y. and Fetters, L. J., *Appl. Polym. Symposium*, 1975, **26**, 113.
297. Owen, M. J. and Thompson, J., *Br. Polym. J.*, 1972, **4**, 297.
298. Robeson, L. M., Noshay, A., Matzner, M. and Merriam, C. N., *Angew. Makromol. Chem.*, 1973, **29**/**30**, 47.
299. McGrath, J. E., Ward, T. C., Shchori, E. and Wnuk, A. J., *Polym. Eng. Sci.*, 1977, **17**, 647.
300. McGrath, J. E., Matzner, M., Robeson, L. M. and Barclay, R., *J. Polym. Sci., Polym. Symp.*, 1977, **60**, 29.
301. Ward, T. C., Wnuk, A. J., Shchori, E., Wiswanathan, R. and McGrath, J. E., in: *Multiphase Copolymers*, eds S. L. Cooper and G. M. Estes, Am. Chem. Soc., Adv. Chem. Ser., No. 176, 1979, p. 293.
302. Farbenfabriken Bayer AG, French Patent 1 577 171, 1969.
303. Stewart, D. D., Peters, E. N., Beard, C. D., Moffitt, R. B., Kwiatkowski, G. T., Bohan, J. J. and Hedaya, E., *J. Appl. Polym. Sci.*, 1979, **24**, 115.
304. McGrath, J. E., Robeson, L. M. and Matzner, M., *Polym. Prep., Am. Chem. Soc. Div. Polym. Chem.*, 1973, **14**, 1032.
305. McGrath, J. E., Robeson, L. M. and Matzner, M., in: *Recent Advances in Polymer Blends, Grafts and Blocks*, ed. L. H. Sperling, Plenum Press, New York, 1974, p. 195.
306. Litt, M. and Herz, J., *Polym. Prepr., Am. Chem. Soc. Div. Polym. Chem.*, 1969, **10**(2), 905.
307. Litt, M. H. and Matsuda, T., in: *Copolymers, Polyblends and Composites*, ed. N. A. J. Platzer, Am. Chem. Soc., Adv. Chem. Ser., No. 142, 1975, 320.
308. Seung, S. L. N. and Young, R. N., *J. Polym. Sci., Polym. Lett. Ed.*, 1980, **18**, 89.
309. Berlin, A. A., Gerasimov, B. G., Ivanov, A. A. and Beregovykh, L. P., *J. Macromol. Sci.-Chem.*, 1977, **A11**, 811.
310. Berlin, A. A., Gerasimov, B. G., Ivanov, A. A., Masliukov, A. P. and Sheludchenko, N. I., *J. Macromol. Sci.-Chem.*, 1977, **A11**, 821.
311. Cowie, J. M. G., Lath, D. and McEwen, I. J., *Macromolecules*, 1979, **12**, 52.
312. Zotteri, L. and Giuliani, G. P., *Polymer*, 1978, **19**, 476.
313. Kennedy, J. P. and Smith, R. A., *J. Polym. Sci., Polym. Chem. Ed.*, 1980, **18**, 1539.
314. Kuo, C. and McIntyre, D., *J. Polym. Sci., Polym. Phys. Ed.*, 1975, **13**, 1543.

315. BOOTH, C. and PICKLES, C. J., *J. Polym. Sci., Polym. Phys. Ed.*, 1973, **11**, 249.
316. GALIN, J. C., SPEGT, P., SUZUKI, S. and SKOULIOS, A., *Makromol. Chem.*, 1974, **175**, 991.
317. FRIDAY, A. and BOOTH, C., *Polymer*, 1978, **19**, 1035.
318. COOPER, D. R., LEUNG, Y-K., HEATLEY, F. and BOOTH, C., *Polymer*, 1978, **19**, 309.
319. TAKAHASHI, Y., SUMITA, I. and TADOKORO, H., *J. Polym. Sci., Polym. Phys. Ed.*, 1973, **11**, 2113.
320. TEITELBAUM, B-YA., *J. Thermal Anal.*, 1975, **8**, 511.
321. MINKE, R. and BLACKWELL, J., *J. Macromol. Sci.-Phys.*, 1979, **B16**, 407.
322. PRUD'HOMME, R. E. and MARCHESSAULT, R. H., *Macromolecules*, 1974, **7**, 541.
323. YOKOUCHI, M., SAKAKIBARA, Y., CHATANI, Y., TADOKORO, H., TANAKA, T. and YODA, K., *Macromolecules*, 1976, **9**, 266.
324. JAKEWAYS, R., WARD, I. M., WILDING, M. A., HALL, I. H., DESBOROUGH, I. J. and PASS, M. G., *J. Polym. Sci., Polym. Phys. Ed.*, 1975, **13**, 799.
325. LOTZ, B. and KOVACS, A. J., *Polym. Prepr., Am. Chem. Soc. Div. Polym. Chem.*, 1969, **10**, 820.
326. CRYSTAL, R. G., O'MALLEY, J. J. and ERHARDT, P. F., *Polym. Prepr., Am. Chem. Soc. Div. Polym. Chem.*, 1969, **10**, 804.
327. GERVAIS, M. and GALLOT, B., *Makromol. Chem.*, 1973, **174**, 157, 193.
328. THOMAS, H. R. and O'MALLEY, J. J., *Macromolecules*, 1979, **12**, 323.
329. NEIDLINGER, H. H., HÖCKER, H. and ZACHMANN, H. G., *Polym. Prep., Am. Chem. Soc. Div. Polym. Chem.*, 1977, **18**(1), 606.
330. O'MALLEY, J. J., THOMAS, H. R. and LEE, G. M., *Macromolecules*, 1979, **12**, 996.
331. SHIMURA, Y. and HATAKEYAMA, T., *J. Polym. Sci., Polym. Phys. Ed.*, 1975, **13**, 653.
332. O'MALLEY, J. J., *J. Polym. Sci., Polym. Symp.*, 1977, **60**, 151.
333. BERGER, G., LEVY, M. and VOFSI, D., *J. Polym. Sci., B (Polym. Lett.)*, 1966, **4**, 183.
334. SEOW, P. K., GALLOT, Y. and SKOULIOS, A., *Makromol. Chem.*, 1976, **177**, 177, 199.
335. GOODMAN, I. and LIW, O. H., unpublished results.
336. GHAFFAR, A., GOODMAN, I. and PETERS, R. H., *Br. Polym. J.*, 1978, **10**, 115.
337. GHAFFAR, A., GOODMAN, I., PETERS, R. H. and SEGERMAN, E., *Br. Polym. J.*, 1978, **10**, 123.
338. RAAB, M., MASAŘ, B., KOLAŘÍK, J. and ČEFELÍN, P., *Intern. J. Polym. Mat.*, 1979, **7**, 219.
339. FELLERS, J. F., LEE, Y-D. and ZDRAHALA, R. J., *J. Polym. Sci., Polym. Symp.*, 1977, **60**, 59.
340. FELLERS, J. F., LEE, Y-D., ZDRAHALA, R. J. and HORNE, S. C., *Polym. Eng. Sci.*, 1977, **17**, 506.
341. SAAM, J. C., GORDON, D. J. and LINDSEY, S., *Macromolecules*, 1970, **3**, 1.
342. GALIN, M. and RUPPRECHT, M. C., *Macromolecules*, 1979, **12**, 506.
343. POCHAN, J. M., PACANSKY, T. J. and HINMAN, D. F., *Polymer*, 1978, **19**, 431.
344. FRITSCHE, A. K. and PRICE, F. P., *Polym. Prepr., Am. Chem. Soc. Div. Polym. Chem.*, 1969, **10**(2), 893.

345. Kojima, M. and Magill, J. H., *J. Polym. Sci., Polym. Phys. Ed.*, 1974, **12**, 317.
346. Kojima, M. and Magill, J. H., *J. Macromol. Sci.-Phys.*, 1974, **B10**, 419.
347. Kojima, M. and Magill, J. H., *J. Appl. Phys.*, 1974, **45**, 4159.
348. Okui, N. and Magill, J. H., *Polymer*, 1977, **18**, 845.
349. Li, H. M. and Magill, J. H., *Polymer*, 1978, **19**, 829.
350. Kojima, M. and Magill, J. H., *J. Macromol. Sci.-Phys.*, 1978, **B15**, 63.
351. LeGrand, D. G., *J. Polym. Sci., B (Polym. Lett.)*, 1969, **7**, 579.
352. Hässlin, H. W., Dröscher, M. and Wegner, G., *Makromol. Chem.*, 1978, **179**, 1373.
353. O'Malley, J. J., *J. Polym. Sci., Polym. Phys. Ed.*, 1975, **13**, 1353.
354. Pochan, J. M. and Hinman, D. F., *J. Polym. Sci., Polym. Phys. Ed.*, 1975, **13**, 1365.
355. Goodman, I. and Sheahan, R. J., unpublished results.
356. Khanderia, J. and Sperling, L. H., *J. Appl. Polym. Sci.*, 1974, **18**, 913.
357. Walton, A. G. and Blackwell, J., *Biopolymers*, Academic Press, New York, 1973.
358. Fraser, R. D. B. and MacRae, T. P., *Conformation in Fibrous Proteins and Related Synthetic Polypeptides*, Academic Press, New York, 1973.
359. Douy, A. and Gallot, B., *Polym. Eng. Sci.*, 1977, **17**, 523.
360. Nakajima, A., Kugo, K. and Hayashi, T., *Macromolecules*, 1979, **12**, 844.
361. Anderson, J. M., Hayashi, T., Wong, M., Barenberg, S., Sigler, G., Geil, P. H., Walton, A. G. and Hiltner, A., *J. Polym. Sci., Polym. Symp.*, 1977, **60**, 77.
362. Hayashi, T., Anderson, J. M. and Hiltner, P. A., *Macromolecules*, 1977, **10**, 352.
363. Seymour, R. W., Estes, G. M. and Cooper, S. L., *Macromolecules*, 1970, **3**, 579.
364. Koutsky, J. A., Hien, N. V. and Cooper, S. L., *J. Polym. Sci., B (Polym. Lett.)*, 1970, **8**, 353.
365. Huh, D. S. and Cooper, S. L., *Polym. Eng. Sci.*, 1971, **11**, 369.
366. Estes, G. M., Seymour, R. W. and Cooper, S. L., *Macromolecules*, 1971, **4**, 452.
367. Seymour, R. W., Allegrezza, A. E. and Cooper, S. L., *Macromolecules*, 1973, **6**, 896.
368. Seefried, C. G., Koleske, J. V. and Critchfield, F. E., *J. Appl. Polym. Sci.*, 1975, **19**, 2493, 2503.
369. Seefried, C. G., Koleske, J. V. and Critchfield, F. E., *J. Appl. Polym. Sci.*, 1975, **19**, 3185.
370. Aitken, R. R. and Jeffs, G. M. F., *Polymer*, 1977, **18**, 197.
371. Schneider, N. S. and Paik Sung, C. S., *Polym. Eng. Sci.*, 1977, **17**, 73.
372. Srichatrapimuk, V. W. and Cooper, S. L., *J. Macromol. Sci.-Phys.*, 1978, **B15**, 267.
373. Schollenberger, C. S., in: *Multiphase Polymers*, eds S. L. Cooper and G. M. Estes, Am. Chem. Soc., Adv. Chem. Ser., No. 176, 1979, p. 83.
374. Aitken, R. R., *Polymer*, 1979, **20**, 1160.
375. Senich, G. A. and MacKnight, W. J., in: *Multiphase Polymers*, eds S. L.

Cooper and G. M. Estes, Am. Chem. Soc., Adv. Chem. Ser., No. 176, 1979, p. 97.
376. Zdrahala, R. J., Gerkin, R. M., Hager, S. L. and Critchfield, F. E., *J. Appl. Polym. Sci.*, 1979, **24**, 2041.
377. Van Bogart, J. W. C., Lilaonitkul, A. and Cooper, S. L., in: *Multiphase Polymers*, eds S. L. Cooper and G. M. Estes, Am. Chem. Soc., Adv. Chem. Ser., No. 176, 1979, p. 3.
378. Chang, A. L. and Thomas, E. L., in: *Multiphase Polymers*, eds S. L. Cooper and G. M. Estes, Am. Chem. Soc., Adv. Chem. Ser., No. 176, 1979, p. 31.
379. Senich, G. A. and MacKnight, W. T., *Macromolecules*, 1980, **13**, 106.
380. Ophir, Z. and Wilkes, G. L., *J. Polym. Sci., Polym. Phys. Ed.*, 1980, **18**, 1469.
381. Kong, E. S. W. and Wilkes, G. L., *J. Polym. Sci., Polym. Lett. Ed.*, 1980, **18**, 369.
382. Schneider, N. S., Paik Sung, C. S., Matton, R. W. and Illinger, J. L., *Macromolecules*, 1975, **8**, 62.
383. Paik Sung, C. S. and Schneider, N. S., *Macromolecules*, 1975, **8**, 68.
384. Wilkes, C. E. and Yusek, C. S., *J. Macromol. Sci.-Phys.*, 1973, **B7**, 157.
385. Nakayama, K., Ino, T. and Matsubara, I., *J. Macromol. Sci.-Chem.*, 1969, **A3**, 1005.
386. Ishihara, J., Kimura, I., Saito, K. and Ono, H., *J. Macromol. Sci.-Phys.*, 1974, **B10**, 591.
387. Kimura, I., Ishihara, H., Ono, H., Yoshihara, N., Nomura, S. and Kawai, H., *Macromolecules*, 1974, **7**, 355.
388. Work, J. L., *Macromolecules*, 1976, **9**, 759.
389. Paik Sung, C. S., Smith, T. W., Hu, C. B. and Sung, N. H., *Macromolecules*, 1979, **12**, 538.
390. Paik Sung, C. S., Hu, C. B. and Wu, C. S., *Macromolecules*, 1980, **13**, 111.
391. Paik Sung, C. S., Smith, T. W. and Sung, N. H., *Macromolecules*, 1980, **13**, 117.
392. Chang, Y-J. P. and Wilkes, G. L., *J. Polym. Sci., Polym. Phys. Ed.*, 1975, **13**, 455.
393. Bonart, R., *J. Macromol. Sci.-Phys.*, 1968, **B2**, 115.
394. Bonart, R., Morbitzer, L. and Müller, E. H., *J. Macromol. Sci.-Phys.*, 1974, **B9**, 447.
395. Bonart, R. and Müller, E. H., *J. Macromol. Sci.-Phys.*, 1974, **B10**, 177, 345.
396. Bonart, R., *Polymer*, 1979, **20**, 1389.
397. Flocke, H. A., *Kunststoffe*, 1966, **56**(5), 328.
398. Onder, K., Peters, R. H. and Spark, L. C., *Polymer*, 1972, **13**, 133.
399. Goodman, I., Ahmad, M., Brazier, A. H., Heard, M. B., Patel, H. G. and Suchiva, K., unpublished results.
400. Clough, S. B. and Schneider, N. S., *J. Macromol. Sci.-Phys.*, 1968, **B2**, 553.
401. Clough, S. B., Schneider, N. S. and King, A. O., *J. Macromol. Sci.-Phys.*, 1968, **B2**, 641.
402. Guise, G. B. and Smith, G. C., *J. Appl. Polym. Sci.*, 1980, **25**, 149.
403. Samuels, S. L. and Wilkes, G. L., *J. Polym. Sci., Polym. Phys. Ed.*, 1973, **11**, 807.

404. NG, H. N., ALLEGREZZA, A. E., SEYMOUR, R. W. and COOPER, S. L., *Polymer*, 1973, **14**, 255.
405. ALLEGREZZA, A. E., SEYMOUR, R. W., NG, H. N. and COOPER, S. L., *Polymer*, 1974, **15**, 433.
406. WILKES, G. L., SAMUELS, S. L. and CRYSTAL, R., *J. Macromol. Sci.-Phys.*, 1974, **B10**, 203.
407. SCHNEIDER, N. S., DESPER, C. R., ILLINGER, J. L., KING, A. O. and BARR, D., *J. Macromol. Sci.-Phys.*, 1975, **B11**, 527.
408. FERGUSON, J., HOURSTON, D. J., MEREDITH, R. and PATSAVOUDIS, D., *Eur. Polym. J.*, 1972, **8**, 369.
409. ILLINGER, J. L., SCHNEIDER, N. S. and KARASZ, F. E., *Polym. Eng. Sci.*, 1972, **12**, 25.
410. BLEIJENBERG, A. C. A., HEIKENS, D., MEIJERS, A., LAMPE, H. G. M. and VON RETH, P. H., *Br. Polym. J.*, 1972, **4**, 125.
411. TURNER-JONES, A. and BUNN, C. W., *Acta Crystallogr.*, 1962, **15**, 105.
412. CHATANI, Y., OKITA, Y., TADOKORO, H. and YAMASHITA, Y., *Polym. J.*, 1970, **1**, 555.
413. IMADA, K., MIJAKAWA, T., CHATANI, Y., TADOKORO, H. and MURAHASHI, S., *Makromol. Chem.*, 1965, **83**, 113.
414. CESARI, M., PEREGO, G. and MAZZEI, A., *Makromol. Chem.*, 1965, **83**, 196.
415. SAITO, Y., NANSAI, S. and KINOSHITA, S., *Polym. J.*, 1972, **3**, 113.
416. FRANK, W. F. X. and STROHMEIER, W., *Progr. Colloid & Polym. Sci.*, 1979, **86**, 205.
417. BLACKWELL, J. and GARDNER, K. H., *Polymer*, 1979, **20**, 13.
418. BLACKWELL, J. and ROSS, M., *J. Polym. Sci., Polym. Lett. Ed.*, 1979, **17**, 447.
419. HOCKER, J. and BORN, L., *J. Polym. Sci., Polym. Lett. Ed.*, 1979, **17**, 723.
420. BLACKWELL, J. and NAGARAJAN, M. R., *Polymer*, 1981, **22**, 202.
421. JACKSON, W. J. and KUHFUSS, H. F., *J. Polym. Sci., Polym. Chem. Ed.*, 1976, **14**, 2043.
422. TAKAYANAGI, M., OGATA, T., MORIKAWA, M. and KAI, T., *J. Macromol. Sci.-Phys.*, 1980, **B17**, 591.
423. JIN, J.-I., ANTOUN, S., OBER, C. and LENZ, R. W., *Br. Polym. J.*, 1980, **12**, 132.
424. JACKSON, W. J., *Br. Polym. J.*, 1980, **12**, 154.
425. MCINTYRE, J. E. and MILBURN, A. H., *Br. Polym. J.*, 1981, **13**, 5.

Chapter 6

MORPHOLOGY AND PROPERTIES OF POLYURETHANE BLOCK COPOLYMERS

P. E. Gibson, M. A. Vallance and S. L. Cooper

Department of Chemical Engineering,
University of Wisconsin, USA

SUMMARY

Typical condensation block copolymers are $[AB]_n$ *type segmented block copolymers which have thermoplastic properties due to phase separation on a microscopic scale. The relationship of their engineering properties to morphology, and to changes in morphology due to thermal or mechanical processing, is a complex one. The characterisation of the solid state morphology of segmented copolymers requires the use of a variety of experimental techniques. The most commonly used methods for investigating the engineering properties and morphology of segmented copolymers are described and illustrative morphological studies of segmented polyurethanes are presented. Although the materials discussed in this chapter are polyurethanes, the experimental methods and general approach to morphological studies are applicable to condensation block copolymers in general.*

1. INTRODUCTION

Block copolymers are synthesised by chemically combining blocks of two dissimilar homopolymers along the chain backbone. If A and B represent two homopolymers, then the possible molecular architecture includes AB diblock structures, ABA triblock polymers and $[AB]_n$ multiblock systems. The nature of the blocks and their sequential arrangement play an important role in determining block copolymer properties.

Many condensation block copolymers are $[AB]_n$ type alternating block copolymers in which the blocks are relatively short and numerous. Depending on the nature of the blocks and the average segment length, the properties of these condensation segmented copolymers may vary from those of random copolymers to thermoplastic elastomers. The former generally has been observed in systems which have either short segment lengths or similar intersegmental and intrasegmental secondary binding forces or both. The solid-state structure of these compatible segmented polymers is relatively homogeneous, with the copolymers displaying properties approximating to those of a weighted average of the homopolymer segments. However, most segmented copolymers exhibit a two-phase structure due to incompatibility of the two dissimilar segment types and are thermoplastic elastomers.

Thermoplastic elastomers are block copolymers that exhibit rubberlike elasticity without requiring chemical crosslinking.[1] For rubber-like solids, the chemical structure of the blocks is chosen such that both a rigid high-modulus phase and a flexible elastomeric phase are formed so that at service temperatures one of the components is viscous or rubbery (soft segment) while the other is of a glassy or semi-crystalline nature (hard segment). Consequently, the hard domains serve as virtual crosslinks by providing junction points for the rubbery chain segments. At the same time, the hard domains also serve as reinforcing filler, an effect similar to that of carbon black in conventional rubbers.

The reinforcing domains of elastomeric block copolymers are rigid because their chain segments are below their glass transition temperature (T_g) or melting point (T_m) when at the temperature of polymer service. Since the hard domains soften at their transition temperature, a thermally reversible elastomeric system results, and rapid thermoplastic processing methods such as injection moulding and extrusion can be applied. These materials are also easily applied as coatings and are used extensively as additives, processing aids and compatibilising agents.

It is now widely accepted that the unusual properties of thermoplastic elastomers are due to their two-phase microstructures. Because the morphology determines to a large extent the engineering properties of thermoplastic elastomers, it is important to determine what factors affect morphology and to characterise the morphology of the phase separated block copolymers. Some of the factors which can affect morphology of

$[AB]_n$ segmented copolymers include hard segment and soft segment type, volume fraction, intermolecular bonding, M.W. distribution, as well as the thermal and mechanical history of the material. The relationship of engineering properties to domain structure and to changes in domain structure due to thermal or mechanical processing is a complex one. Clearly, the original morphology of a thermoplastic elastomer is altered by stretching to high elongations, annealing, or annealing under strain (i.e. heat setting). It is also important to note that the solid-state of these materials is seldom, if ever, an equilibrium state.

The thermodynamics of microphase separation in block polymers has been described by several authors[2–5] and is covered in Chapter 4 of this book. To date, the thermodynamics of microphase separation applicable to short segment $[AB]_n$ block copolymers has not been developed effectively.

The synthesis of condensation-type block copolymers is covered in Chapters 5 and 7 of this book so that only a brief sketch is given here. Segmented copolymers of $[AB]_n$ type are usually synthesised by condensation polymerisation reactions.[6–8] The reaction components consist of a difunctional soft segment, the basic hard segment component and a chain extender for the hard segment.

Soft blocks of segmented polyurethanes are usually derived from linear, α,ω-dihydroxy- polyethers or polyesters (macroglycols) having M.W.'s between 600 and 3000. In a typical procedure leading to a thermoplastic polyurethane elastomer, the macroglycol is end-capped with the full amount of aromatic diisocyanate required to form the final composition. Subsequently, the mixture of end-capped prepolymer and excess diisocyanate is reacted with the required stoichiometric amount of a short-chain diol to complete the reaction. The diol links the prepolymer segments together while excess diol and diisocyanate form short hard-block segments, leading to an $[AB]_n$ structure. Block lengths in $[AB]_n$ polymers are frequently much shorter than those in anionically synthesised ABA block copolymers.

Molecular structure can be varied by changing the chemical composition of the three reactants (macroglycol, diisocyanate and short-chain diol) or by changing the method of polymerisation. All three reactants can be polymerised simultaneously in a one-step[9] reaction or they can be added sequentially[10] after forming an isocyanate-capped prepolymer. From a theoretical standpoint, Peebles[11,12] has shown that a two-step method of block polymer synthesis should lead to a narrower distribution of hard segment lengths as compared with that for a one-step synthesis, providing that reaction of the first isocyanate moiety occurs at a faster rate than that of the second one.

Chemical composition determines many molecular properties such as polarity, hydrogen bonding capability and crystallisability of the blocks. If the short-chain diols are replaced by diamines, highly polar urea linkages are formed in the hard blocks. Polyurethanes also have been synthesised with piperazine replacing the diisocyanate,[13] thereby eliminating all possibility of hydrogen bonding. The synthesis reactions for other segmented copolymers, such as the segmented copolyesters,[14–18] are analogous to those for the urethanes. In the case for the copolyesters, however, the reaction is a melt transesterification which produces methanol as a by-product and requires low-pressure evaporation for removal (see Chapter 7).

Although the materials discussed in this chapter are polyurethanes, the experimental methods and general approach to morphological studies are useful for condensation block copolymers in general.

The characterisation of the solid state morphology of segmented copolymers, as well as the determination of how the morphology of a material changes with processing conditions, requires the use of a variety of experimental techniques. Before illustrative morphological studies are presented, the most commonly used experimental methods for investigating the engineering properties and morphology of segmented elastomers are described.

2. EXPERIMENTAL METHODS FOR MORPHOLOGY PROPERTIES STUDIES

2.1. Small-angle X-ray Scattering

Small-angle X-ray scattering (SAXS) is a technique by which structural features of the order of 20–2000 Å in size may be studied. Virtually any substance can be studied because few materials have large absorption coefficients for wavelengths of electromagnetic radiation less than 2 Å. Typically, the 1·542 Å wavelength Cu-Kα X-rays are used in SAXS experiments. Essentially, SAXS is a probe of a material's electron density distribution. Analysis of the angular distribution of SAXS intensity reveals information about the magnitude and periodicity of the electron distribution.

Morphological information, representing an average of all features within the scattering volume, is obtained from the scattered intensity data by the use of mathematical analyses based on theoretical models.

The intensity of the scattering is related to the extent of phase separation within the system via the invariant. The breadth of the scattering curve is inversely related to the average size of the phases. The correlation function, which is the Fourier transform of the scattering curve, is related to the average phase dimensions and also indicates the average spacing between phases. Finally, the shape of the SAXS curve tail determines the thickness of the interface between phases. The determination of the invariant, correlation function and the interfacial thickness is described in greater detail below.

The intensity of X-ray scattering, $I(s)$, is related to a material's spatial electron density distribution, $\rho(x)$, as follows:

$$I(s) = I_e(s)\left[\int_V \rho(x)\exp(-2\pi i\, s.x)\,\mathrm{d}x\right]^2 \tag{1}$$

where $I_e(s)$ = intensity scattered by a single electron; $s = 2\sin\theta/\lambda$ = scattering vector magnitude; 2θ = scattering angle; V = volume of material exposed to X-rays.

Because the observable function $I(s)$ is related to the squared magnitude of the Fourier transform of $\rho(x)$, phase information is lost and $\rho(x)$ cannot be obtained from $I(s)$. However, methods have been developed to extract information about $\rho(x)$. One such method is the calculation of the correlation function, $\gamma(r)$. If the electron density is written in the form $\rho(x) = \rho_0 + \eta(x)$, where ρ_0 is the average electron density, then $\gamma(r)$ is defined by the relation:

$$\gamma(r) = \frac{1}{\overline{\eta^2}V}\int_V \eta(x)\eta(x+r)\,\mathrm{d}x \tag{2}$$

Equation (1) may be written in terms of $\gamma(r)$ as:

$$I(s) = I_e(s)\overline{\eta^2}V\int_V \gamma(r)\exp(-2\pi i\, s.r)\,\mathrm{d}r \tag{3}$$

The correlation function, γ, represents the probability that the electron density at one point is equal to the electron density at another point a distance $|r|$ away. It has real space dimensions and is an averaged image, although indirect, of the morphology of the sample. The correlation function contains all the morphological information that can possibly be obtained from the SAXS curve. The interpretation of the correlation function is confounded by the fact that it contains intermixed effects resulting from the form of the phases as well as from their respective arrangement in space.[19]

For a two-phase system, the weighted integral of scattering intensity, $\int_0^\infty s^2 I(s)\,\mathrm{d}s$, known as the invariant, is related to the mean square of the electron density fluctuation by the relation:

$$\overline{\eta^2} = \frac{4\pi}{V} \int_0^\infty s^2 \frac{I(s)\,\mathrm{d}s}{I_e(s)} \tag{4}$$

For a two-phase system composed of phase A and phase B with sharp interfaces between the A and B domains, the value for the mean square electron density fluctuation, $\overline{\eta^2}$, is given in terms of the electron density of each phase, ρ_i, and the phase volume fraction, ϕ_i, as follows:

$$\overline{\eta^2} = \phi_{\mathrm{A}}\phi_{\mathrm{B}}(\rho_{\mathrm{A}} - \rho_{\mathrm{B}})^2 \tag{5}$$

The magnitude of $\overline{\eta^2}$ is thus related to the purity of each phase and the degree of phase separation within the system.

A simpler, but less descriptive means of describing the morphology of a two-phase system is through the use of the inhomogeneity length parameters. If a two-phase material is randomly intersected by an infinite number of lines, the average length of all chords lying in the phase A will be given by l_{A}, the phase A inhomogeneity length. The inhomogeneity length in phase A is related to the overall inhomogeneity length, l_p, as follows:

$$l_{\mathrm{A}} = \frac{l_p}{1 - \phi_{\mathrm{A}}} \tag{6}$$

where ϕ_{A} is the volume fraction of phase A. The overall inhomogeneity length is obtained from the relation:

$$l_p = \frac{2}{\mathrm{K}\pi^2} \int_0^\infty s^2 I(s)\,\mathrm{d}s \tag{7}$$

where K is Porod's constant.

Up to this point in the SAXS discussion, we have been concerned with the analysis of two-phase systems with sharp interfaces. Should the system contain density gradients at the interface between phases, the tail of the SAXS curve will fall off more rapidly and theoretically will follow the general form:[20–22]

$$\lim_{s \to \infty} I(s) = \frac{\mathrm{K}}{s^4} \cdot H^2(s) \tag{8}$$

In the above equation, K is Porod's constant and $H^2(s)$ is a smoothing function related to the shape of the electron density gradient at the interface. For a sharp interface, $H^2(s) = 1$. If the interfacial gradient is assumed to have a sigmoidal shape, then $H^2(s)$ will take the form:[20,22]

$$H^2(s) = \exp(-4\pi^2\sigma^2 s^2) \tag{9}$$

The interface in two-phase polymer systems is best approximated by a sigmoidal gradient.[23] It is easier, however, to grasp conceptually the dimensions of an assumed linear (ramp gradient) interface. The thickness, t, of a linear interface is given in terms of the standard deviation of a sigmoidal gradient, σ, as:

$$t = \sqrt{12}\sigma \tag{10}$$

Experimental apparatus and the difficulties involved in the analysis of SAXS data will not be discussed. An excellent, comprehensive review of the applications of scattering to polymers is given by Higgins and Stein.[24] For a more comprehensive treatment of SAXS, the reader is referred to the appropriate texts and reviews.[19,25,26]

2.2. Infrared Dichroism

Infrared dichroism can be used to study the orientation response of the hard and soft segments of segmented copolymer materials subjected to strain. Polarised infrared radiation is used to study the absorbances of a chemical moiety (N—H, C=O or C—H) characteristic of a particular segment. Light polarised parallel and perpendicular to the direction of the sample stretch axis is monitored with differences in the two absorbed intensities, $A_{\parallel}$ and $A_{\perp}$, respectively, related to the orientation of that particular segment.[27]

Polarised monochromatic infrared radiation can be absorbed by a chemical bond only if there is a component of the radiation resolvable into the direction of the changing electrical dipole moment associated with that functional group's vibration. The dipole moment may be characterised by a transition moment vector **M** which species its magnitude and direction with respect to the chain backbone. The absorbance contribution from each functional group depends on the angle (β) between **M** and **E** according to the equation:

$$A = \mathrm{K}(\mathbf{M}.\mathbf{E})^2 = \mathrm{K}(ME)^2 \cos^2\beta \tag{11}$$

where A is the absorbance, **E** is the electric field vector, and K is a proportionality constant. Maximum absorbance thus occurs when the direction of polarisation and the transition moment direction are parallel.

If the absorbing structural units in a macromolecule are randomly oriented then the absorbance will be independent of the direction of incident polarised infrared radiation. However, if these groups are preferentially aligned, e.g. by stretching the polymer film, then the absorbance will vary according to the alignment of transition moments with respect to the polarisation direction of the incident radiation. This phenomenon, called infrared dichroism, can be used to measure the degree of orientation of the polymer chain if the angle between the transition moment and the polymer chain axis is known. It is essential for quantitative work that the angle, α, between the transition vector and the chain backbone axis be known.

It is generally assumed that the transition moment direction is along the chemical bond in such highly localised vibrations as the C—H and N—H stretch.[28,29] This corresponds to an α value of 90°. However, transition moment directions associated with functional groups may be found to deviate from that of the chemical bond direction. For example, the value of α for C=O bonds in polyurethanes and polyamides has been found to be about 79°.[30,31] Transition moment angles can be calculated by assuming a molecular structure corresponding to a particular vibration or they can be found experimentally by combining infrared measurements with other methods of determining molecular orientation such as X-ray diffraction.[32]

Conventionally, infrared dichroism measurements are made by uniaxially orienting a sample of polymer film and determining the absorbance of selected bands with radiation polarised parallel and perpendicular to the stretch direction. From the resulting unequal absorbances ($A_{\parallel}$ and $A_{\perp}$) a dichroic ratio, D, can be calculated as follows:

$$D = A_{\parallel}/A_{\perp} \tag{12}$$

The dichroic ratio can be related to an orientation function, f, through the relation:

$$f = [(D_0 + 2)/(D_0 - 1)][(D - 1)/(D + 2)] \tag{13}$$

D_0 is the dichroic ratio for perfect alignment and is equal to $2\cot^2\alpha$. The orientation function, f, may be related to an average angle of disorientation, θ, by

$$f = [3\cos^2\theta - 1]/2 \tag{14}$$

This function has a value of unity for a sample whose elements are

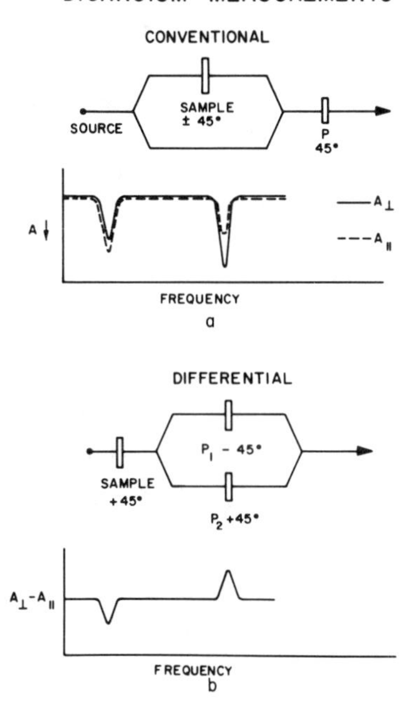

FIG. 1. Schematic representation of techniques for measurement of infrared dichroism. (a) Conventional, two spectra recorded; (b) differential dichroism, single difference spectrum recorded. (From reference 130, with permission John Wiley & Sons, Inc.)

completely oriented in the stretch direction. For perfect orientation transverse to the stretch direction $f = -1/2$ and for random orientation $f = 0$.

Figure 1(a) illustrates the experimental arrangement for data output in a conventional infrared dichroism measurement. A double-beam spectrophotometer is employed with a single polariser in the common beam. In practice the polariser is set at 45° to the slit direction to minimise machine polarisation effects. Two spectra are run with the sample positioned first at a +45° angle and then at −45° so that the sample elongation direction is parallel and then perpendicular to the direction of polarisation. Potential sources of error and correction techniques have been discussed by Zbinden.[28]

Infrared dichroism studies are frequently carried out by monitoring fundamental stretching vibrations which absorb quite strongly. This requires very thin film specimens (thicknesses ranging from 2 μm to

0·1 mm) in order to avoid complete absorption of the radiation at the frequencies of interest. Typically, such films may be prepared by casting from dilute solutions onto mercury surfaces.[33] Alternatively, a spin-casting device has been developed for the preparation of thin uniform films from dilute solutions evaporated in a centrifugal force field. Stretching devices have been constructed in order that infrared dichroism measurements can be carried out on strained films which can be rotated in the infrared beam.[33]

For dichroic effects which are small in magnitude, the conventional experimental procedure diagrammed in Fig. 1(a) is not very adequate. The differential technique illustrated in Fig. 1(b), increases the sensitivity and allows measurements to be made on polymer films while they undergo continuous elongation or relaxation. Two polarisers are used so that the chopped common beam going to the monochrometer contains both $A_{\parallel}$ and $A_{\perp}$ information. The quantity recorded is the dichroic difference, $A_{\perp} - A_{\parallel}$. Thus the output for an unoriented sample will be a straight line while preferential orientation of any vibration will result in an upward or downward peak depending on the relative magnitudes of $A_{\parallel}$ and $A_{\perp}$.

The dichroic difference is related to f, the orientation function. A_0 is the absorbance of the peak in an unstretched specimen and d/d_0 is the stretched to unstretched thickness ratio.

$$f = \left(\frac{D_0 + 2}{D_0 - 1}\right)\left(\frac{A_{\parallel} - A_{\perp}}{3A_0(d/d_0)}\right) \tag{15}$$

A quantitative comparison of the two infrared dichroism techniques has been published by Read *et al.*[34]

The development of Fourier transform infrared (FTIR) spectrometers has made possible the study of phenomena occurring on a time scale as short as 1 s.[122] The advantages of FTIR spectrometers over conventional dispersive infrared spectrometers include higher signal-to-noise ratio, rapid scanning, computer-aided signal analysing and spectral manipulation, and higher energy throughput. Rapid scanning of frequency and rapid data acquisition have allowed the investigation of deformation, chemical reactions, phase transitions and conformational changes of polymers.

Siesler has reported the use of FTIR to study the complex nature of isocyanate reactions in the formation of polyurethane foams.[123] In addition, studies have been made of segmental orientation of polyurethane films undergoing uniaxial deformation. The simultaneous measurement of infrared dichroism, stress and strain in a film were rapidly and effectively carried out using FTIR.

Clearly, the technique of FTIR spectroscopy has opened many new areas of investigation for polymers in general and condensation block copolymers in particular. The recent book by Siesler and Holland-Moritz is an excellent reference for the application of infrared and Raman spectroscopy to the study of polymers.[35]

2.3. Dynamic Mechanical Analysis

If a sinusoidal strain is applied to a solid exhibiting relaxation, then the resultant stress is sinusoidal in the case of linear viscoelastic response and generally leads the strain by some phase angle, δ. This phase relationship results from the time necessary for molecular rearrangements, and is analogous to the time lags observed in creep and relaxation experiments. Expressing strain in complex form

$$\varepsilon = \varepsilon_0 \exp(i\omega t) \tag{16}$$

where ε_0 is the strain amplitude and ω is the angular frequency, the stress response can be described as

$$\sigma = \sigma_0 \exp(i\omega t + \delta) \tag{17}$$

and the stress–strain relationship is given as

$$\sigma = E^* \varepsilon = (E' + iE'')\varepsilon \tag{18}$$

where

$$E^* = (E' + iE'') = (\sigma_0/\varepsilon_0)\exp(i\delta) \tag{19}$$

The most popular mode of dynamic mechanical analysis (DMA) involves measuring $|E^*| = \sigma_0/\varepsilon_0$ and δ or $\tan\delta = E''/E'$ with ω and ε_0 fixed and temperature varying, usually rising at a constant rate. For thermoplastic elastomers, a typical experiment might involve quenching a polymer film to $-150\,°C$ followed by heating at $2°\,min^{-1}$ to $200\,°C$, during which time measurements are made and recorded periodically. Because of the wide range of $|E^*|$ values which a single material can exhibit (several orders of magnitude), it is often necessary to use two or more films with different cross-sections in order to obtain data on the entire temperature range of interest. From a typical temperature scan it is possible to observe glass transitions, crystallisation and melting phenomena, as well as inelastic processes in the glassy state which are usually associated with specific chemical sequences in the polymer chains.[124] If temperature scans are taken at multiple frequencies, activation energies can be associated with the various relaxations.

Thus far it has been assumed that strains are applied uniaxially in a normal mode and that strain is the forcing function while stress is the response. This is the case for instruments such as the Toyo Rheovibron®. However, many variations of DMA are recorded in the literature. The torsion pendulum,[36] which provides very similar spectra, measures shear strains and shear stresses. Flexural deformation can also be used. In some applications resonance characteristics are measured.[37] The dynamic mechanical test is related to creep and stress relaxation experiments.[38] As pointed out by Ferry[39] the complex modulus is easily obtained from the relaxation modulus by Fourier transform of the latter from the temporal to the spectral domain.

2.4. Differential Scanning Calorimetry

In differential scanning calorimetry (DSC) the power consumption necessary to heat a small, typically 20 mg, sample of polymer through a programmed temperature ramp is measured and plotted as a function of temperature. For practicality of experimental design, the quantity monitored is actually the difference between the power consumption of the polymer sample of interest and that of a standard ramped at the same rate. The standard is usually chosen to have a uniform heat capacity throughout the temperature range of interest. In this fashion thermograms can be obtained with temperatures increasing or decreasing. A typical scan might involve quenching the sample to $-150\,°C$, then heating at $20°\ min^{-1}$ to $250\,°C$. From such thermograms it is possible to observe glass transitions, crystallisation and melting. The data are useful for inferring the degree of phase separation in thermoplastic elastomers, as well as the composition-dependence of the transitions. Recent advances in processor interfacing have led to highly quantitative DSC investigations.

2.5. Small-angle Light Scattering

Small-angle light scattering (SALS) is a technique for probing morphological structures 1000–100 000 Å in size in polymer films. This technique is insensitive to the presence of microdomains whose typical dimensions are of the order of 100 Å. On the other hand, spherulitic superstructures, as are found in semi-crystalline systems, are often of sufficient size for characterisation by SALS. SALS provides data on fluctuations in the refractive index of a material. Light scattering also gives information about the orientation of the polymer molecules.

To record a SALS signature for a polymer system, radiation incident on the sample is plane polarised. A monochromatic columnated beam is used,

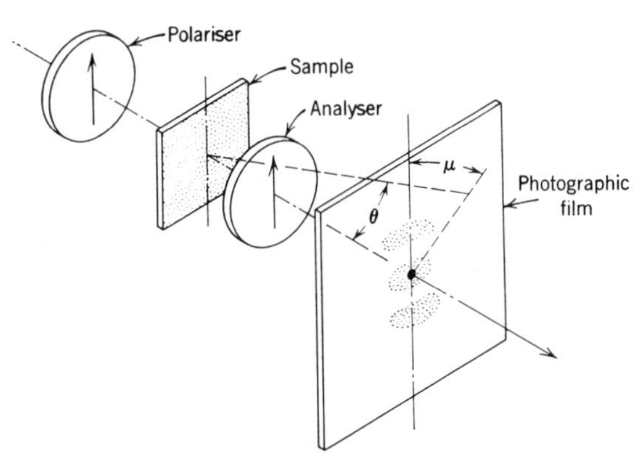

FIG. 2. Experimental arrangement for photographic small-angle light scattering experiments. (From reference 129, with permission John Wiley & Sons, Inc.)

such as that available from a small laser. Samples are thin films, often coated with silicon oil to minimise surface scattering. Scattered radiation is analysed with a second polariser aligned with the first polariser, to obtain the V_v scattering pattern, or rotated 90° from the first polariser to obtain the H_v scattering pattern. Optically anisotropic scattering sources will demonstrate an H_v pattern, whereas isotropic scatterers will not. The patterns are recorded on photographic film or by electronic position-sensitive detectors. The pattern is characterised in terms of relative scattered intensity as a function of the scattering angle θ and the azimuthal angle μ (see Fig. 2).

The interaction of light with matter results in a polarisation of the electronic charge distribution, yielding induced dipoles. For polymers, the polarisability along the chain axis is generally different from that perpendicular to the chain axis, so that a preferential arrangement of the chains will result in optical anisotropy. Spherulites are structures up to several microns in diameter which accompany polymer crystallisation under certain processing conditions. The crystallised chains are thought to lie perpendicularly to the radial direction within a spherulite.[40] The intercrystalline spaces within spherulites are occupied by the amorphous components of the polymer system. It is found that spherulites are anisotropic scatterers. If spherulites are idealised as isolated spherical inclusions with radial polarisability, α_r, different from tangential polarisability, α_t, in an isotropic matrix of polarisability, α_0, then theoretical scattering patterns can be calculated, as has been done by Stein[41] and by Samuels.[42] Figure 3 shows typical calculated H_v and V_v patterns and Fig. 4

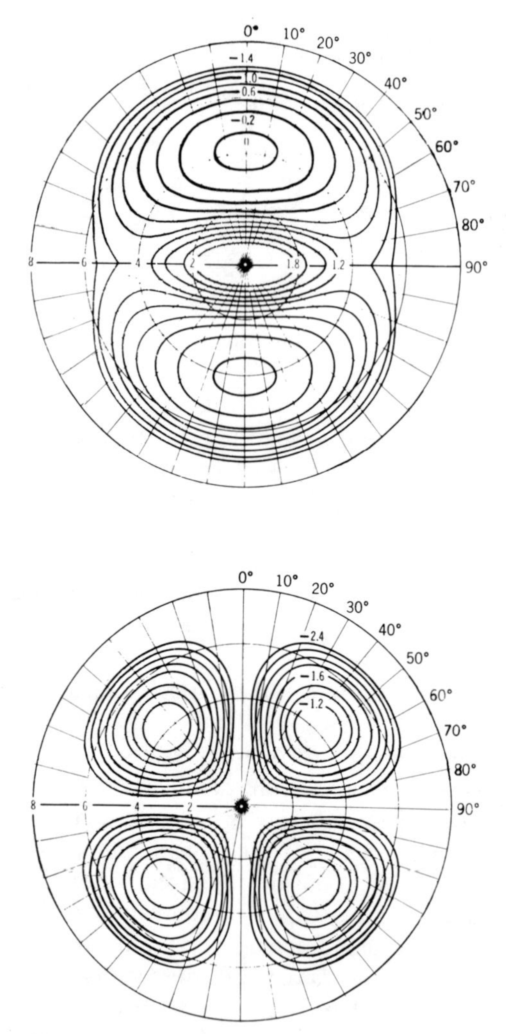

FIG. 3. Typical V_v and H_v calculated scattering patterns from isolated anisotropic spheres. The contour lines correspond to constant levels of logarithmic intensity. (From reference 129, with permission John Wiley & Sons, Inc.)

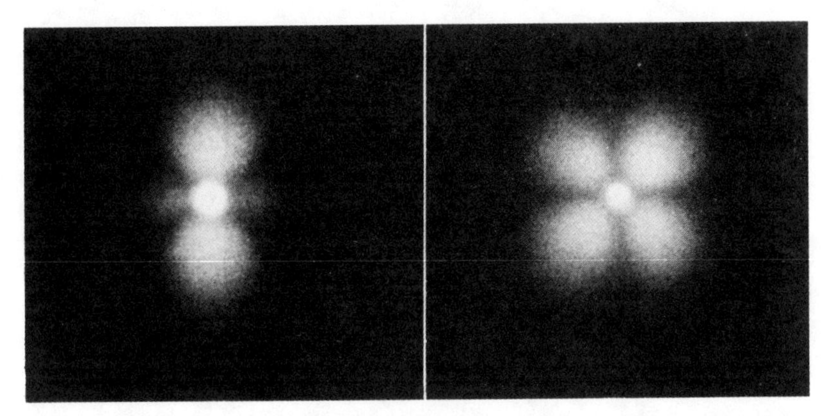

FIG. 4. Typical V_v and H_v scattering patterns for spherulitic texture (polyethylene). (From reference 129, with permission John Wiley & Sons, Inc.)

shows experimental photographs for comparison. The average spherulitic radius can be estimated as

$$R \approx \lambda/\pi \sin(\theta_m/2) \tag{20}$$

where λ is the wavelength of the light and θ_m is the scattering angle at which the intensity maximum occurs in the four symmetrical lobes of the H_v pattern.

2.6. Stress–strain Measurements

There are several forms of mechanical testing used to characterise thermoplastic elastomers, each of which is used to measure a specific set of properties: fatigue strength, toughness, abrasion resistance, etc. The most widely used mechanical test in thermoplastic elastomer morphology studies is the constant strain rate, uniaxial tensile test. In this experiment, engineering stress (force normalised by initial cross-section area) is measured as a function of elongation until rupture occurs. The characteristic measurements include initial modulus, ultimate strength, ultimate elongation and capacity for strain energy dissipation. In some cases the experiment is run in a cyclic fashion in which case strain energy hysteresis and permanent elongation are observed. Such studies yield information regarding stress transfer mechanisms in the microstructure and mechanical disruption and reorganisation of microdomains.

2.7. Miscellaneous Techniques

The technique of wide-angle X-ray diffraction (WAXD) is well-known and is detailed in several texts.[43-8] The reader is referred to these texts for a description of experimental equipment, procedures and analysis of data. In brief, WAXD can be used (1) to determine crystal structure, (2) to determine degree of crystallinity, (3) to measure the size of crystallites, (4) to determine the degree of perfection of crystallites and (5) to determine the orientation of crystallites.[49] The most common use of WAXD in studying short segment block copolymers is in determining the presence of crystallinity in hard or soft domains and in determining how the degree of crystallinity changes as the material undergoes deformation, annealing, heat setting, etc. WAXD has also been used to determine the contribution of WAXD to SAXS curves in the region where SAXS and WAXD overlap. Blackwell and Gardner have used wide-angle X-ray diffraction to determine the crystalline structure of the MDI–butanediol polyurethane which is a common hard segment component of technical block copolyurethanes.[50]

Dielectric spectroscopy is used to measure the complex permittivity of polymer films as a function of both temperature and frequency. Many transitions are observed, including glass transitions and melting.[51] With respect to dipole relaxations observed in the glassy state, calculation of associated activation energies reveals information about the configuration and surrounding environment of the relaxing species. At very low frequencies, multiphase polymers exhibit interfacial polarisations.[52] The observed relaxations involve the polarisation of space charge moving through adjacent regions of different conductivities. Such relaxations can be used to shed light on the nature of microdomain geometry.[53]

In addition to infrared dichroism, SAXS, WAXD and SALS, other techniques which can be used to determine orientation in anisotropic materials include birefringence, sonic velocity measurement, polarised Raman scattering, and polarised fluorescence.[49,54] Some of the techniques are limited because they cannot distinguish orientation due to different components in a multiphase material.

In the last decade, the technique of neutron scattering has been applied to polymers.[24] The application of neutron scattering to segmented block copolymers may shed more light on the role of the interface in phase separated microcomposites.

The application of transmission electron microscopy to the study of microphase morphology in segmented polyurethanes and similar systems is limited by the lack of phase contrast and the supposed small size of the microdomains. Efforts to enhance contrast by defocus techniques[55] have resulted in mixed success, largely due to failures on the side of the investigators to account for artifacts introduced by the microscope system. A review of the theory of phase contrast and the defocus technique for contrast enhancement by Roche and Thomas[56] demonstrates striking similarity in electron micrographs of a multiphase segmented polyurethane film and films of polystyrene and carbon. The authors noted that several previously published micrographs,[57,59] which claimed to demonstrate phase separation in polyurethanes, had features all attributable to spatially filtered noise structure, variation in the intensity of scattering as a function of scattering angle due to defocus and spherical aberration of the microscope system. Among the other difficulties is that of preparing a suitable thin film ($\simeq 100$ Å thick). In the case where such a film can be prepared and treated with selective stains or coatings to enhance contrast, the relationship between the thin film morphology and the bulk morphology is unknown.

3. MORPHOLOGY OF POLYURETHANE BLOCK POLYMERS

Segmented thermoplastic elastomers exhibit structural heterogeneity on the molecular, the domain and in some cases on a larger scale involving periodic or spherulitic texture. Each level of structural organisation is studied by specific methods. Molecular sequence distributions can be studied by chemical methods, such as NMR or infrared spectroscopy. Domain structures are studied directly by electron microscopy or more quantitatively by small-angle X-ray scattering (SAXS) methods which are particularly applicable because of the size range of typical domains. From the SAXS intensity curve important parameters are obtained which can be used to characterise the morphology of a two-phase system.

Electron microscopy,[6,57–9,60,61] on the other hand, can provide direct information on the domain structure under favourable conditions. SALS and polarised microscopy have been used in addition to electron microscopy to study spherulitic structure. These methods are complemented by DSC and by various techniques for studying dynamic mechanical behaviour which can be interpreted to give additional, if somewhat less direct information on domain structure.

Much attention has been directed toward an elucidation of domain morphology in segmented polyurethane copolymers. Direct evidence of a domain structure in polyurethanes was first suggested by Koutsky *et al.*,[57] using transmission electron microscopy (TEM). The domain structure is

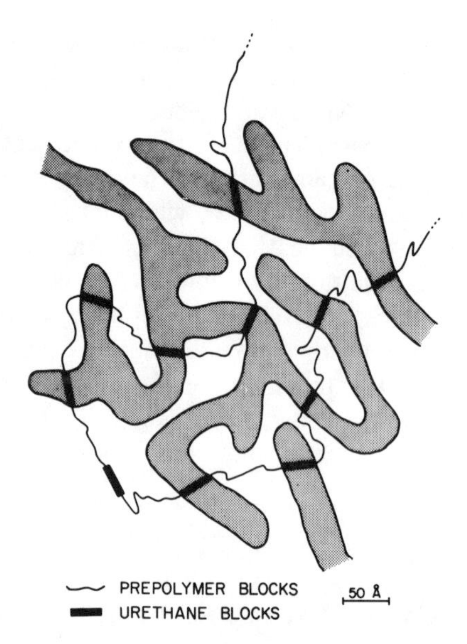

FIG. 5. Representation of domain structures in a segmented copolymer. (From reference 65.)

not as clearly observable as in most styrene–diene–styrene block copolymers because of difficulties of phase staining and the smaller domain sizes involved.

Subsequent investigations by Allport[6] and Thomas[59] also revealed a heterogeneous morphology. It cannot be assumed on the basis of the evidence, however, that complete phase separation occurs. In fact, there is evidence to suggest that appreciable hydrogen bonding occurs between hard and soft blocks,[62–4] which implies incomplete phase separation.

An illustration of the domain structure in a segmented polyurethane proposed by Estes[65] is shown in Fig. 5 in which the shaded areas are the hard domains. Both phases are represented as being continuous and interpenetrating. The model also presumes that phase separation is not complete; some urethane blocks are dispersed in the rubbery matrix.

The presence of hard and soft domains in segmented polyurethanes has also been confirmed by experimental results obtained using pulsed NMR and low-frequency dielectric measurements. Assink[66] has recently shown that the nuclear-magnetic, free-induction decay of these thermoplastic elastomers consists of a fast Gaussian component attributable to the glassy hard domains and a slow exponential component associated with the rubbery domains. Furthermore, the NMR technique can also be used to determine the relative amounts of material in each domain.

Dielectric relaxation study of two-phase microstructures in segmented copolymers was attempted by North and his co-workers.[67-70] Dielectric measurements down to 10^{-5} Hz were made on MDI-based segmented polyether- and polyesterurethanes using a Fourier transform technique.[70] These materials displayed large, low-frequency dielectric absorptions which were temperature dependent and had the characteristics of Maxwell-Wagner interfacial polarisation. The study led to the conclusion that the occluded hard domains were non-spherical with diffuse phase boundaries.

Crystallisation of either segment of thermoplastic elastomers provides a driving force towards phase separation. As well as promoting phase separation, crystallisation is often associated with the presence of a superstructure organisation which can be detected by light scattering. The morphology depends on several factors, such as the nature and concentration of the crystallisable component, casting solvent and thermal history, etc. Wilkes and Samuels[71] reported spherulitic morphology in a segmented piperazine polyurethane cast from chloroform whereas a similar series of materials investigated by Cooper *et al.*[27] showed no superstructure when using methylene chloride as solvent. Spherulites have also been observed in segmented polyurethaneureas,[72] segmented polycaprolactone-urethanes,[60] segmented poly(oxypropylene)urethanes,[59] and segmented polyetherester thermoplastic elastomers.[61,73-7] The effect of different spherulitic structures on the mechanical properties of segmented polyether-esters has been reported recently.[76,77]

3.1. Thermal Analysis Studies

The study of transition behaviour by various thermoanalytical techniques (DTA, DSC, thermal expansion measurement, and thermomechanical analysis) has been important to the understanding of morphology and intermolecular bonding in segmented copolymers. Several transitions can be observed in a DSC thermogram, depending on the nature of the solid-state structure of the sample. These include the glass transition(s) and endothermic disruptions of crystalline and paracrystalline domains. The variation of T_g of the soft matrix in segmented polyurethanes as a function of composition or segmental chemical structure is used as an indicator of the degree of microphase separation.

Typical DSC thermograms of segmented polyurethanes are shown in Fig. 6. The samples are based on MDI-4G hard segments combined with either poly(oxytetramethylene) (POTM) or poly(tetramethylene adipate)

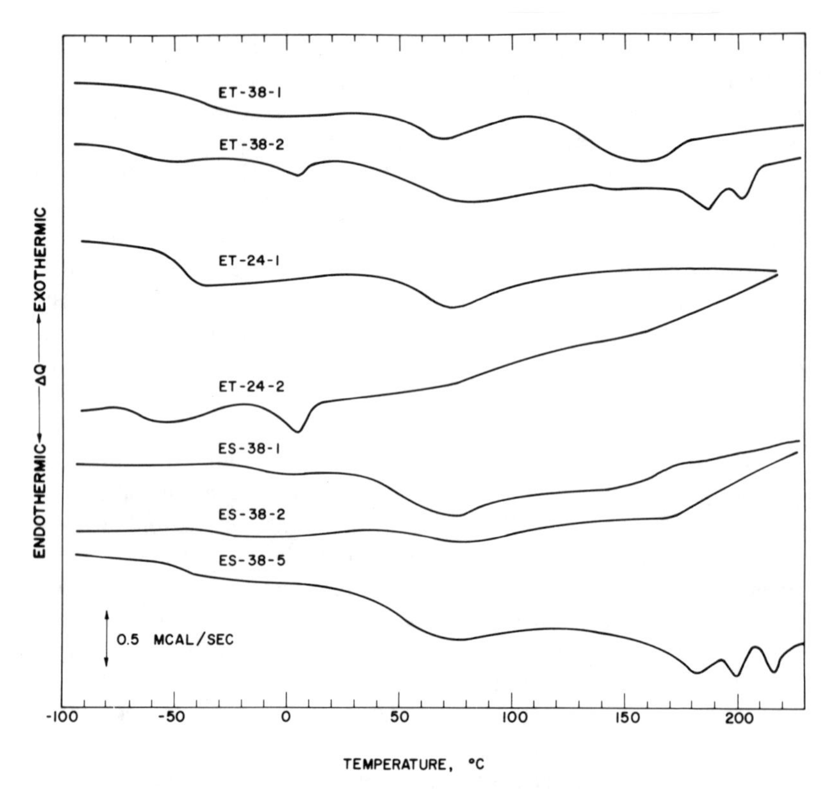

FIG. 6. DSC curves for ET-24, ET-38 and ES-38 series of segmented polyurethanes. (From reference 64, with permission Marcel Dekker, Inc.)

(PTMA) as soft segment. The nomenclature has been established whereby the two figures following ET (POTM/4G/MDI copolymer) or ES (PTMA/4G/MDI copolymer) indicate the wt % diisocyanate and the soft segment molecular weight in thousands, respectively. The shift in base line in the region of −60– −10 °C corresponds to the T_g of the soft segment. ET polyurethanes having a soft segment molecular weight of 1000 show a constant soft segment T_g at about −44 °C. In ES polyurethanes, the soft segment T_g generally shows an increase with increasing MDI content from −30 °C at 24 % to −10 °C at 38 %. Both ES and ET polymers show a decrease in soft segment T_g with increasing soft segment molecular weight. The relatively constant T_g values observed as a function of composition exhibited by the ET elastomers of 1000 M.W. soft segments indicate that the penetration of isolated hard segments into the soft phase is limited. The

ES samples, on the other hand, have a greater tendency for the hard segment units to be trapped in the soft matrix. This is probably because of the greater polarity of the polyester segment[62,64] rather than the formation of hard segment–soft segment hydrogen bonding which also may take place in polyetherurethanes.[78] Because of phase mixing in polyester-urethanes at 1000 M.W. soft segment, the T_g of the elastomeric phase increases as the hard segment content is raised.

Analogous studies on two series of segmented poly(oxytetramethylene) (M.W. = 1000) polyurethanes, having either a symmetric 2,6-toluene diisocyanate (TDI) or an asymmetric 2,4-TDI-based hard segment (butanediol chain-extended), show similar results.[79,80] The 2,6-TDI specimens, having crystalline hard domains which restrict phase separation, exhibit a soft segment T_g which is relatively independent of hard segment content. The 2,4-TDI systems, on the other hand, give soft segment T_g values which increase with increasing hard segment content indicative of considerable phase mixing allowed by the amorphous 2,4-TDI-based hard domains. Paik Sung *et al.* have thoroughly investigated segmented polyurethane–polyureas based on 2,4-toluene diisocyanate hard segment, ethylene diamine and butanediol chain extenders and poly(oxytetramethylene) or poly(butylene adipate) soft segments. It was found that the extent of phase segregation improved significantly when ethylene diamine was used instead of butanediol as the chain extender.[125,126]

The DSC spectra shown in Fig. 6 exhibit several endotherms associated with disordering processes which occur in the urethane domains.[64] Early studies[81–3] assumed that these endotherms were attributable to hydrogen-bond disruption, e.g. an endotherm of about 80 °C for dissociation of hard-soft segment hydrogen bonds and an endotherm around 150–170 °C for interurethane hydrogen bond dissociation. More recent studies have shown that the intermediate DSC transitions are not attributable to hydrogen bond dissociation.[84]

Seymour and Cooper[85,86] studied the effect of annealing temperature (up to 200 °C) and time on the thermal responses of polyester– and polyether–polyurethanes using DSC analysis. Transition behaviour depended strongly upon thermal history as shown in Fig. 7. Three characteristic endothermic transitions were observed: (I) an endotherm centred at approximately 70 °C which was attributed to the disruption of domains with limited short-range order; (II) a transition at 120–190 °C which represented the dissociation of domains containing long-range order; and (III) a transition above 200 °C which was attributed to the melting of microcrystallites of the hard segments. Schematic structures for lesser (A)

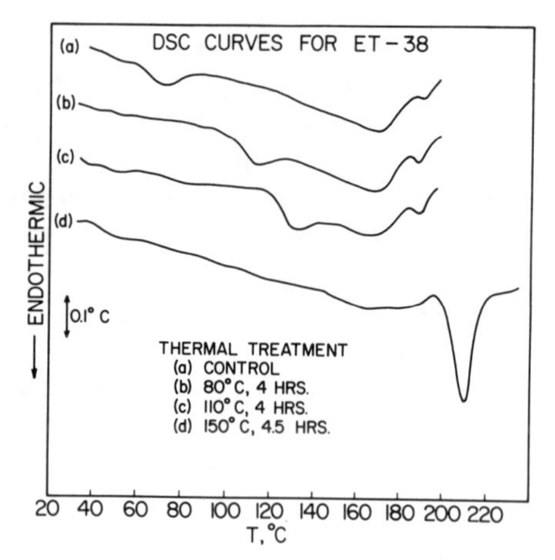

FIG. 7. Effect of annealing on the DSC curves for a MDI/4G/POTM polyurethane. (From reference 130, with permission John Wiley & Sons, Inc.)

and greater (B) ordered non-crystalline domains and a highly ordered, crystalline domain (C) are shown in Fig. 8 as proposed by Hesketh *et al.*[87] Each hard segment will have a zig-zag conformation because of the angle at the centre of the unit derived from MDI (see Fig. 9 in Chapter 5).[50] Hard segments are represented as being linear in Fig. 8, however, for simplicity. Seymour *et al.*[85,86] demonstrated that short-range ordering could be improved continuously by annealing as evidenced by a consistent upward temperature shift of endotherm (I) until its merger with endotherm (II). Greater shifts were correlated with higher annealing temperatures. Endotherm (II) could be shifted into the region of endotherm (III) by severe annealing at long times if the hard block was of a sufficient length to produce a microcrystalline domain.

Jacques[88] has studied the effect of high temperature annealing (150–250 °C) on the morphology of polyester–polyurethane elastomers. DSC analysis of semi-crystalline polyurethanes showed that annealing of control samples below 200 °C resulted at best in long-range ordered hard segment domains (endotherm at 210 °C), whereas annealing at 200 °C and above led to crystalline domains (melting at 249 °C) with an associated dramatic increase in Young's modulus (up to 300 %).

Hesketh *et al.*[87] studied the morphological changes in segmented elastomers induced by annealing and quenching using DSC. As a result of annealing, endotherms attributed to the disruption of long-range ordering in the hard segment domains were observed at a temperature slightly

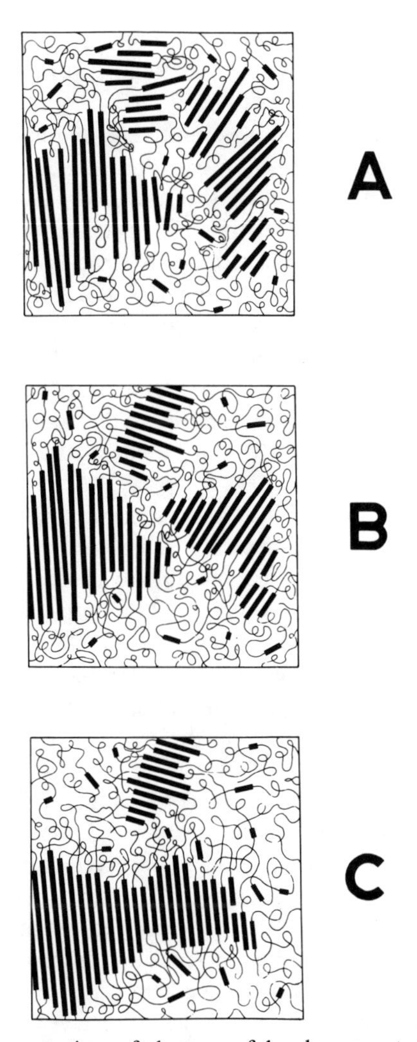

Fig. 8. Schematic representation of degrees of hard segment domain order. (Broad line, hard segment; thin line, soft segment. Distribution of hard segment lengths is based on the most probable distribution with an average length of six units.) (A) Lesser ordered, non-crystalline hard segment domains; (B) greater ordered, non-crystalline hard segment domains; (C) microcrystalline hard segment domains. (From reference 87, with permission Society of Plastics Engineers.)

(10–20 °) above that of the annealing temperature. Furthermore, these endotherms could be shifted to higher temperatures and their peak heights increased by use of higher annealing temperatures. On the other hand, the temperatures corresponding to positions of crystalline melting endotherms were independent of the annealing/quenching conditions investigated. It was also found that during storage at room temperature following sample quenching, an endotherm, in the range 50–80 °C, appeared which was observed to shift to higher temperatures and grow in size with longer aging times. This endotherm was attributed to structure developed in the material due to room temperature annealing. Hesketh *et al.*[87] suggested that during annealing, hard segments coming from aggregations which were disrupted by the annealing conditions were reformed into long-range ordered domains which dissociated at a temperature just above the annealing temperature. The size of this endotherm depended on the amount of hard segments which dissociated at temperatures below the annealing temperature.

Recently, Van Bogart *et al.* used thermal analysis to study annealing-induced ordering in segmented elastomers.[127] Twelve segmented elastomers were studied each having approximately 50 % by weight of hard segment content. Seven general classes of materials were examined including polyether– and polyester–polyurethanes, polyether–polyurethane-urea, and polyether–polyester. The materials were slowly cooled (–10 °C min^{-1}) from the melt to particular temperatures (–10 °, 20 °, 60 °, 90 °, or 120 °C) where they were annealed (16, 12, 8, 6, or 4 days, respectively). Annealing was followed by slow cooling (–10 °C min^{-1}) to –120 °C after which a DSC experiment was run. In general, annealing caused an endothermic peak at a temperature 20–50 °C above that of the temperature of annealing. This phenomenon was observed in both semi-crystalline and amorphous materials. The closer the annealing endotherm was to a crystalline endotherm without exceeding it in temperature, the larger the size of the endotherm. Annealing endotherms resulted from hard segment ordering in some cases and soft segment ordering in other cases. Only one annealing endotherm was observed for a given annealing history, even though in some materials hard and soft segments could exhibit annealing-induced morphological changes. Hard segment homopolymers were studied yielding results similar to those from block polymers containing shorter sequences of the same material. The results suggest that annealing-induced ordering is an intradomain phenomenon not associated with the interphase between domains, or necessarily dependent on the chain architecture of segmented elastomers.

3.2. Dynamic Mechanical Property Studies

Dynamic mechanical properties[89] provide information about first- and second-order transitions (T_m and T_g, respectively), phase separation and the mechanical behaviour of polymers. Below T_g the glassy state prevails with modulus values in the order of 10^{10} dyne cm^{-2} for unfilled polymers. A rapid decrease in modulus is seen as the temperature is increased through the glass transition region. A linear amorphous polymer which has not been crosslinked shows a short rubbery plateau region followed by a continued rapid drop in modulus. Crosslinking causes the modulus to remain nearly constant with increasing temperature at about two or three decades below that of the glassy state. In block copolymers such as segmented polyurethanes, an enhanced rubbery plateau region appears where the modulus changes little with increasing temperature. Another rapid drop in modulus occurs when the temperature is increased through the hard segment transition point. In contrast, a semi-crystalline polymer maintains high modulus through the glass transition region and up to the crystalline melting point where the structural integrity of the crystallites is destroyed.

In segmented polyurethanes, the low temperature transition can be related to the glass transition in the amorphous soft segment rich phase. When this phase is relatively pure in soft segments, the correlation between this storage modulus transition and the glass transition of the soft segment homopolymer is very good. The high temperature transition correlates well with the temperature at which hard segment-rich domains disrupt. Again, high purity in these domains shifts the transition to higher temperatures, approaching the softening temperature of the hard segment homopolymer. Sample composition, segmental length, inherent interseg-ment solubility and sample preparation method influence the degree of phase separation and thereby the shape and temperature location of the dynamic mechanical transition points. Phase mixing between domains is indicated by a decreased slope in storage-modulus transitions and by broadened loss peaks.

Extensive dynamic mechanical property studies have been carried out on hydrogen-bonded[90] and non-hydrogen-bonded[71,91] polyurethanes. Several secondary relaxations were found in the tan δ curve in addition to the major hard and soft segment transitions. Molecular mechanisms could be assigned to each of these. A low temperature γ transition ($\simeq -125\,^{\circ}$C) was attributed to localised motion in polyether sequences. Similar γ transitions have been found in other block copolymers.[61,73–5] In poly-urethanes with long soft segments (M.W. $\gtrsim$2000), a soft segment melting transition was detected. The T_g loss peak occurred at lower temperatures when soft segment length was increased. Since the longer segments are expected to produce better-ordered and larger domains, some soft segments can exist in regions well removed from the domain interface and hard domain interactions, so that their motion can be relatively unrestricted by the hard domains. There is also better microphase separation in

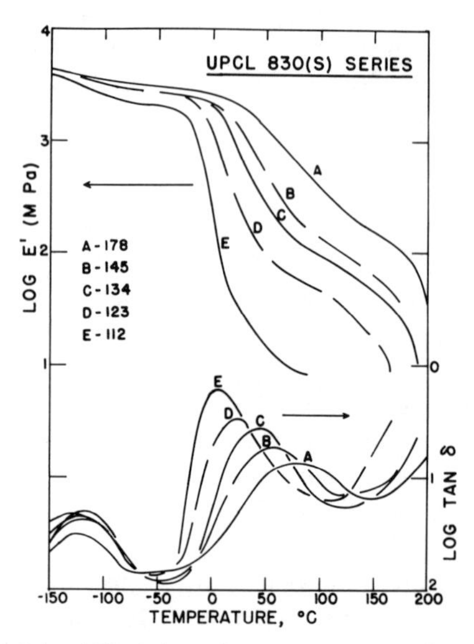

FIG. 9. Storage modulus (E') and tan δ curves for the UPCL 830 series materials. (A) 830-178; (B) 830-145; (C) 830-134; (D) 830-123; (E) 830-112. (From reference 77, with permission Marcel Dekker, Inc.)

these systems and therefore less hard segment material dissolved in the soft segment phase. Soft segment T_g values were lower in materials that were not hydrogen-bonded than in hydrogen-bonded samples with equivalent hard segment content. Presumably, this was primarily attributable to the influence of hydrogen bonding interactions between hard and soft segments. Since this bonding persists even above the hard segment T_g,[62] it is evident that the lower temperature, soft segment T_g is not accompanied by a marked disruption of these bonds.

As an example, the dynamic mechanical responses of UPCL 830 and UPCL 2000 systems are shown in Figs 9 and 10, respectively. The UPCL designation means that the hard segments are based on MDI/4G and the soft segments are polycaprolactone. For example, UPCL 830-178 has a soft segment M.W. of 830 and the molar ratios of soft segment glycol to hard segment chain extender to hard segment used are 1:7:8, respectively. The secondary transition at approximately −130 °C results from the local mode motion of the methylene sequences in the polycaprolactone soft segment. The transition between −50 and 0 °C corresponds to the soft segment glass transition β_s.

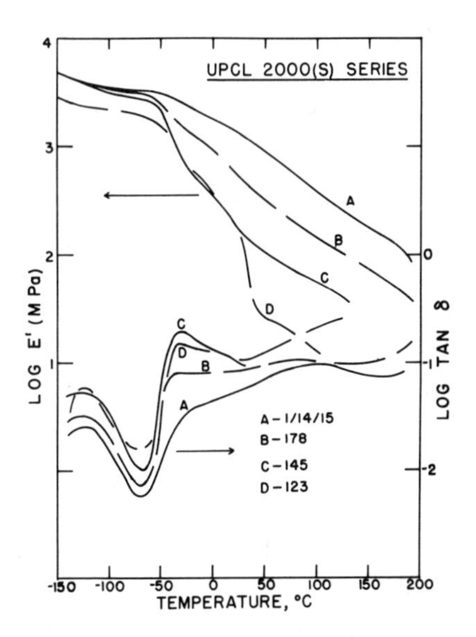

FIG. 10. Storage modulus (E') and tan δ curves for the UPCL 2000 series materials: (A) 2000-1/14/15; (B) 2000-178; (C) 2000-145; (D) 2000-123. (From reference 77, with permission Marcel Dekker, Inc.)

The UPCL 830 series shown in Fig. 9 exhibits a shift in the β_s transition to higher temperatures as the hard segment content is increased, which is characteristic of a compatible system. The UPCL 2000 series shown in Fig. 10 exhibits a relatively constant β_s while the slope of the storage modulus curve (E') at temperatures above the β_s transition temperature, increases with increasing hard segment content. This is a characteristic of an incompatible system whereby the hard segment domains behave as filler particles and act as reinforcement. These findings are consistent with earlier dynamic mechanical data reported by Koleske[92–5] on similar PCL/MDI/4G segmented copolymer systems.

The E' curve for sample UPCL 2000-123 exhibits a small plateau after the β_s transition, followed by a sharp drop in modulus at approximately 40 °C. The polycaprolactone soft segments in this system are partially crystalline and help to reinforce the material above the soft segment T_g until they melt at approximately 40 °C.

Koleske *et al.*[92–5] made a number of important observations concerning the influence of segment length and structure on morphology and dynamic mechanical properties for a series of polycaprolactone-based polyurethanes with MDI/4G and TDI/4G hard segments. The shift in the soft segment domain T_g with varying hard segment content was found to be related directly to the extent of phase separation.[94] At low soft segment M.W.'s ($\simeq$830), T_g increased with increased hard segment content,[92] indicating segment mixing in the soft matrix phase. At a fixed hard segment length, an increasing soft segment length (M.W. $\gtrsim$830) led to a lower T_g.[95] Hard segment symmetry and composition was found to be a major factor in determining the extent of phase segregation.[93,94] The extent of domain segregation in thermoplastic urethane elastomers was concluded to have a dominant effect on resultant mechanical properties.[93]

3.3 Stress–Strain Studies

Smith[96–8] has studied the relationship between ultimate properties and morphology in segmented copolymers. In general, the behaviour of a strained system depends on the size and concentration of the hard segment domains,[99–101] the resistance of hard segment domains to plastic deformation, the ability of the segments to orient in the direction of stretch and the ability of the soft segments to crystallise under strain.

Studies of two polyurethanes (POTM/MDI/4G) of approximately 50 % by weight hard segment content each demonstrate the effect of differences in domain size. At low strains, infrared dichroism[31,102] showed that hard segments in ET-38-2 (2000 M.W. soft segment) orient transversely to the direction of stress. Parallel orientation is achieved at higher strains. Furthermore, at a fixed strain, reorientation of the segments is time-dependent. For ET-38-1 having block lengths which are half that of ET-38-2, however, orientation of the hard segments is always parallel to the stretch direction and much less relaxation is observed. Because of the shorter length, hard segment domains for ET-38-1 are non-crystalline, whereas those for ET-38-2 are crystalline, which results in the differences observed. ET-38-1 domains can be deformed readily and disrupted quickly, allowing rapid relaxation and parallel orientation unlike the crystalline ET-38-2 which tends to oppose domain disruption and reorientation.

Stress–strain curves[98] for ET-38-2 are higher than those for ET-38-1 by virtue of a greater degree of phase separation in the former. Furthermore, the moduli at a fixed strain for these samples decrease with increasing temperature, a characteristic attributed to their thermoplastic nature. Samples with a urea-urethane hard segment (based on 4,4′-methylene-bis-(2-chloroaniline) and capable of forming rigid, highly crystalline hard segments) exhibit stress–strain curves which are basically independent of temperature up to 150 °C.

Hysteresis is prominent in segmented copolymers because of the disruption of hard segments at high strain.[103,104] Unlike Young's modulus which depends on the rigidity and morphology of hard segment domains, elastoplastic behaviour at high strains is a function of domain restructuring and ductility and the nature of the mixed hard and soft segment interfacial regions.[98,105] In segmented polyurethanes, hard segment crystallisation has been found to increase the hysteresis, permanent set and tensile strength. Heat build-up in polyurethanes attributable to their high hysteresis losses has limited their suitability in applications such as high-speed tyres.

Smith[98] studied both strain and true stress at break as a function of temperature for four segmented elastomers (all with POTM soft segments). Comparing ET-38-1 and ET-38-2, the former was superior in strength and elongation. The smaller, more numerous domains in ET-38-1 were apparently more efficient at stopping catastrophic crack growth than those for ET-38-2 (both had approximately the same hard segment content), resulting in a larger tensile strength. Furthermore, the more readily deformable domains in ET-38-1 permitted higher elongation which increased with increasing temperature before fracture. ET-24-2 exhibited a drastic change in properties near 40 °C. Below 40 °C, the soft segment of M.W. 2000 partially crystallised and strengthened the specimen.[96] Above 40 °C, however, the soft segments were no longer crystalline, and the hard segment domains could not retard crack growth because of insufficient cohesive strength at these temperatures. If another hard segment characterised by high cohesive energy was used, however, as in the case of the urea-urethane specimen (similar to ET-24-2 except for the nature of the hard segment), higher tensile strengths were observed. The urea-urethane sample was unique in that elongation at break and tensile strength began to improve above 120 °C. Smith[98] offered the explanation that the hard segment domains become more deformable above this temperature, permitting reorganisation into a more fibrous structure than is possible near 100 °C.

3.4. Studies of Microphase Separation Kinetics

Recent work[87,106–10] has revealed that the morphology of segmented copolymers following thermal treatment is time-dependent. Raising the temperature of a polymer system induces phase mixing. Subsequent cooling causes phase separation, giving the original morphology. However, because of kinetic and viscous effects, a finite amount of time is required to produce a given change in morphology.

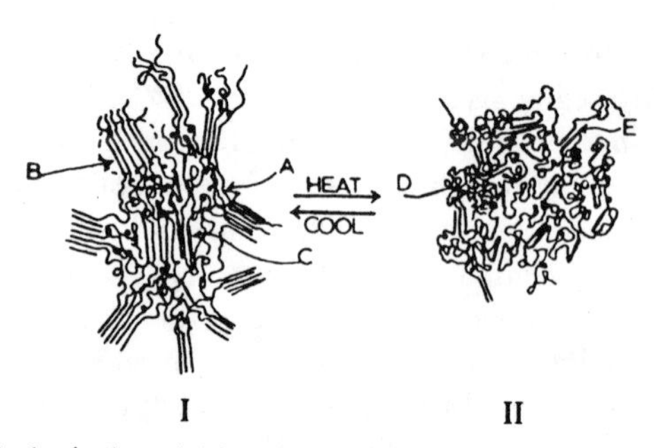

FIG. 11. Morphological model for phase mixing/demixing. (A) Partially extended soft segment; (B) Hard segment domain; (C) hard segment; (D) relaxed soft segment; (E) lower order hard segment domain. (From reference 108, with permission American Institute of Physics.)

Wilkes *et al.*[106–10] have investigated the kinetics of domain formation in segmented polyurethanes. Samples which were heated briefly (5 min) at high temperature (170 °C) and then quenched exhibited changes in soft segment glass transition temperature, a degree of phase separation (as evidenced by the SAXS intensity at a fixed angle) and Young's modulus as a function of time after quench. To explain this phenomenon, Wilkes[106,107] has proposed the following model, shown schematically in Fig. 11. Initially the well-aged segmented elastomer consists of two phases, (1) hard domains which are reasonably pure and (2) soft domains comprised of soft segments in a partially extended conformation plus some trapped hard segments—particularly those hard segments with the shortest block lengths. Upon heating, the extended soft segments contract or relax, causing additional hard segments to be pulled out of the hard domains and phase mixing to occur. Subsequent rapid cooling creates a thermodynamic driving force for the phases to separate again, but the high viscosity requires that the demixing (and domain purification) occur slowly over a period of time.

Hesketh *et al.*[87] have performed similar time-dependent T_g studies on a number of segmented copolymers, the samples here being annealed for 4 h at various temperatures (120 °, 150 °, 170 °, or 190 °C). All the samples had similar hard segment contents ($\simeq 50\%$ by weight) and a POTM soft segment of M.W. 1000, and differed only in hard segment composition. DSC measurements taken immediately after quenching from the annealing temperature showed that the soft segment T_g was higher than that of the unannealed control material. The displacement from equilibrium at a given time was greater for higher annealing temperatures. Thus it was suggested that the applied thermal history promoted increased mixing of the hard and the soft segments. It was further noted[87] that the shift in T_g was smaller for those materials having crystalline hard segments. As time passed following quenching, the soft segment T_g values decreased and approached an 'equilibrium' value.

Ophir *et al.*[110] have also studied a series of polyesterurethanes (PTMA/MDI/4G) crosslinked to various degrees by reaction with a peroxide. For well-aged samples at room temperature, the samples with lower degrees of crosslinking showed better phase separation as revealed by higher intensity SAXS curves. Upon thermal treatment and quenching to room temperature, the greater displacements from steady state (as monitored by SAXS intensity at a fixed angle and storage modulus) were exhibited by the more lightly crosslinked specimens. These samples also came to a steady state in a much shorter time than the more highly crosslinked systems. As well as inhibiting phase separation[104,111] and lowering the rubbery modulus, crosslinking restricts the transient response of the morphology to temperature change.

3.5. Small-angle X-ray Studies

Microphase separation in segmented polyurethane copolymers was first detected by small-angle X-ray scattering by Clough *et al.*[112] Polyether- and polyester–polyurethanes having an MDI/4G hard segment were studied. On the basis of densitometer measurements of SAXS photographs, the intensity of sample scattering was correlated with the size of the endotherm indicated by DSC at about 150 °C. These results were consistent with an earlier interpretation[81] that the designated endotherm was due to the disruption of urethane domains. Comparison of scattering intensities also led to two important conclusions: (1) microphase separation is greater for polyether-based polyurethanes than for the equivalent polyester–polyurethanes, and (2) separation into domains is enhanced by longer segment lengths.

Bonart[113] has made more quantitative X-ray studies on the structure of domains. Polyether- and mixed-polyester–polyurethaneurea copolymers

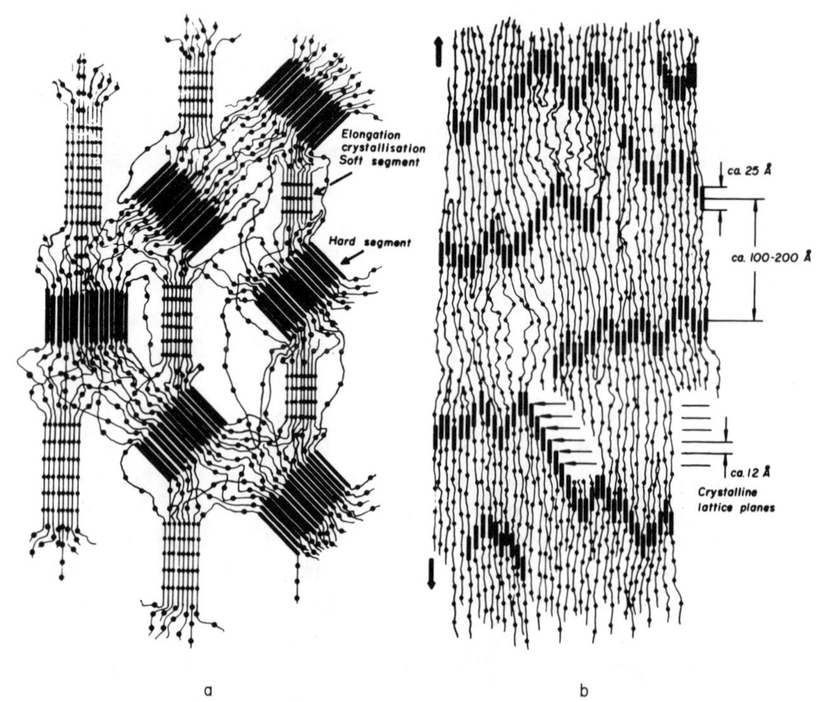

FIG. 12. Morphological model for extended polyurethane elastomer domain structure. (a) Structure at 200% elongation; (b) structure of material annealed at 500% elongation. (From reference 113, with permission Marcel Dekker, Inc.)

were studied. Wide-angle X-ray diffraction (WAXD) of unstretched samples showed amorphous halos which were attributed to the distances between adjacent hard segments in paracrystalline (non-crystalline, but ordered) regions and to the distance by which adjacent hard segments are staggered (see Fig. 12), respectively. Staggering of the hard segments occurs in such a way as to produce a maximum number of hydrogen bonds between the N—H and C=O groups of adjacent segments. SAXS showed a diffuse interference corresponding to the spacing between hard segment domains. Upon stretching, the polyester samples underwent stress-crystallisation above 150% elongation as indicated by the presence of WAXD reflections. Crystalline soft segments were oriented in the stretch direction as indicated by an equatorial interference pattern. Elongations up to 500% showed no appreciable improvement in orientation. The above was also true for the polyether-based samples, except that they exhibited paracrystallinity (diffuse WAXD reflections) instead of true crystallinity.

More complex orientation behaviour was observed for the hard segments as indicated by shifts in one of the reflections, that related to the distance between adjacent hard segment units, with elongation. At elongations of about 200% the reflection was positioned on the equator of the WAXD patterns indicating that the hard segments were oriented transversely to the direction of stretch. At 500% elongation, however, this interference shifted to a meridional position, indicating that the hard segments had reoriented into the stretch direction. Annealing at 500% elongation led to disruption of the soft segment ordering while the hard segments remained oriented. SAXS showed an isotropic amorphous ring for the unstretched samples. At 100% elongation, a four-point pattern was observed. At 500% elongation, the scattering intensities were greatest at meridional positions.[113]

Bonart[113] proposed the model shown in Fig. 12 to explain the observed phenomenon. At an elongation of 200% (Fig. 12(a)), soft segments stress-crystallise to produce 'force strands'. These force strands apply growing shear stresses to hard segments more and more oriented in the direction of elongation. This restructuring of the hard segments causes a homogenisation of soft segment tensions as well. Upon annealing, this process is accelerated, producing the structure shown in Fig. 12(b). This morphology exhibits hard segment orientations in the direction of elongation and stress-relaxed, disoriented soft segments. The restructuring of the hard segments was believed to explain the stress hysteresis phenomenon characteristics of segmented polyurethanes.[113]

In a subsequent paper, Bonart *et al.*[114] studied differences between paracrystalline and crystalline polyurethanes of identical composition and structure. The samples were polyester–polyurethanes having an MDI/4G hard segment. A sample composed of 120 Å long hard blocks could exhibit either paracrystallinity or crystallinity of the hard segment with adequate heat treatment. SAXS results showed a greater scattering intensity for the crystalline sample, which was attributed to its better segregation of phases. Consistent with this finding, DSC analysis indicated a greater degree of hard segment aggregation for the crystalline sample which had a 230 °C endotherm as compared to 170 °C for the paracrystalline one. Other SAXS analysis showed increased scattering intensity with increasing hard segment content. WAXD of the crystalline sample showed that, upon stretching, most crystalline domains were disrupted to form paracrystalline structures with orientation parallel to elongation. At 400% elongation, however, some residual crystalline domains existed with a transverse orientation. Bonart concluded that crystallised hard segments participate in the

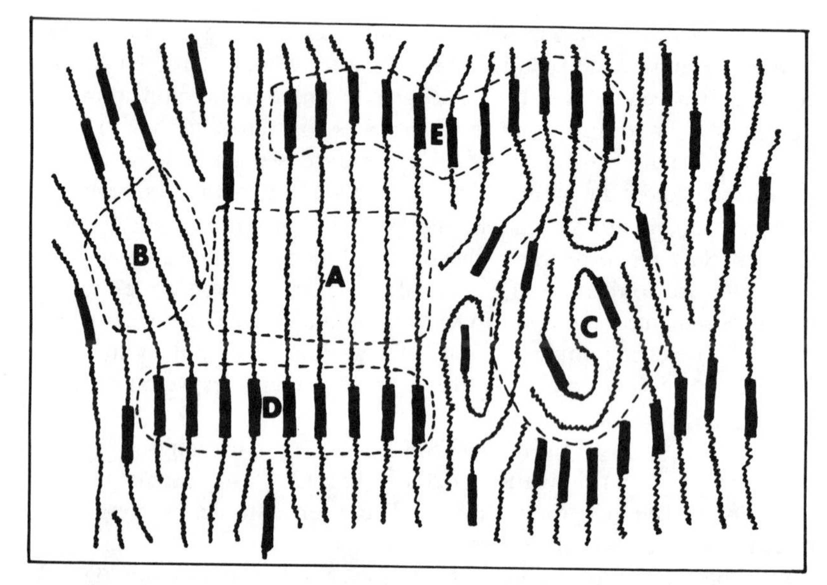

FIG. 13. Morphological model of segmented polyurethanes. (A) Stress-crystallised soft segment; (B) paracrystalline soft segment; (C) amorphous 'solution' of hard and soft segments; (D) crystalline hard segment domains; (E) paracrystalline hard segment domains. (From reference 115, with permission Marcel Dekker, Inc.)

elongation mechanism only very indirectly. Bonart did not deal explicitly with the structural differences between paracrystalline and crystalline domain structures.

Wilkes and Yusek[115] studied domain formation in polyester–polyurethanes by means of SAXS and WAXD. They concluded that the domains are lamellar in shape with an average separation of 100–250 Å. Higher spacings were given by the samples with longer soft segments and higher urethane contents. The hard segments are believed to serve as crosslinks, preventing relaxation upon stress and inducing stress-crystallisation of the soft segments, giving a higher tensile strength. Upon heat setting, the hard segments break up and reform,[113] relieving stress while maintaining considerable hard segment orientation. Wilkes and Yusek[115] presented the morphological model shown in Fig. 13 for a segmented polyurethane which has been stretched 200–400 % and annealed at 60–100 °C. It represents a summary of the morphological features determined from structural studies conducted before 1973.

Bonart *et al.*[22,116] studied the state of segregation in polyurethane

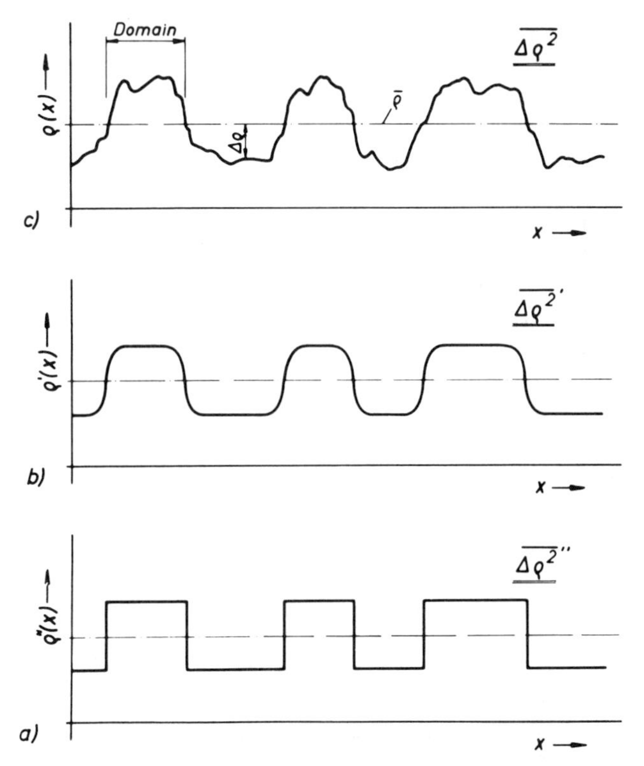

FIG. 14. Schematic electron density distributions for segmented polyurethanes. (a) idealised system $\rho''(x)$ with sharp boundaries between phases; (b) system $\rho'(x)$ with diffuse interfaces and no statistical density fluctuations; (c) real system $\rho(x)$ with statistical density fluctuations and diffuse boundaries. (From reference 22, with permission Marcel Dekker, Inc.)

elastomers as characterised by (1) the size distribution of domains, (2) the diffuseness of the phase boundaries, and (3) the degree to which hard segments dissolve in the soft segment matrix, and vice versa. Bonart *et al.*[116] described theoretical methods for obtaining quantitative information about domain morphologies from SAXS data based on Porod's theories[117–19] of high-angle scattering limits. Bonart claims that domain boundaries are more or less diffuse as consequences of varying hard segment lengths and longitudinal segmental shifts (see Fig. 12). A method was proposed to determine boundary diffuseness on the basis of the model shown in Fig. 14 giving spatial electron density distributions along a line within the sample. A comparison of the mean square variances in electron density for the real system (Fig. 14(b)) corrected for statistical density fluctuations with a model for ideal segregation (Fig. 14(c)) indicates the degree of interfacial mixing.

Application of the theory discussed above to experimental SAXS data was also carried out by Bonart *et al.*[22] Polyether- and polyester–polyurethanes having approximately equal soft segment M.W.'s and hard segment contents (15 wt %) were studied. The scattering intensity for the material containing polyester segments was approximately half that with polyether, implying better phase separation in the latter. Analysis of the data, however, showed that there were no distinct differences in the degree of segregation within the samples assuming that excess MDI/4G units dissolved in the soft segment matrix. The results indicated that there were no preferential affinities of the hard segment for either soft segment. The width of the transition zone between domains was calculated to be of the order of 20 Å corresponding qualitatively to the length of a hard segment structural unit.

Bonart *et al.*[22] used WAXD and SAXS to study the role of hydrogen bonding in hard segment domain organisation. For MDI/4G hard segments, the theoretical arrangement satisfying all hydrogen bonds (lattice planes separated by 9·2 Å at an angle of 30° to the segments) proved to be the arrangement observed in reality by WAXD.

A recent study of polyether- and polyester–polyurethanes by Ophir[120] showed that those with polyester soft segments are characterised by a greater degree of phase mixing at the interface between domains than with polyethers. Mixing at the interface in a POTM/MDI/4G polyether–polyurethane and a PTMA/MDI/4G polyester–polyurethane accounted for 18 % and 31 % reductions in the degree of phase separation, respectively, relative to a material with no interfacial mixing. Values for the interfacial thickness, assuming a linear density gradient model, were given as 5 Å and 10 Å for the polyether- and polyester-containing materials, respectively.

Chang and Wilkes[121] have studied the effects of soft segment length and structure on the domain morphology of polyether–polyurethanes with a urethane-urea hard segment. Samples of poly(oxyethylene) (POE) polyurethanes were varied in soft segment M.W. (200–4000) while maintaining a relatively constant hard segment length. WAXD curves showed that crystallisation of the soft segment occurred in the sample containing the highest M.W. POE units. A corresponding DSC endotherm was present in this sample. SAXS curves showed the presence of domains in the crystalline sample exhibiting a Bragg spacing of 180 Å. SAXS curve invariants showed

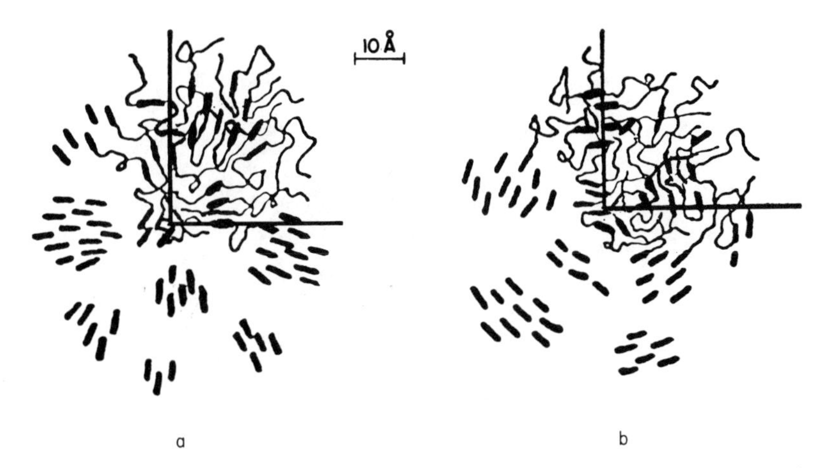

FIG. 15. Morphological model for the superstructure of semi-crystalline polyurethane copolymers. (a) Radial hard segment orientation; (b) tangential hard segment orientation: straight line, hard segment; wavy line, soft segment. (From reference 121, with permission John Wiley & Sons, Inc.)

a systematic loss in value with increasing soft segment length due to increased interfacial mixing. Investigations on poly(oxypropylene) (POP) polyurethanes led to similar conclusions. However, the preferential ordering of hard segments in the POP-containing samples was greater than in the corresponding POE samples. These observations were rationalised on the basis of polymer incompatibility. Apparently, the lower affinity of the hard segments for POP led to enhanced phase separation.

Based on electron microscopy findings, Chang and Wilkes[121] proposed a model for the superstructure of non-crystalline segmented polyurethanes as shown in Fig. 15. The paracrystalline hard segment domains are thought to have a disc-like shape with an average thickness of 10–20 Å (from WAXD studies), and a diameter of 100–150 Å (from SAXS studies) was suggested.[120]

Studies of the superstructure in polyether–polyurethanes using SAXS, WAXD and electron microscopy have also been made by Schneider *et al.*[58] Qualitative agreement was seen between the calculated hard segment domain spacing and experimental SAXS results. The authors suggested a radial orientation of hard segments, as shown in Fig. 15(a), in concentric pseudo-spherical lamellae. More recent electron microscopy studies by Fridman and Thomas,[59] however, suggested a tangential orientation of hard segments in a spherulitic lamellar structure as shown in Fig. 16.

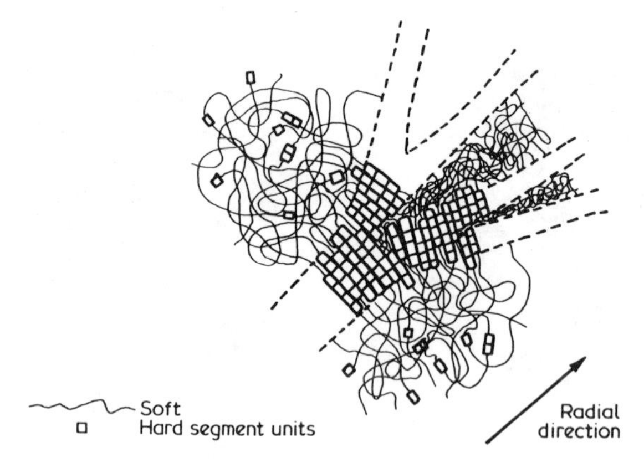

FIG. 16. Morphological model for the structure of hard segment domains in semi-crystalline polyurethane copolymers. (Reproduced from Fridman, I. D. and Thomas, E. L., *Polymer*, 1981, **22**, 333; by permission of the publishers, IPC Business Press, Ltd. ©)

The recent work of Van Bogart *et al.* illustrates the use of a variety of techniques to study structure–property relationships of segmented co-polymers.[128] Polycaprolactone-based segmented polyurethanes were studied using DSC, SAXS, WAXD, dynamic mechanical testing and stress–strain testing. The materials studied varied in hard segment type (diphenylmethane-4,4′-diisocyanate/butanediol [MDI/4G] or di-cyclohexylmethane-4,4′-diisocyanate/butanediol [H_{12}MDI/4G]), soft segment M.W. (830 or 2000, polycaprolactone), hard segment content (23–77 wt %) and thermal history. On the basis of DSC and WAXD results, it was concluded that the materials with aromatic (MDI/4G) hard segments had semi-crystalline hard segment domains while the materials with cycloaliphatic (H_{12}MDI/4G) hard segments had mostly amorphous domains. Materials with the shorter polycaprolactone soft segment (830 M.W.) exhibited thermal and mechanical behaviour which indicated hard and soft segment compatibility, while the materials with 2000 M.W. polycaprolactone soft segment exhibited thermal and mechanical behaviour indicative of microphase separation. SAXS results revealed both systems to have a well-defined two-phase microstructure, however. Using SAXS, the morphology of the hard segment domains in the MDI/4G extended polyurethanes was found to be that of cylindrical or parallelepiped-shaped hard segment crystallites roughly 40–60 Å in length composed of chain folded MDI/4G segments surrounded by amorphous hard segments. The H_{12}MDI/4G-based materials had spherical hard segment domains 30–80 Å in size containing coiled, amorphous H_{12}MDI/4G segments. Analysis of the SAXS data showed that interfacial thicknesses in the (MDI/4G)-based and (H_{12}MDI/4G)-based materials were in the range 20–36 Å and 8–14 Å, respectively. Hard domains in the (H_{12}MDI/4G)-based materials were smaller than those in corresponding MDI/4G materials. Greater values for ultimate strength were found for the smaller domain (H_{12}MDI/4G)-based materials.

REFERENCES

1. TRELOAR, L. R. G., *Rubb. Chem. Technol.*, 1974, **47**(3), 625.
2. KRAUSE, S. and REISMILLER, P. A., *J. Polym. Sci., A-2*, 1975, **13**, 1975.
3. HELFAND, E. and WASSERMAN, Z., *Polym. Eng. Sci.*, 1977, **17**, 582.
4. MEIER, D. J., *J. Appl. Polym. Symp.*, 1974, **24**, 67.
5. LEGRAND, A. D., VITALE, G. G. and LEGRAND, D. G., *Polym. Eng. Sci.*, 1977, **17**, 582.
6. ALLPORT, D. C. and JANES, W. H. (Eds.), *Block Copolymers*, Wiley, New York, 1973.
7. NOSHAY, A. and MCGRATH, J. E. (Eds.), *Block Copolymers, Overview and Critical Survey*, Academic Press, New York, 1977.
8. SAUNDERS, J. H. and FRISCH, K. C., *Polyurethanes, Chemistry and Technology, Part I, Chemistry*, Interscience, New York, 1962.
9. SCHOLLENBERGER, C. S., US Patent 2 871 218, 1955.
10. CARVEY, R. M. and WITENHAFER, D. E., British Patent 1 087 743, 1965.
11. PEEBLES, L. H., JR., *Macromolecules*, 1974, **7**, 872.
12. PEEBLES, L. H., JR., *Macromolecules*, 1976, **9**, 58.
13. HARRELL, L. L., JR., *Macromolecules*, 1969, **2**, 607.
14. WITSIEPE, W. K., *Advan. Chem.*, 1973, **129**, 39.
15. HOESCHELE, G. K. and WITSIEPE, W. K., *Angew. Makromol. Chem.*, 1973, **29**/**30**, 267.
16. HOESCHELE, G. K., *Polym. Eng. Sci.*, 1974, **14**, 848.
17. WOLFE, J. R., JR., *Rubb. Chem. Technol.*, 1977, **50**, 688.
18. WOLFE, J. R., JR., *Polym. Preprints*, 1978, **19**(1).
19. GUINIER, A. and FOURNET, G., *Small-angle Scattering of X-rays*, Wiley, New York, 1955.
20. RULAND, W., *J. Appl. Cryst.*, 1971, **4**, 70.
21. KOBERSTEIN, J. T., MORRA, B. and STEIN, R. S., *J. Appl. Cryst.*, 1980, **13**, 34.
22. BONART, R. and MÜLLER, E. H., *J. Macromol. Sci.-Phys.*, 1974, **B10**(1), 177.
23. HELFAND, E., *Accounts of Chemical Research*, 1975, **8**, 295.
24. HIGGINS, J. S. and STEIN, R. S., *J. Appl. Cryst.*, 1978, **11**, 346.
25. KAKUDO, M. and KASAI, N., *X-ray Diffraction by Polymers*, Elsevier, New York, 1972.
26. SCHELTEN, J. and HENDRICKS, R. W., *J. Appl. Cryst.*, 1978, **11**, 297.

27. ALLEGREZZA, A. E., JR., SEYMOUR, R. W., NG, H. N. and COOPER, S. L., *Polymer*, 1974, **15**, 433.
28. ZBINDEN, R., *Infrared Spectroscopy of High Polymers*, Academic Press, New York, 1964.
29. FRASER, R. D. B., *J. Chem. Phys.*, 1953, **21**, 1511.
30. BELLAMY, L. J., *The Infrared Spectra of Complex Molecules*, Methuen, London, 1962.
31. SEYMOUR, R. W., ALLEGREZZA, A. E., JR. and COOPER, S. L., *Macromolecules*, 1973, **6**(6), 896.
32. SAMUELS, R. J., *Structured Polymer Properties*, Wiley, New York, 1974.
33. SEYMOUR, R. W., ESTES, G. M. and COOPER, S. L., *Macromolecules*, 1975, **8**, 68.
34. READ, B. E., HUGHES, D. A., BARNES, D. C. and DRURY, F. W. M., *Polymer*, 1972, **13**, 485.
35. SIESLER, H. W. and HOLLAND-MORITZ, K., *Infrared and Raman Spectoscopy of Polymers*, Marcel Dekker, New York, 1980.
36. NIELSEN, L. E., *Rev. Sci. Instr.*, 1951, **22**, 690.
37. FOERSTER, F., *Z. Metallk.*, 1955, **96**, 297.
38. TOBOLSKY, A. V., *Properties and Structure of Polymers*, Wiley, New York, 1960.
39. FERRY, J. D., *Viscoelastic Properties of Polymers*, 3rd ed., Wiley, New York, 1980.
40. BASSET, S. C., *Principles of Polymer Morphology*, Cambridge University Press, Cambridge, 1981.
41. STEIN, R. S. and RHODES, M. B., *J. Appl. Phys.*, 1960, **31**, 1873.
42. SAMUELS, R. J., *J. Polym. Sci., A-2*, 1971, **9**, 2165.
43. ALEXANDER, L. E., *X-ray Diffraction Methods in Polymer Science*, Wiley, New York, 1969.
44. KLUG, H. P. and ALEXANDER, L. E., *X-ray Diffraction Procedures for Polycrystalline and Amorphous Materials*, Wiley, New York, 1974.
45. GUINIER, A., *X-ray Diffraction*, W. H. Freeman, San Francisco, 1963.
46. KAELBLE, E. F. (Ed.), *Handbook of X-rays*, McGraw-Hill, New York, 1967.
47. HOSEMANN, R. and BAGCHI, S. N., *Direct Analysis of Diffraction by Matter*, North Holland Publishing, Amsterdam, 1962.
48. TADOKORO, H., *Structure of Crystalline Polymers*, Wiley-Interscience, New York, 1979.
49. WILKES, G. L., *J. Macromol. Sci.-Revs. Macromol. Chem.*, 1974, **C10**(2), 149–261.
50. BLACKWELL, J. and GARDNER, K. H., *Polymer*, 1979, **20**, 13–17.
51. HILL, N., VAUGHAN, W. E., PRICE, A. H. and DAVIES, M., *Dielectric Properties and Molecular Behavior*, Van Nostrand, New York, 1969.
52. VALLANCE, M. A., FAITH, D. C., III and COOPER, S. L., *Rev. Sci. Instrum.*, 1980, **51**, 1338.
53. VAN BEEK, L. K. H., *Prog. in Dielec.*, 1967, **7**, 69.
54. STEIN, R. S. and WILKES, G. L., in: *Structure and Properties of Oriented Polymers*, ed. I. M. Ward, Halsted Press, New York, 1975.
55. MILES, M. J. and PETERMAN, J., *J. Macromol. Sci.-Phys.*, 1979, **B16**, 243.
56. ROCHE, E. J. and THOMAS, E. L., *Polymer*, 1981, **22**, 333.

57. KOUTSKY, J. A., HIEN, N. V. and COOPER, S. L., *J. Polym. Sci.*, *B*, 1970, **8**, 353.
58. SCHNEIDER, N. S., DESPER, C. R., ILLINGER, J. L., KING, A. O. and BARR, D., *J. Macromol. Sci.-Phys.*, 1975, **B11**(4), 527.
59. FRIDMAN, I. D. and THOMAS, E. L., *Polymer*, 1980, **21**, 388.
60. CHANG, A. L. and THOMAS, E. L., *Adv. Chem. Series*, 1979, **176**, 31.
61. SHEN, M., MEHRA, U., NIEINOMI, M., KOBERSTEIN, J. K. and COOPER, S. L., *J. Appl. Phys.*, 1974, **45**(10), 4182.
62. SEYMOUR, R. W., ESTES, G. M. and COOPER, S. L., *Macromolecules*, 1970, **3**, 579.
63. TANAKA, T., YOKOYAMA, T. and YAMAGUCHI, Y., *J. Polym. Sci.*, *A-1*, 1968, **6**, 2137.
64. SRICHATRAPIMUK, V. W. and COOPER, S. L., *J. Macromol. Sci.-Phys.*, 1978, **B15**(2), 267.
65. ESTES, G. M., *Ph.D. Thesis*, Department of Chemical Engineering, University of Wisconsin, 1971.
66. ASSINK, R. A., *J. Polym. Sci.*, 1977, **15**, 59.
67. NORTH, A. M., *J. Polym. Sci.*, *C*, 1975, **50**, 345.
68. NORTH, A. M., REID, J. C. and SHORTALL, J. B., *Europ. Polym. J.*, 1969, **5**, 565.
69. NORTH, A. M. and REID, J. C., *Europ. Polym. J.*, 1972, **8**, 1129.
70. DEV, S. B., NORTH, A. M. and REID, J. C., in: *Dielectric Properties of Polymers*, ed. F. E. Karasz, Plenum, New York, 1972.
71. SAMUELS, S. L. and WILKES, G. L., *J. Polym. Sci.*, *C*, 1973, **43**, 149.
72. KIMURA, I., IGHIHARA, H., ONO, H., YOSHIHARA, N., NOMURA, S. and KAWAI, H., *Macromolecules*, 1974, **7**, 355.
73. SEYMOUR, R. W., OVERTON, J. R. and CORLEY, L. S., *Macromolecules*, 1975, **8**, 331.
74. LILAONITKUL, A., WEST, J. C. and COOPER, S. L., *J. Macromol. Sci.-Phys.*, 1976, **B12**(4), 563.
75. LILAONITKUL, A. and COOPER, S. L., *Rubb. Chem. Technol.*, 1977, **50**, 1.
76. LILAONITKUL, A., ESTES, G. M. and COOPER, S. L., *Polym. Preprints*, 1977, **18**(2), 500.
77. VAN BOGART, J. W. C., LILAONITKUL, A., LERNER, L. E. and COOPER, S. L., *J. Macromol. Sci.-Phys.*, 1980, **B17**(2), 267.
78. SCHNEIDER, N. S. and PAIK SUNG, C. S., *Polym. Eng. Sci.*, 1977, **17**(2), 73.
79. SCHNEIDER, N. S., PAIK SUNG, C. S., MATTON, R. W. and ILLINGER, J. L., *Macromolecules*, 1975, **8**, 62.
80. PAIK SUNG, C. S. and SCHNEIDER, N. S., in: *Polymer Alloys*, eds. D. Klempner and K. C. FRISCH, Plenum, New York, 1977, p. 261.
81. CLOUGH, S. B. and SCHNEIDER, N. S., *J. Macromol. Sci.-Phys.*, 1968, **B2**, 553.
82. MILLER, G. W. and SAUNDERS, J. H., *J. Polym. Sci.*, 1970, **A1**(8), 1923.
83. VROUENRAETS, C. M. F., *Polym. Preprints*, 1972, **13**(1), 529.
84. SCHOLLENBERGER, C. S. and HEWITT, L. E., *A.C.S. Polymer Preprints*, 1978, **19**(1), 17.
85. SEYMOUR, R. W. and COOPER, S. L., *J. Polym. Lett.*, 1971, **9**, 689.
86. SEYMOUR, R. W. and COOPER, S. L., *Macromolecules*, 1973, **6**, 48.
87. HESKETH, T. R., VAN BOGART, J. W. C. and COOPER, S. L., *Polym. Eng. Sci.*, 1980, **20**(3), 190.

88. JACQUES, C. H. M., in: *Polymer Alloys*, eds. D. Klempner and K. Krisch, Plenum, New York, 1977.
89. NIELSEN, L., *Mechanical Properties of Polymers*, Reinhold, New York, 1961.
90. HUH, D. S. and COOPER, S. L., *Polym. Eng. Sci.*, 1971, **11**(5), 369.
91. NG, H. N., ALLEGREZZA, A. E., SEYMOUR, R. W. and COOPER, S. L., *Polymer*, 1973, **14**, 255.
92. SEEFRIED, C. G., JR., KOLESKE, J. V. and CRITCHFIELD, F. E., *J. Appl. Polym. Sci.*, 1975, **19**, 2493.
93. SEEFRIED, C. G., JR., KOLESKE, J. V., CRITCHFIELD, F. E. and DODD, J. L., *Polym. Eng. Sci.*, 1975, **15**(9), 646.
94. SEEFRIED, C. G., JR., KOLESKE, J. V. and CRITCHFIELD, F. E., *J. Appl. Polym. Sci.*, 1975, **19**, 3185.
95. SEEFRIED, C. G., JR., KOLESKE, J. V. and CRITCHFIELD, F. E., *J. Appl. Polym. Sci.*, 1975, **19**, 2503.
96. SMITH, T. L., *J. Polym. Sci.-Phys.*, 1974, **2**, 1825.
97. SMITH, T. L., *Polym. Eng. Sci.*, 1977, **17**(3), 129.
98. SMITH, T. L., *IBM J. Res. Develop.*, 1977, **21**(2), 154.
99. GUTH, E., *J. Appl. Phys.*, 1945, **16**, 20.
100. AGGARWAL, S. L., LIVIGNI, R. A., MARKER, L. F. and DUDEK, T. J., in: *Block and Graft Copolymers*, eds. J. J. Burke and V. Weiss, Syracuse University Press, Syracuse, 1973, p. 157.
101. NIELSEN, L. E., *Rheol. Acta*, 1974, **13**, 86.
102. SEYMOUR, R. W. and COOPER, S. L., *Rubb. Chem. Technol.*, 1974, **47**, 19.
103. TRICK, G. S., *J. Appl. Polym. Sci.*, 1960, **3**, 252.
104. COOPER, S. L., HUH, D. S. and MORRIS, W. J., *Ind. Eng. Chem., Prod. Res. Dev.*, 1968, **7**, 248.
105. BONART, R., *J. Macromol. Sci.-Phys.*, 1968, **B2**(1), 115.
106. WILKES, G. L., BAGRODIA, S., HUMPHRIES, W. and WILDNAUER, R., *J. Poly. Sci., Lett.*, 1975, **13**, 321.
107. WILKES, G. L. and WILDNAUER, R., *J. Appl. Phys.*, 1975, **46**, 4148.
108. WILKES, G. L. and EMERSON, J. A., *J. Appl. Phys.*, 1976, **47**, 4261.
109. ASSINK, R. A. and WILKES, G. L., *Polym. Eng. Sci.*, 1977, **17**, 603.
110. OPHIR, Z. H. and WILKES, G. L., *Adv. Chem. Series*, 1979, **176**, 53.
111. COOPER, S. L. and TOBOLSKY, A. V., *J. Appl. Polym. Sci.*, 1967, **11**, 1361.
112. CLOUGH, S. B., SCHNEIDER, N. S. and KING, A. O., *J. Macromol. Sci.-Phys.*, 1968, **B2**(4), 641.
113. BONART, R., *J. Macromol. Sci.-Phys.*, 1968, **B2**(1), 115.
114. BONART, R., MORBITZER, L. and HENTZE, G., *J. Macromol. Sci.-Phys.*, 1969, **B3**(2), 337.
115. WILKES, C. E. and YUSEK, C. S., *J. Macromol. Sci.-Phys.*, 1973, **B7**(1), 157.
116. BONART, R., MORBITZER, L. and MÜLLER, E. H., *J. Macromol. Sci.-Phys.*, 1974, **B9**(3), 447.
117. POROD, G., *Kolloid-Z.*, 1951, **124**(2), 83.
118. POROD, G., *Kolloid-Z.*, 1952, **125**(1), 51.
119. POROD, G., *Kolloid-Z.*, 1952, **125**(2), 108.
120. OPHIR, Z. and WILKES, G. L., *Polym. Preprints*, 1979, **20**, 503.
121. CHANG, Y. J. P. and WILKES, GARTH, *J. Polym. Sci., Polymer Phys. Ed.*, 1975, **13**, 455.

122. HOLLAND-MORITZ, K., STACH, W. and HOLLAND-MORITZ, I., *Progr. Colloid and Polymer Sci.*, 1980, **67**, 161.
123. SIESLER, H. W., in: *Proc. 5th European Symp. on Polymer Spectroscopy*, ed. D. O. Hummel, Verlag Chemie, Basel, 1979.
124. MCCRUM, N. G., READ, B. E. and WILLIAMS, G., *Anelastic and Dielectric Effects in Polymeric Solids*, Wiley, New York, 1967.
125. PAIK SUNG, C. S., HU, C. B. and WU, C. S., *Macromolecules*, 1980, **13**, 111.
126. PAIK SUNG, C. S., SMITH, T. W. and SUNG, N. H., *Macromolecules*, 1980, **13**, 117.
127. VAN BOGART, J. W. C., BLUEMKE, D. A. and COOPER, S. L., *Polymer*, accepted for publication.
128. VAN BOGART, J. W. C., GIBSON, P. E. and COOPER, S. L., *J. Polymer Sci., Polymer Phys. Ed.*, submitted for publication.
129. STEIN, R. S., in: *Newer Methods of Polymer Characterization*, ed. B. Ke, Wiley-Interscience, New York, 1964.
130. WEST, J. C. and COOPER, S. L., *J. Polym. Sci., Polym. Symp.*, 1977, **60**, 127.

Chapter 7

DEVELOPMENTS IN POLYETHER-ESTER AND POLYESTER-ESTER SEGMENTED COPOLYMERS

R. W. M. van Berkel, S. A. G. de Graaf, F. J. Huntjens and C. M. F. Vrouenraets

Enka BV, Research Institute Arnhem, The Netherlands

SUMMARY

This chapter reviews developments in the preparation, properties and applications of thermoplastic copolyether-ester and copolyester-ester elastomers.

First of all a historical outline is given of the development of these elastomers and their position amongst other elastomers and plastics. The second section is concerned with the chemistry of the two types of segmented elastomeric copolyesters, special attention being given to the technologically important types, which have a poly(butylene terephthalate) hard segment. The third and the fourth sections compare the two types of the title elastomers with respect to mechanical and physico-chemical properties and their relationship to thermoplastic polyurethanes in particular cases. Rheological behaviour and its effect on processing techniques are discussed in the fifth section. The final section is concerned with the major fields of application.

1. INTRODUCTION

This chapter deals with thermoplastic elastomers composed of segmented copolyether-esters and segmented copolyester-esters.

The copolyether-esters are essentially linear in that they are normally produced by polycondensation from difunctional components. Their composition can be described as an $\text{-}[A\text{—}B]_p\text{-}$ structure, where A represents a polyether segment and B a polyester segment. In most cases the polyether segment A is derived from a hydroxyl-terminated aliphatic polyether: poly(oxyethylene)-, poly(oxypropylene)- or poly(oxytetramethylene)-α,ω-diol (POE, POP and POTM, respectively). The polyester segment B is usually a rigid high melting aromatic polyester derived from terephthalic acid. Elastomers based upon naphthalene dicarboxylic acid[1,2] or bis(4-carboxyphenoxy)ethane[3] also have been described. Like other segmented thermoplastic elastomers (e.g. thermoplastic polyurethanes (TPU)) the copolyether-esters possess a continuous two-phase structure[4] in which the soft segments provide flexibility because of their low glass transition temperatures. The polyester segments provide physical cross-linking as a result of their strong, high-melting, crystalline structure. Segmented copolyether-esters have been known for almost 30 years from publications by Coleman[5,6] and du Pont;[7] see also Chapter 5. An overview by Noshay and McGrath[8] shows that most of the literature on copolyether-esters was published after 1967.

The developments led, in 1972, to the introduction of commercial copolyether-esters by du Pont (Hytrel®) and Toyobo (Pelprene®) and in 1975 by Akzo (Arnitel®).

All three elastomers are based on poly(oxytetramethylene) terephthalate (POTM-T) as the amorphous soft segment and poly(butylene terephthalate) (represented as 4G.T in the ciphering system of Chapter 5 but abbreviated here as PBTP), or a combination of tere- and isophthalates, which was also used in the case of the early Hytrels, as the hard and crystalline segment. These copolyether-esters, which are available in a hardness range 40–74 Shore D, find an ever increasing acceptance because many of their mechanical properties are between those of conventional rubbers and thermoplastic homopolymers, as shown in Fig. 1.

Copolyester-ester thermoplastic elastomers built up of amorphous soft segments of aliphatic polyesters and hard segments of high melting crystalline polyester have attracted considerably less interest. The main reason for this is the instability displayed by the segmented copolymer both when prepared in a melt process and during further processing, due to ester-interchange between the two polyester segments. Although no commercial segmented copolyester-esters are known, the increasing amount of patent literature published in the last few years points to a growing interest in this class of elastomers.

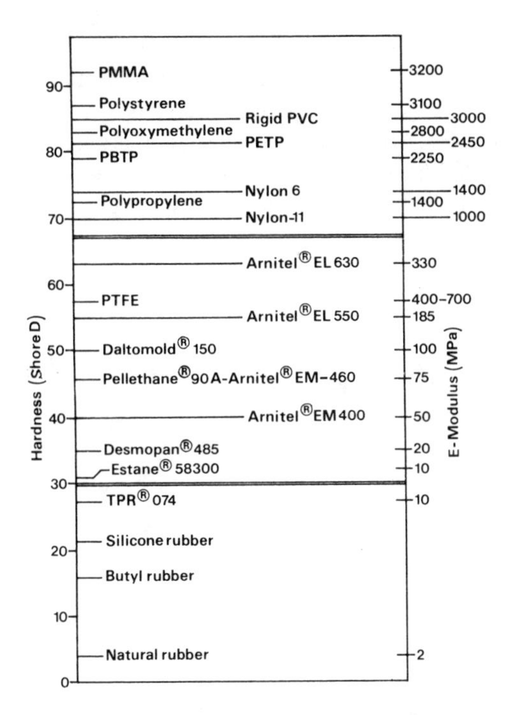

FIG. 1. Hardness and E modulus of plastics and elastomers. Registered Trade Marks belong to the following companies: Akzo (Arnitel), ICI (Daltomold), Upjohn (Pellethane), B. F. Goodrich (Estane), Uniroyal (TPR), Bayer (Desmopan).

Various techniques of coupling different polyester segments have been described: coupling of hydroxyl-terminated oligomers by diisocyanate,[9,18] by bis-*N*-acyl-lactams,[61] condensation of acyl chloride-terminated and hydroxyl-terminated oligomers,[10] transesterification,[11,12] limited transesterification of two high molecular weight homopolymers in the melt[13] or crystallisation-induced transesterification.[14,15] Various recent patents[16,17,18,19] disclose methods of coupling an aliphatic oligomer having a low glass transition temperature to a high melting aromatic polyester using a diisocyanate. These methods make use of a batch or continuous process in which the two polyester oligomers are reacted in the melt with a diisocyanate, or one oligomer is first extended with diisocyanate and subsequently reacted with the second polyester.

As all these patents use PBTP as the high melting component, these copolyester-ester elastomers contain fewer urethane groups than polyester-urethanes of comparable hardness. They are therefore physically cross-linked in the same way as copolyether-esters, as a result of the crystalline PBTP structure. When the two polyester segments are coupled by an excess of diisocyanate, additional chemical crosslinks are obtained in the solid state due to the formation of allophanate structures.

2. CHEMISTRY

2.1. Segmented Copolyether-esters

2.1.1. Polymer Preparation

The preparation of segmented copolyether-esters has been described at great length in the literature. The polymers have the following general structure:

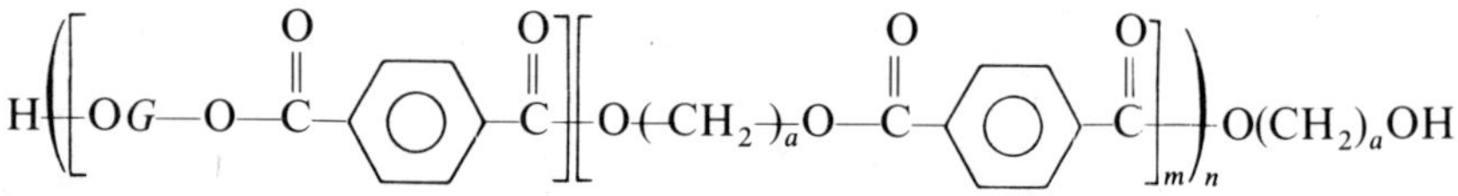

$$\mathrm{H}\!\left(\left[\mathrm{O}G\!-\!\mathrm{O}\!-\!\overset{\overset{\displaystyle\mathrm{O}}{\|}}{\mathrm{C}}\!-\!C_6H_4\!-\!\overset{\overset{\displaystyle\mathrm{O}}{\|}}{\mathrm{C}}\right]\left[\mathrm{O}\!\left(\mathrm{CH_2}\right)_a\!\mathrm{O}\!-\!\overset{\overset{\displaystyle\mathrm{O}}{\|}}{\mathrm{C}}\!-\!C_6H_4\!-\!\overset{\overset{\displaystyle\mathrm{O}}{\|}}{\mathrm{C}}\right]_m\right)_n\mathrm{O(CH_2)}_a\mathrm{OH}$$

soft segment G = POE. POP, POTM

hard segment $a = 2, 4, 6$

In 1954 Coleman[5] described segmented copolyether-esters composed of soft segments from POE-α,ω-diols and hard segments of poly(ethylene terephthalate) (PETP). In 1962 Shivers[20] broadened this class of segmented copolymers by including other poly(oxyalkylene)-α,ω-diols, e.g. POTM-α,ω-diol, and other polyester hard segments where a is between 2 and 10.

The preparation of segmented copolyether-esters is based on the ester-interchange reaction, as is well-known from the preparation of poly(ethylene terephthalate) and poly(butylene terephthalate). In the first reaction step, dimethyl terephthalate is heated to 150–210 °C with an excess of ethylene glycol or butylene glycol in the presence of a catalyst. In the second step, the polycondensation reaction, the product of the first reaction together with a poly(oxyalkylene)-α,ω-diol is heated to 240 °C or 265 °C, depending on the use of butylene glycol or ethylene glycol in the first step. During the polycondensation reaction the pressure is continuously lowered to below 100 Pa to remove excess low molecular weight glycol and volatile by products.

Step 1, ester interchange:

$$H_3COC(=O)-C_6H_4-C(=O)OCH_3 + HO\text{+}CH_2\text{+}_aOH \xrightarrow[\text{catalyst; (1) } Mn(OAc)_2\text{, (2) } Ti(OBu)_4]{\text{excess; (1) } a=2\text{, (2) } a=4} HO\text{+}CH_2\text{+}_aOC(=O)-C_6H_4-C(=O)O\text{+}CH_2\text{+}_aOH + 2CH_3OH \quad (1)$$

Step 2, polycondensation:

$$HO\text{+}CH_2\text{+}_aO-C(=O)-C_6H_4-C(=O)-O\text{+}CH_2\text{+}_aOH + HO-G-OH \xrightarrow{\text{catalyst: (1) } Sb_2O_3 \text{ at } 265\,°C\text{, (2) } Ti(OBu)_4 \text{ at } 240\,°C} H\left(\left[O-G-OC(=O)-C_6H_4-C(=O)\right]\left[O\text{+}CH_2\text{+}_aOC(=O)-C_6H_4-C(=O)\right]_m\right)_n O\text{+}CH_2\text{+}_aOH + HO\text{+}CH_2\text{+}_aOH \quad (2)$$

The composition of the polymer is governed by the ratio of the starting monomers; as there is only a loss of ethylene glycol or butylene glycol, which must take place in order to complete the polycondensation, the weight percentage and the molecular weight of HO—*G*—OH determine the value of *m*. The flow diagram of the process is shown in Fig. 2.

2.1.2. Materials

All the monomers are commercially available. Suitable molecular weights of POTM-α,ω-diol are 1000 and 2000 and for POE-α,ω-diol 600–4000. Use has also been made of POP-α,ω-diols or ethyleneglycol-capped POP diols. Due to the lesser thermal stability of the POP-based diols it is advantageous to carry out the polycondensation at about 230 °C. The ether segments in these copolymers, being sensitive to light (ultraviolet (UV)) and liable to oxidative degradation, require adequate protection. In most cases use is made of about 0·5 wt % antioxidant, especially those of the sterically

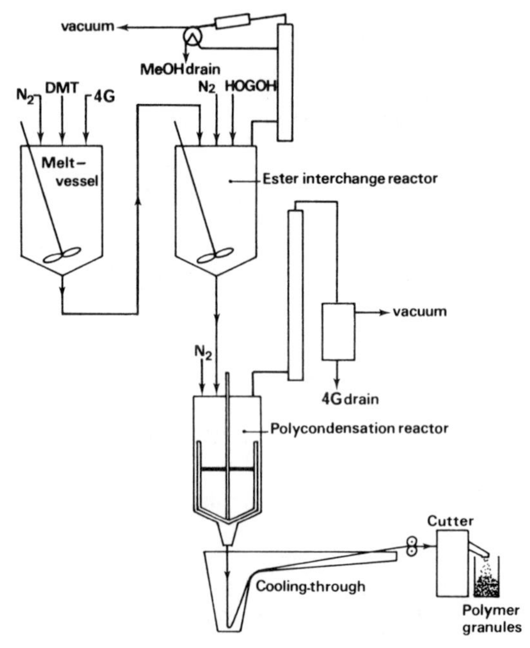

FIG. 2. Flow diagram of the process of preparing a segmented copolyether-ester.

hindered phenol-type, e.g. Irganox 1330 or Irganox 1098 (Ciba-Geigy). The antioxidant may be added before the start of the polycondensation reaction; however, for practical reasons it may be incorporated with the starting monomers.

2.1.3. *Post-condensation in the Solid Phase*

The segmented copolyether-esters prepared as described above, have a suitable melt viscosity for injection moulding. For optimum extrusion quality, however, the melt viscosity needs to be raised. As these high melt viscosities lead to unacceptably long discharging times of the polycondensation reactor and to an unacceptably high hold-up level, the polymer granules must be post-condensed in the solid phase.

It is known that polyesters[21] and polyamides[22] can be post-condensed in the solid state in a convenient way at temperatures below the melting point. By reaction of a COOH end group with an OH end group, or reaction between two glycol end groups in a polyester such as poly(ethylene terephthalate), the molecular weight will increase. The water molecule (or glycol molecule) formed in the amorphous part of the polymer, diffuses from the inside outwardly, especially when this process is carried out under low pressure.

The same process has been applied to segmented copolyether-esters. A sufficient rate of solid phase condensation is obtained at a temperature of 190 °C or higher. A segmented copolymer consisting of 65 % m/m (mass/mass) PBTP and 35 % m/m POTM (M.W. 1000) and having a melt viscosity suitable for injection moulding, was post-condensed at 190 °C to 195 °C for 24 h at a pressure of 100 Pa to a melt viscosity suitable for extrusion.

As the harder grades of segmented copolyether-esters show melting points above this temperature, no difficulties are encountered in post-condensation. However, softer segmented copolyether-esters, e.g. consisting of 40 % m/m PBTP and 60 % m/m POTM (M.W. = 1000) have a relatively low melting point, viz. 165 °C. As the rate of polycondensation at 150 °C is too low for practical purposes, this polymer cannot be post-condensed. For various reasons use is made of POTM having a M.W. of 2000, which results in a melting point of about 196 °C, permitting the polymer to be post-condensed in the solid phase.

The concentration of COOH end groups also plays an important role in post-condensation. Generally, the concentration should not exceed 40 meq/kg, in order to maintain a sufficiently high rate of post-condensation.

Thermal oxidative degradation of $O(CH_2)_4OH$ end groups during polycondensation is responsible for a high COOH end group concentration. Therefore the polycondensation must be carried out by the shortest possible route (i.e. to an acceptable level of carboxyl end groups and M.W.) and with avoidance of leakage of oxygen.

2.2. Segmented Copolyester-esters

In thermoplastic polyurethane elastomers the soft segments are formed either from an aliphatic polyether or an aliphatic polyester. The commercially available elastomeric segmented copolyether-esters possess a POTM unit. Replacing the polyether unit in a copolyether-ester by an aliphatic polyester unit having a low melting point and a low glass transition temperature may be done for obvious reasons. The main advantages of polyester over polyether are the lesser susceptibility to oxidation and UV degradation, and that they are in general, lower priced starting materials. The disadvantages however are not inconsiderable: there is generally less flexibility at lower temperatures due to the higher T_g, a higher sensitivity to hydrolysis and, especially when combined with a high melting polyester hard segment, the possibility of undesirable chemical interaction. This interaction, the ester interchange reaction between the two polyesters, may be a valuable tool in the preparation of the segmented copolyester-ester. On the other hand, however, it presents the risk of leading to too short segments (or even complete randomisation), resulting in a complete loss of elastomeric properties.

When, in the procedure of preparing copolyether-esters as described in Section 2.1, the POTM-α,ω-diol is replaced with a polyester-α,ω-diol, severe randomisation occurs and a non-elastomeric product is obtained. To circumvent this problem, the preparation of segmented copolyester-esters has been carried out by two different routes:

(a) coupling two preformed polyester blocks by a controlled ester-interchange reaction or controlled ester-interchange of two high M.W. polyester homopolymers;
(b) coupling two preformed OH-terminated polyester segments by diisocyanates.

2.2.1. Coupling by Ester-interchange

A method of synthesising segmented copolyester-esters with alternately soft (flexible) and hard (rigid) segments has been developed by Teijin Ltd.[13] An aliphatic and an aromatic polyester, both of high M.W., are mixed at 230–70 °C in a nitrogen atmosphere. The mixture of the two polymers changes gradually into a segmented copolymer by ester-interchange reactions. Increasing the reaction time decreases the average M.W. of the resulting segments. To obtain a segmented copolymer within a reasonable time, the ester-interchange reaction must be catalysed.

When PBTP is used as the rigid aromatic polyester, the catalyst is already present as a titanium compound used in the preparation of the homopolymer. A suitable segment length is attained as soon as the two-phase melt becomes transparent. The reaction is then discontinued, because further heating will randomise the segmented copolymer with deterioration of elastomeric properties.

The influence of the heating time on the properties of a product obtained by heating a mixture of poly(ethylene adipate) (PEA) and poly(butylene terephthalate) (50/50 m/m) at 240 °C in a nitrogen atmosphere is shown in Table 1. The decreasing melting points (PBTP melts at 225 °C) show that after 180 min severe randomisation has taken place. Judged from the elongation at break, the product shows elastomeric properties only after 60 min reaction.

TABLE 1
INFLUENCE OF HEATING TIME AT 240 °C ON THE PROPERTIES OF A PEA/PBTP MIXTURE[13]

Property	*Value after heating for:*		
	30 min	*60 min*	*180 min*
Tensile strength (MPa)	12·1	17·0	—
Tensile elongation (%)	65	600	—
Melting point (°C)	174	189	118
Appearance of the melt	opaque	clear	clear
Appearance of the injection-moulded specimen	opaque	opaque	clear

To stop further randomisation at this point, a phosphorus compound was added to deactivate the titanium catalyst. Suitable deactivation compounds include orthophosphoric acid, phosphorous acid, triphenyl phosphite and triphenyl phosphate.[23] As with other polyesters, the M.W. of the resulting copolyester-ester can be raised by solid state post-condensation. In that case the catalyst has to be deactivated in a separate step.[13]

Especially with high M.W. polyesters, the rate of ester-interchange is highly dependent on end group concentrations (OH, COOH and OCH_3) and on the concentration of catalyst residue from homopolymer preparation. For these reasons the ester-interchange method may give irreproducible results.

2.2.2. *Coupling of Polyester Segments by Diisocyanate*

An alternative method for the preparation of a segmented copolyester-ester elastomer comprises coupling an aliphatic polyester-α,ω-diol and a high melting aromatic polyester-α,ω-diol by a diisocyanate, as represented schematically by eqn (3).

$$\mathrm{HO{\sim\sim}OH + 2\,OCN{-}R{-}NCO + HO{=}OH \longrightarrow}$$

$$\left[\mathrm{O{\sim\sim}O{-}\underset{\underset{O}{\|}}{C}{-}\underset{H}{N}{-}R{-}\underset{H}{N}{-}\underset{\underset{O}{\|}}{C}{-}O{=}O{-}\underset{\underset{O}{\|}}{C}{-}\overset{H}{N}{-}R{-}\overset{H}{N}{-}\underset{\underset{O}{\|}}{C}}\right]_n \quad (3)$$

HO〰OH, the aliphatic polyester-diol
HO=OH, the aromatic polyester-diol

As the coupling takes place at elevated temperature—because of solubility reasons or because of the high melting temperature of the aromatic polyester—ester interchange might also occur. Here, too, it depends on reaction temperature, time and the presence of catalyst.

Iwakura[9] describes the coupling of aromatic and aliphatic OH-terminated polyester segments by tetramethylene diisocyanate at 170 °C in nitrobenzene solution. Judged from decreasing melting points, randomisation took place at 250 °C. However, the non-randomised copolyester-ester from poly(ethylene terephthalate) and poly(ethylene adipate) showed elastomeric properties.[9] Since about 1975 a number of companies have developed techniques of preparing segmented copolyester-ester elastomers by coupling through diisocyanates. Dainippon[19] have claimed a method in which an aliphatic polyester oligomer is converted into an isocyanate capped prepolymer. This prepolymer is subsequently coupled to an aromatic polyester of relatively low M.W. in an extruder or kneader for 10 min beyond the melt temperature of the aromatic polyester. The method was used to upgrade the M.W. of the PBTP and at the same time to obtain a higher impact strength. Monsanto[16] also describe a melt process in which a slurry containing solid aromatic polyester-diol (e.g. PBTP), a liquid aliphatic polyester-diol (e.g. poly(ethylene-*co*-butylene adipate)) and a diisocyanate, is fed into a continuous reactor heated to 250–280 °C.

Due to the short reaction time (<15 min) transesterification can largely be avoided, making it possible to obtain an elastomer having a low glass transition temperature and a high melting point. In a patent publication[17] Bayer discloses a batch process in which an aliphatic and an aromatic polyester oligomer are melt-blended and subsequently mixed with MDI. These segmented polyester-esters, containing about 60 wt % of PBTP and MDI, show a relatively low hardness, a low stress build-up and a low tensile strength.

A European patent application[18] of Akzo discloses a batch process in which the transesterification catalyst, used in the preparation of the PBTP-α,ω-diol, is deactivated prior to the addition of the diisocyanate-terminated aliphatic polyester prepolymer at 240 °C. These elastomers show the high melting point of PBTP segments and the low glass transition temperature of the soft segment. Little evidence of transesterification was found between the two polyesters.

Deactivation of the catalyst (e.g. a titanic orthoester) is accomplished by heating the PBTP-α,ω-diol, with a phosphorus compound for $\frac{1}{2}$ h at 240 °C. Deactivation was demonstrated in an experiment in which 1 mol TDI-capped poly(butylene adipate), M.W. 1600, was mixed with 1 mol PBTP-

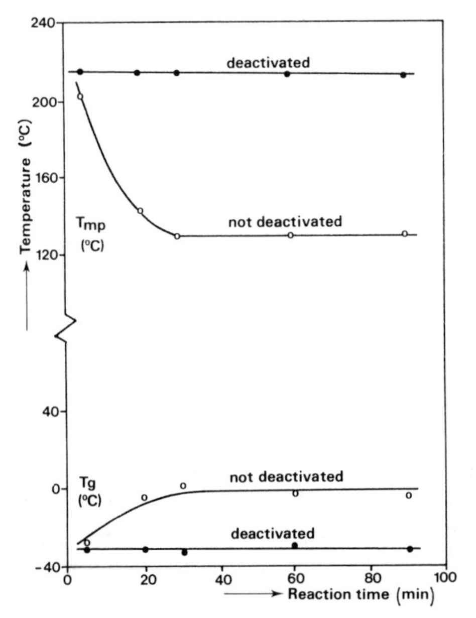

FIG. 3. Influence of deactivation of the catalyst on randomisation.

α,ω-diol, M.W. = 2000 and heated for 90 min at 245 °C. The PBTP-diol contained 375 ppm $Ti(OBu)_4$. Samples were drawn after 5, 20, 30, 60 and 90 min.

A similar experiment was carried out in which the Ti catalyst was deactivated by adding 1500 ppm carbethoxymethyldiethyl phosphonate (CMDP). The samples were analysed by DTA to determine the average glass transition temperature of the soft segment (T_g) and the temperature of the melting peak of the hard segment (T_{mp}).

The effect of deactivation is demonstrated clearly in Fig. 3, when compared with samples containing a catalyst that is still active, which results in an increased T_g and a decreased T_{mp}.

The deactivated poly(butylene terephthalate)-α,ω-diol having a M.W. of about 2000–2500 can be obtained in a reproducible way by the ester-interchange of dimethyl terephthalate (DMT) with 1,4-butanediol (4G). To ensure complete removal of OCH_3 end groups a 4G/DMT molar ratio of 1·5 was used in the presence of 440 ppm $Ti(OBu)_4$. The ester-interchange was followed by a polycondensation reaction for 30 min at a temperature of 240 °C and a pressure of 2600 Pa. Deactivation was effected by adding 1500 ppm of CMDP at 240 °C for another 30 min in a nitrogen atmosphere, which resulted in obtaining the aromatic α,ω-diol corresponding to eqn (4).

$$CH_3O\overset{O}{\overset{\|}{C}}-C_6H_4-\overset{O}{\overset{\|}{C}}OCH_3 + \underset{\text{excess}}{HO\!\left(CH_2\right)_4\!OH} \xrightarrow[\substack{10^5\text{–}2600\ \text{Pa} \\ 440\ \text{ppm Ti(OBu)}_4}]{160\text{–}240\ °C}$$

$$\underset{\text{PBTP-}\alpha,\omega\text{-diol}}{H\!\left[O(CH_2)_4O\overset{O}{\overset{\|}{C}}-C_6H_4-\overset{O}{\overset{\|}{C}}\right]_{9-10}\!O\!\left(CH_2\right)_4\!OH} + CH_3OH + HO\!\left(CH_2\right)_4\!OH$$

$$\xrightarrow[1500\ \text{ppm CMDP}]{30\ \text{min},\ 240\ °C} \text{deactivated PBTP-}\alpha,\omega\text{-diol} \qquad (4)$$

For the preparation of a segmented copolyester-ester, a diisocyanate-terminated aliphatic polyester was added to the deactivated PBTP-α,ω-diol at 240 °C and the mixture was heated with stirring for 30 min. The polymer was then extruded into strands, followed by chopping and drying. For this process the same flow diagram can be used as for the preparation of segmented copolyether-esters, shown in Fig. 2; an additional reaction vessel with stirrer is needed for the preparation of the diisocyanate-capped aliphatic prepolymer. This prepolymer vessel is connected to the polycondensation reactor. The aliphatic polyester-diols that were used include: poly(ethylene adipate)-α,ω-diol (PEA), poly(butylene adipate)-α,ω-diol (PBA), polycaprolactone-α,ω-diol (PCL), copolyester-α,ω-diols and mixtures, having M.W.'s 1000 and/or 2000. The diisocyanates TDI and MDI were used.

In order to obtain optimum mechanical properties the reaction was carried out at an overall NCO/OH ratio in excess of 1·0. Especially preferred is a ratio in the range 1·1–1·3. The use of an excess of diisocyanate leads to the formation of allophanate structures upon cooling the polymer melt to below 190 °C (eqn (5)).

$$\underset{\text{excess}}{-N{=}C{=}O} + \underset{\text{urethane}}{-NH-\overset{O}{\overset{\|}{C}}-O-} \rightleftharpoons \underset{\text{allophanate structure}}{-N(-C{=}O-NH-)-\overset{O}{\overset{\|}{C}}-O-} \qquad (5)$$

TABLE 2
COMPOSITION OF TWO TYPICAL COPOLYESTER-ESTER ELASTOMERS. (KILOGRAMMES OF STARTING MATERIAL NEEDED FOR 1000 kg OF POLYMER)

Constituent	*38 Shore D*	*50 Shore D*
Dimethyl terephthalate	225·2 kg	404·4 kg
1,4-Butanediol	156·7 kg	281·4 kg
Tetrabutyl titanate	0·112 kg	0·200 kg
CMDP	0·900 kg	1·617 kg
Resulting PBTP ester-interchange product	266·3 kg	478·3 kg
Poly(butylene adipate-α,ω-diol)		
M.W. = 1000	550·5 kg	86·4 kg
M.W. = 2000	—	313·5 kg
Diphenylmethane diisocyanate (MDI)	183·1 kg	121·8 kg
M.W. PBTP	2400	2400
NCO/OH in prepolymer	1·32	2·0
NCO/OH in copolymer	1·10	1·10
%PBA (m/m) in copolymer	55·0	40·0

These allophanate crosslinks improve the mechanical properties and thermal processing remains possible since above 190 °C the allophanate structures dissociate, as can be demonstrated by applying infrared techniques to a compression-moulded film. Thus, above 190 °C an absorption band assigned to the —NCO group appeared at 2280 cm^{-1}, the intensity of which increased with increasing temperature. During cooling the absorption slowly disappeared and was no longer visible below 180 °C. Upon re-heating the same phenomenon was observed.

The composition of two typical segmented copolyester-ester elastomers is given in Table 2.

3. MECHANICAL PROPERTIES

The mechanical properties of copolyether-esters based on POTM and PBTP have been described in numerous publications[24–7] and brochures from polymer producers. A very detailed description of Hytrel elastomers is given by Wells.[28] Properties of copolyether-esters made up of different soft and/or hard segments have been described by Hoeschele,[29] Wolfe[30,36] Goodman[31] and Goldberg.[59]

Properties of copolyester-ester elastomers obtained by coupling through diisocyanates were described in the early 1960s. These polymers were prepared by solution polymerisation or by coupling the low melting polyester oligomers in bulk.[9,32–4] Methods of preparation and properties of diisocyanate-coupled copolyester-esters, based on high melting aromatic polyester segments and produced by melt processing, are mainly described in the patent literature published during the mid 1970s.[14–16,19] In the following sections diisocyanate-coupled copolyester-esters will be referred to as copolyester-ester, unless otherwise indicated.

3.1. Stress–Strain Behaviour

The proportion of combined high melting aromatic polyester and the chemical compositions of the soft and the hard segments have a very strong influence on the stress–strain behaviour of these elastomers. Although any composition ranging from 1 % to 99 % aromatic polyester can be prepared, commercial compositions available range from about 40 % to about 90 % PBTP. The latter composition, although showing a greatly reduced modulus compared with the PBTP homopolymer, shows the behaviour of a ductile polymer but can hardly be regarded as an elastomer.

In tensile tests (specimens stamped from injection-moulded slabs) the

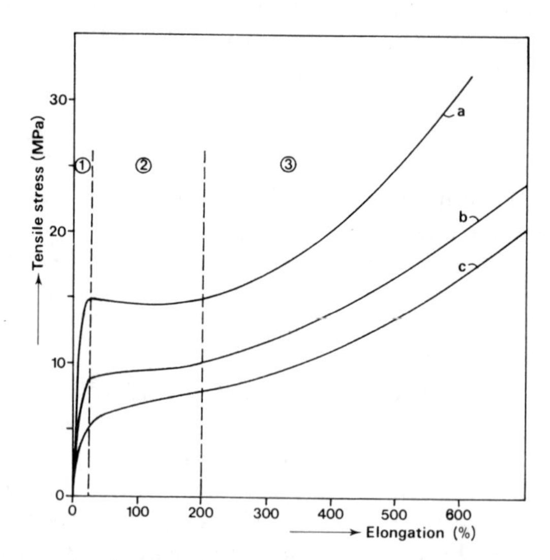

FIG. 4. Stress–strain diagram of copolyether-esters based on POTM and PBTP (in accordance with DIN 53504). (a) 35 % m/m POTM, M.W. = 1000; 55 Shore D. (b) 50 % m/m POTM, M.W. = 1000; 46 Shore D. (c) 60 % m/m POTM, M.W. = 2000; 38 Shore D.

copolyether-esters show a behaviour analogous to that of the copolyester-esters, reflecting their morphological structure. The stress–strain curve can be divided into three regions (see Fig. 4). In the first region (up to about 25 % elongation) the hard crystalline matrix is deformed. Up to a few per cent the deformation is entirely reversible. At higher elongations the crystalline matrix is progressively disrupted until at about 25 % elongation the original crystalline matrix has completely disintegrated.

The second region, between 25 and 200 % elongation, shows the yielding of the polymer, due to orientation and rearrangement of crystallites under the influence of shearing forces. In the third region, beyond 200 %, stress is transmitted through the soft segments up to the point of rupture at their maximum extension.

The polymers of lower hardness show a less pronounced yield point due to the weaker hard segment matrix and probably less perfect crystallites.

The stress–strain behaviour of the copolyester-ester described in Section 2.2.2 resembles that of a polyester-urethane rather than a copolyether-ester, as shown in Fig. 5.

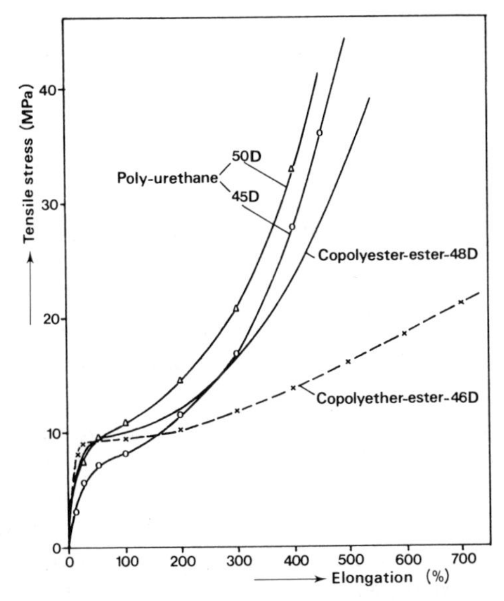

FIG. 5. Stress-strain diagrams of 45–50 Shore D elastomers (DIN 53504). (i) Polyurethanes based on 4G, MDI and PBA_{1000}*; (ii) copolyester-ester based on $PBTP_{2400}$, MDI and a mixture of PBA_{1000} and PBA_{2000}; (iii) copolyether-ester based on $POTM_{1000}$ and PBTP. (* Subscripts denote the M.W.'s of soft segment diols.)

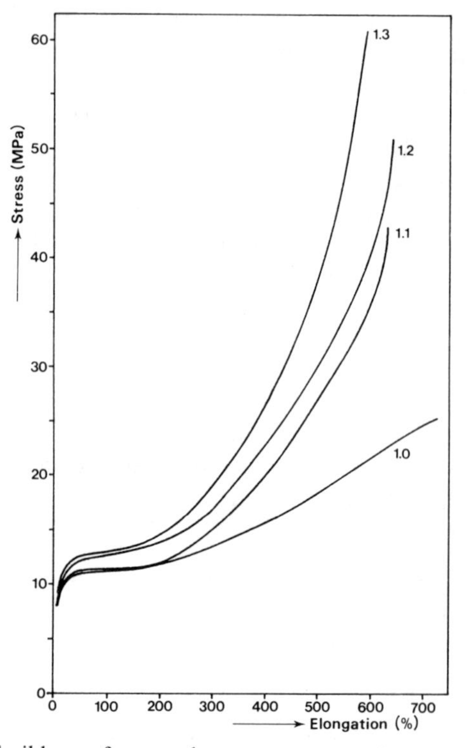

FIG. 6. Stress build-up of a copolyester-ester at various NCO/OH ratios (52 Shore D). Composition: mixture of PBA_{1000} and PBA_{2000}, extended with TDI and $PBTP_{2640}$. Polymer contains 40% m/m PBA.

The higher stress build-up at high elongation of the copolyester-ester, compared with the copolyether-ester, is probably to be attributed to partial differences in degree and the nature of the crosslinks, the crosslinking in copolyether-esters being of an exclusively physical nature as a result of the presence of interconnected PBTP crystals, serving as tie points. The copolyester-ester, prepared at an NCO/OH ratio of 1·1, contains additional chemical tie points formed by allophanate structures. These chemical crosslinks lead to a more uniform stress distribution thoughout the sample, thus preventing plastic flow of the strongest PBTP crystals and leading to stress-transmission through soft segments at lower elongation. A more uniform stress distribution in combination with increasing chemical crosslink density (i.e. increasing NCO/OH ratio) also leads to the tensile strength increasing with decreasing elongation (see Fig. 6).

TABLE 3

THE INFLUENCE OF SOFT AND HARD SEGMENT COMPOSITIONS ON THE PROPERTIES OF SOME COPOLYETHER-ESTERS AND COPOLYESTER-ESTERS

Property	*Unit*	*Test method*	*Copolyether-ester*				*Copolyester-ester*[f]		
			POE_{1000}[a] $PBTP$ – (35)[e]	POP_{2400}[c] $PBTP$ – (35)	$POTM_{1000}$ $PBTP$ – (35)	$POTM_{1000}$ $PETP$ – (37)	PBA_{1000} $PBTP_{3000}$ MDI (36)	PBA_{2000} $PBTP_{3000}$ MDI (42)	PCL_{2000} $PBTP_{3000}$ MDI (40)
Hardness	Shore D	ASTM-D-2240	56	52	55	53	54	45	48
σ (10)[b]	MPa	DIN 53504	12·5	10·4	10·5	11·0	6·2	4·1	5·2
σ (20)	MPa	DIN 53504	14·5	13·1	14·6	14·3	9·7	6·3	8·1
σ (30)	MPa	DIN 53504	15·0	13·8	15·4	15·6	12·1	7·4	9·2
σ (50)	MPa	DIN 53504	14·0	14·1	15·1	15·5	13·3	8·2	10·1
σ (100)	MPa	DIN 53504	15·0	14·2	14·4	15·2	14·1	9·2	10·9
σ (200)	MPa	DIN 53504	17·2	15·3	14·9	17·5	16·5	11·9	13·2
σ (300)	MPa	DIN 53504	21·1	17·8	17·2	20·8	25·6	16·9	18·0
σ (400)	MPa	DIN 53504	25·0	22·0	21·2	24·5	37·0	25·2	26·4
σ (break)	MPa	DIN 53504	27·6	40·5	31·4	31·0	47·0	50·5	46·7
ε (break)	%	DIN 53504	470	680	575	540	450	550	550
Tensile set[d]	%	DIN 53518	51	50	52	48	23	29	22

[a] Figure denotes $\bar{M}_n$.
[b] Stress at indicated strain level.
[c] POP capped with OE.
[d] 23 °C; 24 h at 100 % elongation, 0·5 h relaxation.
[e] Soft segment content expressed as % m/m.
[f] NCO/OH = 1·10.

The stress–strain properties of copolyether-esters are not much influenced by the nature of the polyether component, as is shown by Table 3. Other properties, such as water absorption in the case of POE/PBTP, sensitivity to oxidation with POP/PBTP, or too low a rate of crystallisation of the system POTM/PETP, have prevented these copolymers from becoming of significant commercial interest. The copolymer based on POTM and PBTP combines the advantages of the PBTP homopolymer (high rate of crystallisation, high melting point) and the favourable properties of the POTM segments (low glass transition temperature, low water absorption and moderate sensitivity to oxidation and UV degradation).

Compared with the copolyether-esters, the copolyester-esters show lower initial moduli at the same hardness value. This phenomenon is probably due to a less perfect crystalline PBTP matrix caused by urethane linkages which have a distorting effect on the crystallisability.[35]

3.2. Set Properties

A significant difference between the copolyether-esters and copolyester-esters shown in Table 3 is the much lower tensile set of the latter, at nearly

TABLE 4

INFLUENCE OF SEGMENT M.W. AND COMPOSITION ON TENSILE SET AND ELASTIC RECOVERY OF COPOLYETHER-ESTERS AND COPOLYESTER-ESTERS

Soft segment	POTM	POTM	PBA	PBA
Soft segment M.W.	1 000	2 000	1 000	2 000
PBTP segment M.W.	666	1 333	2 500	3 000
Soft segment content (% m/m)	60	60	54	59
Hardness Shore D	40	39	39	35
Tensile set (23°) (100% elongation, 24 h)	44	26	18	61
Elongation 100%[a]	100	100	100	100
ε_E	68	80	89	66
ε_F	20	11	6	28
ε_P	12	9	5	6
Elongation 200%[a]	200	200	200	200
ε_E	118	150	164	116
ε_F	55	30	27	74
ε_P	27	20	9	10

[a] Strain maintained for 15 min, followed by relaxation at 23 °C for 15 min and subsequent heating for 15 min at 100 °C.

the same stress level at 100% elongation, which imparts to these copolymers a rather polyurethane-like character. The greater tendency of polyester soft segments to crystallisation with increasing M.W. is apparent from Table 3 when comparing the tensile set values of the samples based upon PBA_{1000} and PBA_{2000}. The effect is pronounced when polyester and polyether soft segments are compared in 40 Shore D materials, as shown in Table 4. The high tensile set of the copolyether-ester based on $POTM_{1000}$ is not caused by crystallisation of the soft segment, but is merely due to the low M.W. of the hard segment. This low M.W. results in smaller and less perfect crystalline structures, which, compared to its $POTM_{2000}$ counterpart, in their turn lead to increased plastic flow upon being stressed.

By subjecting samples to a given strain (e.g. 100% and 200%) the elastic recovery (ε_E), retraction due to entropic forces upon heating (ε_F) and permanent deformation (ε_P) can be determined.

Permanent deformation is clearly related to the M.W. of the hard segment. Strain-induced crystallisation, which leads to high tension set and high ε_F values, is found especially with PBA_{2000}. The higher ε_F for $POTM_{1000}$, compared with $POTM_{2000}$, is apparently related to the weaker crystalline matrix, which is partly re-established by entropic forces, developed within the soft segment upon heating.

3.3. Hardness and Tear Strength

As do tensile properties, hardness and tear strength depend on the proportion of soft segment. For copolyether-esters a practically linear relationship exists between hardness and the proportion of soft segment, independent of the polyether segment composition and regardless of whether the soft segment is incorporated by direct polymerisation or by melt-blending PBTP homopolymer with a soft grade copolyether-ester based on PBTP.

The copolyester-esters are somewhat below the line of the copolyether-esters (see Fig. 7), which is again indicative of a slightly disturbed crystallisation of the PBTP segment in the former elastomers.

The tear strength (determined in accordance with ASTM-D 624 die C) is also clearly related to the proportion of hard segment in the elastomer. In Fig. 8 the tear strength is plotted against the proportion of PBTP in copolyether-esters. For copolyester-esters the hard segment content is calculated as PBTP plus the amount of MDI connecting the hard and the soft segments. The data in Fig. 8 include values for copolyester-esters based on PBTP, MDI and aliphatic polyester-diols such as PEA, PBA, PCL and a copolyester from hexamethylene glycol, neopentyl glycol and adipic acid.

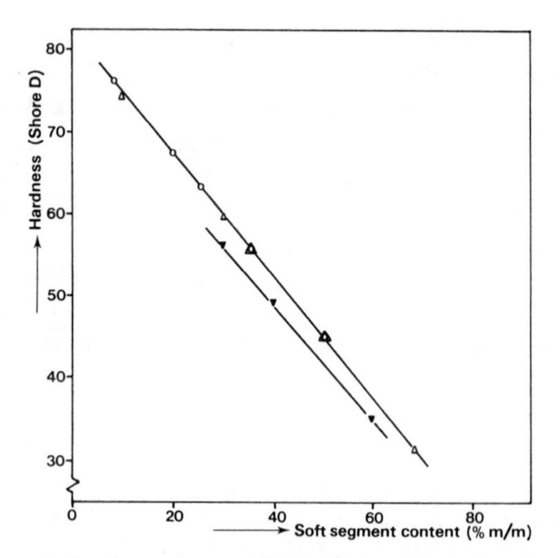

FIG. 7. Shore D hardness versus soft segment content. △, Copolyether-esters based on PBTP and POE, POP and POTM; ○ blends of PBTP and copolyether-esters; ▼, copolyether-esters based on PBTP, MDI and PBA.

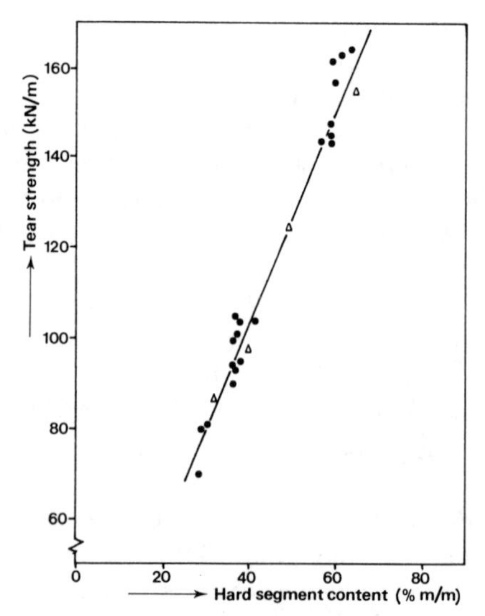

FIG. 8. Tear strength versus hard segment content. △, Copolyether-ester; ●, copolyester-ester.

3.4. Dynamic Mechanical Properties

Dynamic mechanical measurements on copolyether-esters have been reported by various authors.[37-9] Apart from the patent literature,[16] few data exist on segmented copolyester-esters. Dynamic modulus measurements are a very useful tool in establishing the temperature range in which segmented elastomers can be used (see also Chapter 6). The modulus-temperature behaviour, hysteresis, set properties and low and high temperature behaviour all depend on the degree of phase separation between soft and hard segments.

The dynamic mechanical properties of a typical segmented copolyester-ester, (PBA–MDI–PBTP–MDI), of a copolyether-ester (POTM–PBTP), and of polyurethanes with PCL and with POTM as soft segments, are characterised by a single broad glass transition around 0 °C (tan δ_{max}) as demonstrated in Fig. 9. As the soft (PBA) and hard domains (PBTP) are relatively small, the copolyester-ester does not show the separate soft and hard segment glass transitions which are characteristic of a copolymer

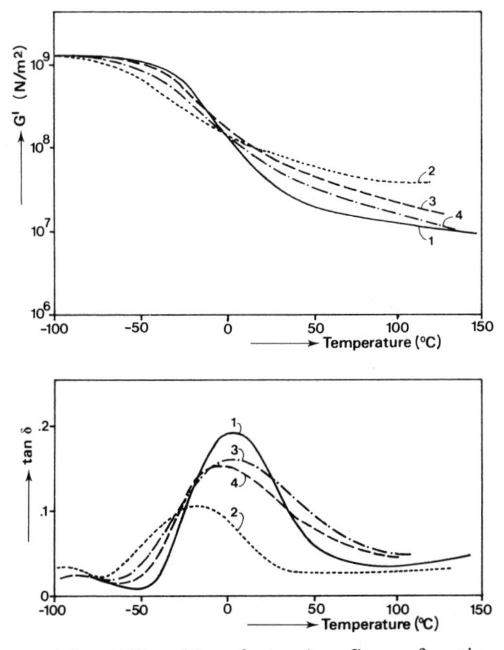

FIG. 9. Shear modulus (G') and loss factor (tan δ) as a function of temperature. (1) Copolyester-ester (50 Shore D); (2) copolyether-ester (55 Shore D); (3) polyesterurethane (55 Shore D); (4) polyetherurethane (55 Shore D).

having much larger domains (e.g. SBS). Instead, a broad glass transition region is observed. The same phenomena are observed for medium hardness copolyether-esters and thermoplastic polyurethanes. The mobility of the soft segment chains (M.W. $\simeq$1500 for PBA and M.W. $\simeq$1000 for POTM) is very much obstructed by the hard segment chains to which they are coupled.

Phase separation, normally found to a lesser extent in copolyester-esters than in copolyether-esters, also leads to reduced soft segment mobility. As a result, the initial soft segment T_g ($\simeq -68\,°C$ for PBA and $\simeq -80\,°C$ for POTM) is *raised*. In both polymer types the M.W. of the hard segment is relatively small (about 1500 in the copolyether-ester and about 2500 in the copolyester-ester). This implies that the T_g of the hard segment will be *lower* than the value obtained for the homopolymer (PBTP) having a high M.W. ($T_g \simeq +50\,°C$). The shifts of the two T_g values and the fact that both polymers are randomly segmented copolymers, result in a single broad glass transition. Thus, the shear modulus does not show two distinct drops, as might be expected with complete phase separation, but varies gradually over a broad temperature range in the transition from the glassy to the rubbery state.

In comparing the four elastomers in Fig. 9 it appears that the copolyester-ester shows more resemblance to the polyurethanes than to the copolyether-ester. An important advantage, however, of the copolyether-ester as well as of the copolyester-ester over the polyurethanes is their slower decrease in modulus level at higher temperatures. In the case of the polyurethanes this is caused by the dissociation of the hydrogen bonds within the urethane hard domains. The copolyester hard segments, however, are thermally stable up to their melting points which lie in the range 200–225 °C, approximately, depending on the segment M.W.'s.

As shown in the lower part of Fig. 9, the tan δ maximum strongly depends on the composition and the high damping of the copolyester-ester compared with that of the copolyether-ester is reflected in the rebound values (35 % versus 55 % at 20 °C).

In thermoplastic elastomers, hysteresis is strongly related to such phenomena as soft segment crystallisation and the yielding and orientation of hard segment crystals. It leads to a permanent set, especially at high elongations.

Comparison of the first five hysteresis cycles of three elastomers (47–50 Shore D) reveals that in all cases the hysteresis loss in the first cycle is substantial and amounts to about 70 %. The next cycles show a strong Mullins effect (Fig. 10).

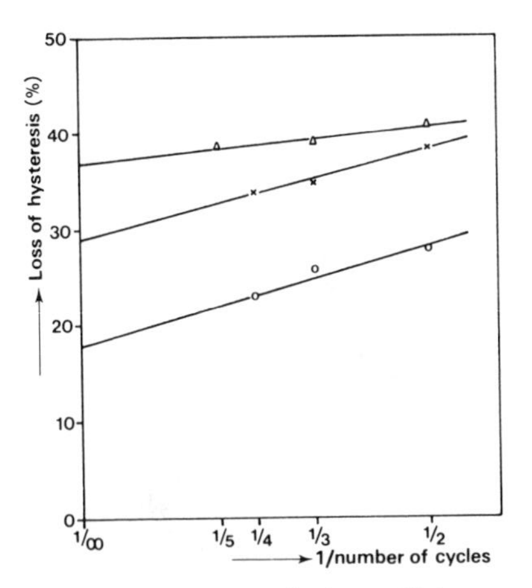

FIG. 10. Hysteresis loss (100% elongation). △, Polyesterurethane (46 Shore D); ×, copolyester-ester (48 Shore D); ○, copolyether-ester (46 Shore D).

In the second and following cycles the copolyether-ester displays the least hysteresis loss, followed by the copolyester-ester and the polyester-urethane (see Fig. 11). In the ultimate permanent deformation, however, after five cycles, the reverse order is found.

3.5. Low-temperature Impact Behaviour

The low-temperature brittleness of an elastomer is governed by the glass transition range of the soft segment. As shown below, the glass transition temperatures of soft segment homopolymers are far below 0 °C:[41]

POE	−67 °C
POP	−73– −61 °C
POTM	−88– −79 °C
PEA	−70– −50 °C
PBA	−68– −60 °C
PCL	−60 °C

Because of the shorter length of the polyether or polyester soft segments, as used in a copolyether-ester or copolyester-ester elastomer, the glass transition tends to lower temperatures.[35] This effect is counteracted, however, by coupling the soft segments to the hard segments, resulting in reduced mobility of the soft segments. A further reduction of soft segment mobility is brought about by incomplete phase separation, leading to increased obstruction of free movement of the soft segments.

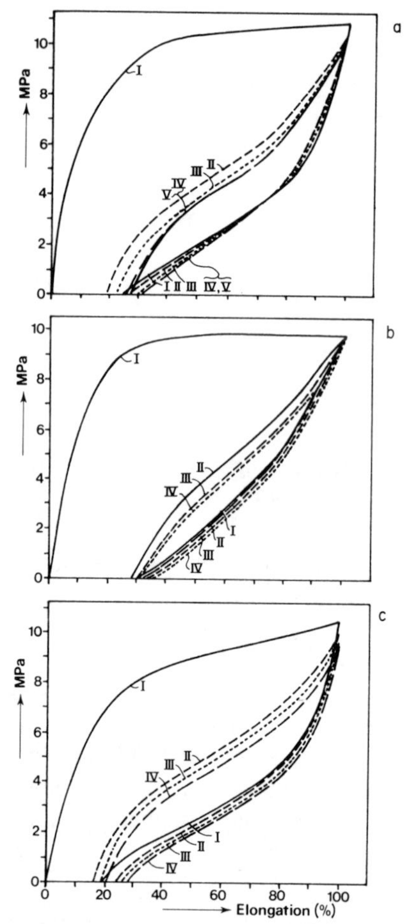

FIG. 11. Hysteresis of elastomers (ISO II bar; 100 mm min^{-1}). (a) Copolyester-ester (48 Shore D); (b) copolyether-ester (46 Shore D); (c) polyesterurethane (46 Shore D).

In the case of copolyester-esters two other important features contribute to further reductions in mobility, thus leading to increased glass transition temperatures: allophanate crosslinks within the diisocyanate-extended aliphatic polyester and randomisation between aliphatic and aromatic polyester sequences, giving rise to segments displaying greatly reduced chain mobility.

Recognising that polyether soft segments give rise to a more complete phase separation than polyester soft segments,[40] the net result leads to

TABLE 5
IMPACT RESISTANCE OF SEGMENTED ELASTOMERS (Charpy METHOD WITH MOULDED U-SHAPED NOTCH IN ACCORDANCE WITH ISO R 179)

Elastomer	*Shore D hardness*	*Test temperature*		
		−34°C	*−43°C*	*−50°C*
Copolyether-ester:				
POTM-PBTP	40	n.b.[a]	n.b.	n.b.
	50	n.b.	n.b.	n.b.
	55	n.b.	30[b]	25
	63	—	3	—
Copolyester-ester:				
PBA-MDI-PBTP-MDI; NCO/OH ratio 1·10	39	n.b.	n.b.	18
	49	n.b.	20	7
Polyesterurethanes:				
Pellethane 2102-80A	37	n.b.	n.b.	n.b.
Pellethane 2102-90A	47	n.b.	20	—
Desmopan 485	36	n.b.	n.b.	n.b.
Desmopan 150-S	50	24	6	—

[a] No break.
[b] Break energy expressed in $kJ\,m^{-2}$.

lower glass transition temperatures for segmented copolyether-esters than for copolyester-esters. For both types of elastomer the soft segment T_g is found to increase with increasing hard segment content, obviously because of an increasing portion of the soft segments being incorporated into an interlamellar compatible amorphous phase along with the aromatic polyester component.[37]

In Table 5 the low-temperature impact behaviour of various segmented elastomers is compared in the Charpy impact test, using specimens with a moulded notch. The copolyether-esters are clearly superior to copolyester-esters and polyesterurethanes. Down to −43 °C these two last-mentioned elastomers display identical impact behaviour and at even lower temperatures the soft polyesterurethanes show a better impact-resistance than the copolyester-esters.

3.6. Properties at Elevated Temperatures

As already pointed out in Section 3.4, segmented copolymers show several main transition ranges in their passage from low to high temperatures. A very useful technique of studying these transitions is DSC. At temperatures

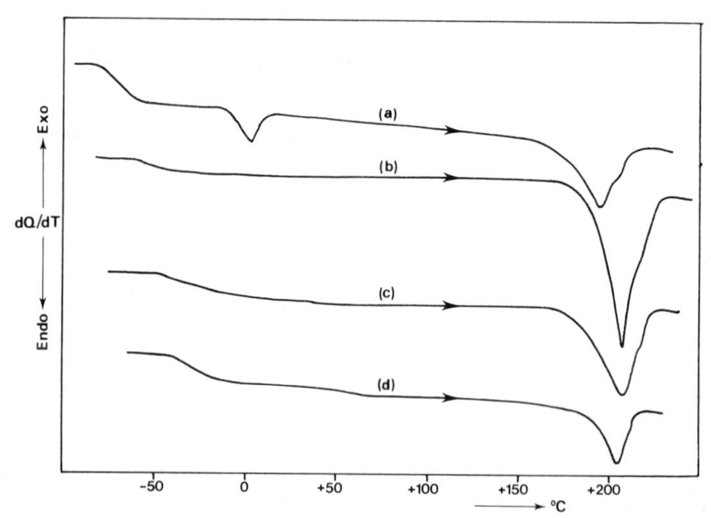

FIG. 12. DSC curves for injection-moulded specimens. du Pont 990 TA, size 10 mg; heating-rate, 20°C min^{-1}. (a) POTM–PBTP, 40 Shore D; (b) POTM–PBTP, 55 Shore D; (c) PBA–MDI–PBTP, 50 Shore D; (d) PBA–MDI–PBTP, 38 Shore D.

below 0 °C, DSC measurements show the soft segment transition from the glassy to the rubbery state. The magnitude and position of this transition is influenced by the nature and the properties of the soft segment, by the degree of phase separation between the soft and hard segments, and—in the case of copolyester-esters—by the additional chemical (i.e. allophanate) crosslink density.[60]

In the case of copolyether-esters and copolyester-esters based on PBTP, the T_g of these hard segments is hardly perceptible. At still higher temperatures (150–220 °C) the melting endotherm of the hard PBTP segments is observed. This is also highly dependent on phase separation and composition. A narrow well-defined endotherm points to good phase separation and to relatively few disturbed crystallites, as in the case of 55 Shore D (or harder) copolyether-esters based on POTM and PBTP, as shown in Fig. 12. The endotherm found around 0 °C in copolyether-ester 40 Shore D (Fig. 12(a)) is caused by crystalline POTM blocks having a M.W. of 2000. POTM blocks having a M.W. of 1000 do not show crystallisation, as can be seen from curve b.

In the case of copolyester-esters based on PBTP, MDI and PBA a somewhat broader melting range is observed, which is indicative of less perfect crystallites due to the disturbing influence of the bulky MDI-based

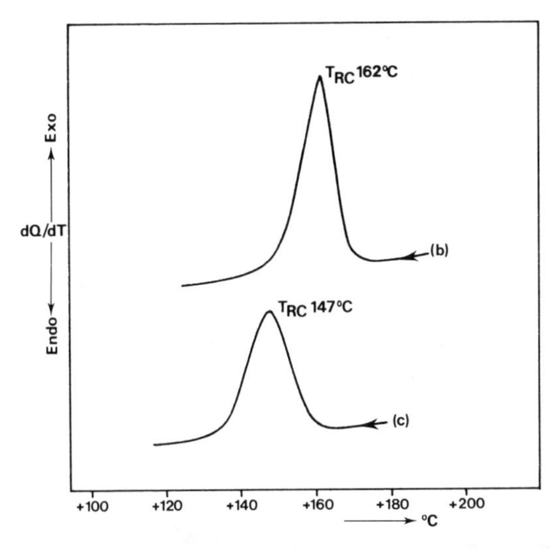

FIG. 13. DSC curves obtained on cooling from the melt; cooling rate, 20 °C min^{-1}.

group. Lower hardness grades show a shift of the melting peak to lower temperatures. In the case of the copolyether-esters this is caused mainly by the lower M.W. of the hard segment, which depends on the composition.[37] A 38 Shore D copolyester-ester sample showed an increased T_g compared with the 50 Shore D material. This may be due to allophanate crosslinks within the soft segment, leading to decreased chain mobility, thus raising the T_g. (Lower hardness copolyester-ester samples were prepared using an NCO/OH ratio of 1·32 for the aliphatic prepolymer (see Table 2).)

Figure 13 demonstrates the crystallisation behaviour of a copolyether-ester (POTM-PBTP) and a copolyester-ester (PBA-MDI-PBTP) of comparable hardness.

It is obvious that the decline in mechanical properties at high temperatures depends mainly on the behaviour of the hard segment domains and additional chemical crosslinks. As in both copolyester-esters and copolyether-esters the hard segment consists of PBTP, a similar behaviour may be expected.

The thermal resistance of the hard segment in thermoplastic polyurethanes is governed by hydrogen bonding of the hard segments. A comparison of three segmented elastomers is given in Table 6, showing the heat distortion temperatures. The heat sag resistance (Fig. 14) of various elastomers also shows the better thermal resistance of copolyether-esters

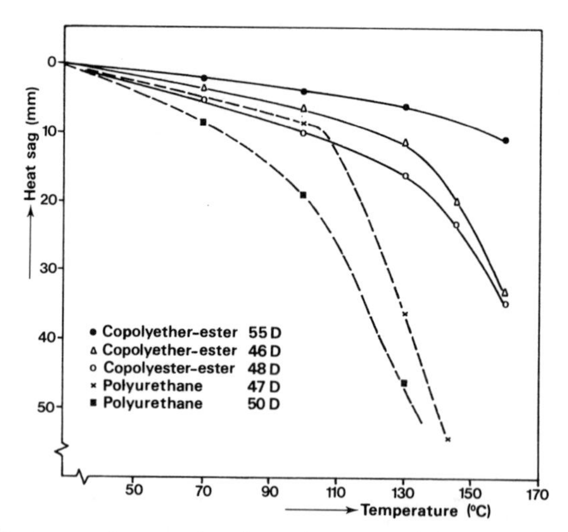

FIG. 14. Heat sag resistance of various elastomers. A 2-mm thick strip is clamped on one side, in such a manner that the free side extends horizontally over 100 mm. The clamped strip is heated for 1 h at the appropriate temperature, followed by determination of the sag.

and copolyester-esters based on PBTP as the hard segment, compared with a polyesterurethane having a hard segment based on MDI and 1,4-butanediol.

4. PHYSICAL AND CHEMICAL PROPERTIES

4.1. Burning Behaviour

Polyether-esters made from aliphatic polyethers having a C/O ratio of 2:4 and with poly(butylene terephthalate) hard segments all show severe burning in air.

The intrinsically high inflammability of the polyether component increases the inflammability of the copolyether-esters as shown by the LOI index. When tested in conformity with the UL-94 test, all polymers of the copolyether-ester series, as well as the PBTP homopolymer, are rated as UL-94 HB (Table 7).

Despite the inflammability shown by the LOI and UL 94 test, all copolyether-ester grades (without flame retardant) are said to satisfy the Federal Motor Vehicle Safety Standard FMVSS-302.[42]

To meet flame retarding requirements, the use of various additives is

TABLE 6
Vicat TEMPERATURES OF THREE THERMOPLASTIC ELASTOMERS
(ASTM D-1525; load = 10·2 N)

Elastomer	*Vicat temperature*
Copolyether-ester: (POTM-PBTP; 55 Shore D)	180 °C
Copolyester-ester: (PBA-MDI-PBTP; 52 Shore D)	167 °C
Polyesterurethane: (PCL-MDI-4G; 55 Shore D)	135 °C

described in the literature. These additives are mostly based on the conventional system of a mixture of halogenated organic compounds and antimony trioxide as a catalyst. The resulting flame retardancy varies with the amount and composition of the flame retardant used.

du Pont report an LOI value of 25·4–27·4 % and a non-burning rating (in accordance with ASTM D-635) when 100 parts of Hytrel 55 Shore D are compounded with 15 parts Sb_2O_3 and 10 parts of chlorinated paraffin, tetrabromophthalic anhydride or hexabromocyclododecane. A still better flame retardancy is claimed[42] when 15 parts of a masterbatch (containing Sb_2O_3 and a brominated compound in 40 Shore D copolyether-ester) were mixed with 100 parts of 55 Shore D material. This composition is reported to give an LOI of 30 %.

Compared with the copolyether-esters and polyurethanes, the copolyester-esters based on PBA, MDI and PBTP show a lower inflammability. In the UL-94 test, 38 and 50 Shore D materials are rated as V-2. The LOI values for copolyester-esters of the above mentioned composition are higher than those for polyurethanes and copolyether-esters.

TABLE 7
LOI AND UL-94 RATINGS OF COPOLYETHER-ESTERS BASED ON POTM AND PBTP

Shore D (ASTM D-2240)	*LOI (%) (ASTM D-2863)*	*UL 94*
40	19·5	HB
46	20·0	HB
55	21·0	HB
63	21·0	HB
72	21·5	HB
PBTP homopolymer	23·0	HB

TABLE 8
BURNING RATES OF VARIOUS THERMOPLASTIC ELASTOMERS (INJECTION-MOULDED SAMPLES, 100 × 6 × 2 mm)

Elastomer	*Shore hardness*	*LOI*	*Burning time (s)*	*Burning distance (mm)*
Copolyether-ester (POTM-PBTP)	40 D	19·5	140	100
Polyurethane (polyester type)[a]	36 D	21	52	40
Polyurethane (PCL-MDI-4G)	37 D	21	30	25
Copolyester-ester (PBA-MDI-PBTP)	39 D	25	0	0
Copolyester-ester (PBA-MDI-PBTP)	50 D	26	0	0

[a] Composition unknown.

The lower inflammability of the copolyester-esters is also demonstrated in a horizontal burning test (Table 8). All samples showed heavy dripping, but whereas the copolyester-esters cease burning immediately after removal of the ignition flame (10 s), the copolyether-ester remains burning.

4.2. Ageing

4.2.1. Hydrolysis

Segmented copolyether-esters and copolyester-esters, which both contain aromatic polyester (PBTP) segments, are sensitive to hydrolysis and this sensitivity is greater in the copolyester-esters which contain aliphatic polyester segments. As is well-known, polyester segments in elastomers can be stabilised by the addition of carbodiimides.[43] Thus, the addition of Stabaxol-P* increases the half-life in tensile strength and elongation to a considerable extent.

The rate of hydrolysis is affected by the concentration of carboxyl end groups (e_c) in the elastomer. It has been found that addition of as little as 1·5 % of Stabaxol-P to copolyether-esters causes the carboxyl content to be reduced from about 25 to about 1–2 meq/kg of polymer. This reduction causes a considerable increase in the half-life in boiling water as shown in Table 9. As not all of the carbodiimide added is consumed in reducing the carboxyl content in the mixing procedure, the excess is available as an effective scavenger for carboxyl groups arising from hydrolysis during end use.

As may be expected, copolyester-esters based on aliphatic and aromatic polyester segments coupled by MDI show a hydrolytic behaviour which is far inferior to that of copolyether-esters. The tensile strength half-life

* Poly(carbodiimide), a product of Bayer AG, W. Germany.

TABLE 9
TENSILE ELONGATION HALF-LIFE (DAYS) OF COPOLYETHER-ESTERS (POTM-PBTP) IMMERSED IN BOILING WATER (ISO II BARS)

% POTM	*% (m/m) Poly(carbodiimide) added*			
	0	*0·75*	*1·0*	*1·5*
25	17	28	40	>119
35	26	40	98	168
60	51	>119	>168	>168

(immersion in boiling water) for copolyester-esters of 50 Shore D varies between 45 and 100 h, compared with 45–180 h for commercial ester-based polyurethanes and over 500 h for copolyether-esters. The decline in tensile properties has been measured for a copolyester-ester based on PCL_{1860}, $PBTP_{3060}$ and MDI. The elastomer contained 40% PCL and had an NCO/OH ratio of 1·15. As shown in Fig. 15, the tensile strength decreased whilst the tensile elongation remained constant or slightly increased. This phenomenon can be ascribed to the hydrolysis of allophanate crosslinks, the breakdown of which soon leads to a polymer having a lower stress build-up. Thereafter, hydrolysis of the aliphatic polyester segments becomes predominant and soon results in a fast decrease in tensile strength and in elongation at break.

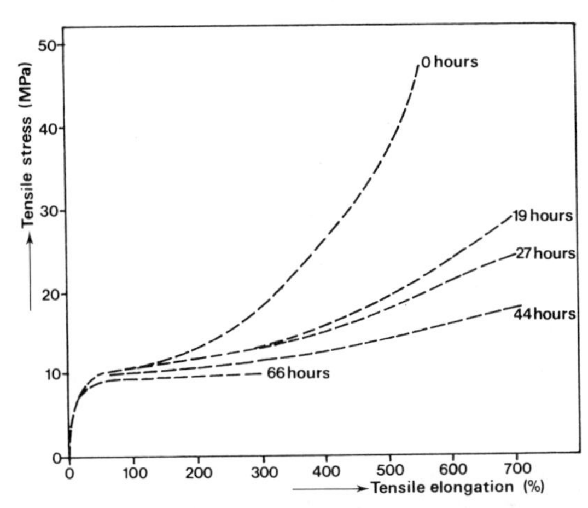

FIG. 15. Change in tensile behaviour of a PCL_{1860}–MDI–$PBTP_{3060}$ copolyester-ester (50 Shore D) on immersion in boiling water.

As with copolyether-esters, melt-blending of these segmented copolyester-esters with 1·5 % Stabaxol P enhances the hydrolytic stability about four-fold.

4.2.2. Oxidative Degradation

Copolyether-esters are generally stabilised with a single antioxidant serving both as processing and end-use antioxidant. Extensive experiments with a number of commercial antioxidants led to the conclusion that the addition of 0·5 % of a hindered phenolic antioxidant during polymerisation (e.g. Irganox 1098) resulted in satisfactory stabilisation for normal use. For high temperature use (up to 150 °C), however, the addition of a thiodipropionate-based co-antioxidant is necessary to maintain prolonged service life. The choice of a particular antioxidant system (with or without synergistic components) depends on its behaviour during polymer preparation (effectivity, volatility) and in end use (polymer colour, blooming, volatility and ability to interrupt oxidation reactions). It is obvious that thermal oxidative stability depends on the amount and type of polyether present in the elastomer. In this respect the life-time of the POP soft segment-based copolyether-esters is about half that of elastomers based on POE or POTM.

Copolyester-esters have an inherently better stability against oxidative

TABLE 10

DECREASE IN TENSILE STRENGTH (MPa) AND TENSILE ELONGATION (%) OF COPOLYETHER-ESTER AND COPOLYESTER-ESTER AT 140 °C IN AIR (INJECTION-MOULDED TESTBARS: ISO II, CROSS HEAD SPEED 200 mm/min)

Time (h)	*Copolyether-ester*				*Copolyester-ester*			
	1[a]		*2*[b]		*3*[c]		*4*[d]	
	MPa	%	*MPa*	%	*MPa*	%	*MPa*	%
0	27	500	27	500	40	550	43	580
120	26	475	27	500	38	680	35	650
360	22	415	27	490	35	660	33	700
700	11	200	25	450	33	670	29	670
1 000	—	—	23	400	31	640	30	680

[a] $POTM_{1000}$-PBTP; 55 Shore D; 0·5 % hindered phenolic antioxidant.
[b] As above, 0·5 % dilauryl thiodipropionate added.
[c] PBA_{2000}-MDI-$PBTP_{3060}$; NCO/OH = 1·15; 50 Shore D.
[d] PCL_{1860}-MDI-$PBTP_{3060}$; NCO/OH = 1·15; 48 Shore D.

degradation and therefore do not require protection by antioxidants. From the figures in Table 10 it is clear that an unstabilised copolyester-ester still functions as an elastomer after 1000 h in air at 140 °C.

The decrease in tensile strength observed in these polymers is probably caused by hydrolysis of allophanate crosslinks.

4.2.3. Resistance to Ultraviolet Radiation

The polyether segment in copolyether-esters is liable to UV radiation-induced oxidation.

As shown in Fig. 16 the results of degradation depend on the type of antioxidant used. The stability against UV-initiated degradation can be improved by incorporating a suitable UV stabiliser. Both the mechanical properties and the colour of the polymer are affected by UV radiation. The amine-based antioxidant (curve a in Fig. 16) causes the test specimen to turn a dark brown. The phenolic antioxidant with or without UV stabiliser causes the polymer to turn a light yellow after 1000 h exposure.

The mechanical properties of copolyester-esters show only slight degradation after 1000 h UV exposure, the colour changing from white to yellowish-brown. In this respect they show a marked resemblance to polyester-based polyurethanes.

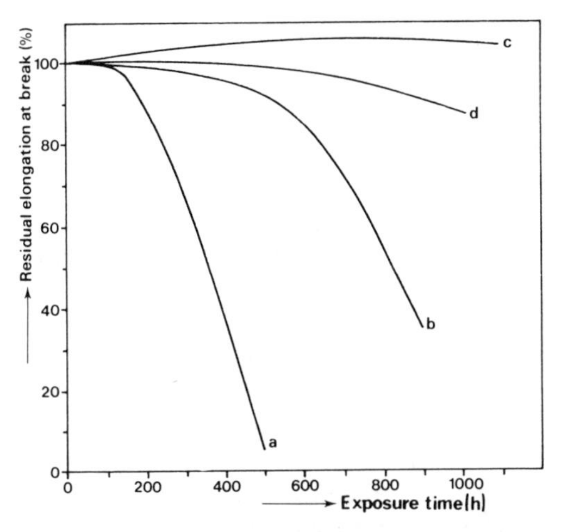

Fig. 16. Resistance to UV radiation (140–160 klux) of a POTM–PBTP copolymer containing 35 % (m/m) POTM. (a) Stabilised with amine-based antioxidant; (b) stabilised with hindered phenol antioxidant; (c), as (b), except that Tinuvin 770 was added; (d), as (b), except that Tinuvin 327 was added.

5. PROCESSING

5.1. Introduction

Segmented copolyether-esters and copolyester-esters can be processed by injection moulding, extrusion and rotational moulding. The choice depends on rheological data obtained by measuring the melt viscosity as a function of the shear rate. Such measurements are carried out in the authors' laboratory with the aid of a Göttfert high pressure capillary melt viscometer.

Ranges for the rheological data required for the various processing techniques are summarised in Table 11. DSC measurements provide supplementary information about processing parameters. Thus, the melting range together with the shape and height of the melt peak determine the minimum processing temperature. The recrystallisation peak temperature and range are indicative of the mould temperature to be used.

TABLE 11
PROCESSING TECHNIQUES AND RHEOLOGICAL DATA

Processing technique	*Shear rate* (s^{-1})	*Viscosity* (*Pa.s*)
Injection moulding	10^3–10^4	5×10^1–5×10^2
Extrusion	10^2–10^3	5×10^2–5×10^3
Calendering	10^1–10^2	5×10^3–5×10^4
Compression moulding	10^0–10^1	5×10^4–5×10^5
Rotational moulding	$<10^0$	$<10^3$

5.2. Melt Viscosity

5.2.1. Copolyether-esters

Figure 17 shows the melt viscosity of a copolyether-ester (POTM–PBTP, containing 35% m/m POTM, 55 Shore D) as a function of shear rate at various temperatures.

The characteristics of an injection moulding type (Arnitel-EL 550) having a relative viscosity of 2·60 (1% m/m, in *m*-cresol at 25 °C) are given in Fig. 17(a). Solid phase post-condensation of this sample (20 h at 195 °C in vacuum) is attended with increasing M.W. which provides an extrusion type (Arnitel-EM 550, $\eta_{rel} = 3{\cdot}20$) having the rheological properties shown in Fig. 17(b). The curves of the melt viscosity between 220 °C and 240 °C show that post-condensation causes the viscosity to increase by a factor of two. Comparing the results in Fig. 17 with the rheological data (Table 11), it is evident that these copolyether-esters can be processed by injection moulding, extrusion or rotational moulding.

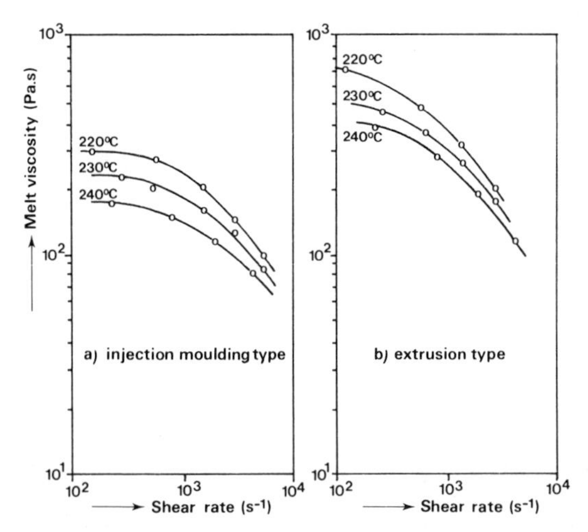

FIG. 17. Melt viscosity of a segmented copolyether-ester (55 Shore D).

5.2.2. Copolyester-esters

DSC measurements show that copolyester-esters have a lower melting range and hence lower processing temperature than copolyether-esters.

The melt viscosity of these diisocyanate-coupled copolyester-esters depends strongly on the type of diisocyanate used. Figure 18 shows the melt viscosity versus shear rate curves of PBA-PBTP copolyester-ester elastomers coupled by TDI and by MDI.

It is evident that the MDI-containing copolyester-ester has a much higher melt viscosity than the TDI-coupled analogue. Therefore the MDI-based polymer can be processed both by injection moulding and extrusion, whereas the TDI-based product is less suitable for extrusion because of too low a melt viscosity. This different rheological behaviour is probably attributable to the different dissociation temperatures of urethane groups originating from MDI and TDI.

$$R_1\text{—NH—}\overset{\overset{\displaystyle O}{\|}}{C}\text{—O—}R_2 \rightleftharpoons R_1\text{—NCO} + \text{HO}R_2$$

Some clue to this phenomenon is found in TGA measurements (heating rate = 10 °C min^{-1}) in which the loss of weight of the MDI product starts at 270 °C whereas with the TDI-coupled copolyester-ester it commences at 240 °C. As the processing temperatures for both types are almost the same, the greater stability of the MDI polymer might well account for its different viscosity behaviour.

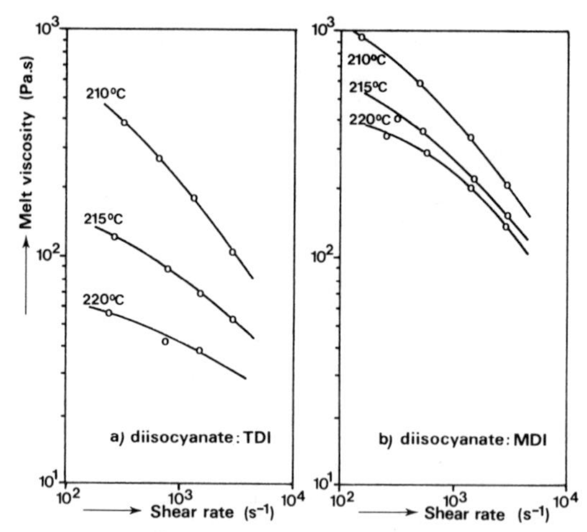

FIG. 18. Melt viscosity of segmented copolyester-esters coupled with (a) TDI and (b) MDI. (NCO/OH = 1·10.)

The melt viscosity of a TDI-based copolyester-ester can be increased by incorporation of a trifunctional compound, e.g. trimethylolpropane (TMP). Adding 0·5 % m/m TMP to the TDI-based copolyester-ester causes the melt viscosity to increase to twice its original value, leading to a product which can be extruded into film, tube, hose, etc.

As mentioned before, coupling with an excess of diisocyanate introduces allophanate structures. The influence of excess diisocyanate on the melt viscosity has been investigated on TDI-based copolyester-esters having NCO/OH ratios ranging from 1·1 to 1·3. The results, summarised in Table 12, show that the melt viscosity decreases only slightly with increasing NCO/OH ratio.

Summarising, it can be concluded that both TDI- and MDI-coupled

TABLE 12
INFLUENCE OF THE NCO/OH RATIO ON THE MELT VISCOSITY (MELT VISCOSITY IN Pa . s; SHEAR RATE = $10^3\,s^{-1}$)

Temperature (°C)	*Melt viscosity at NCO/OH ratio of:*		
	1·1	*1·2*	*1·3*
205	$4{\cdot}6 \times 10^2$	$3{\cdot}7 \times 10^2$	$2{\cdot}8 \times 10^2$
210	$2{\cdot}2 \times 10^2$	$1{\cdot}7 \times 10^2$	$1{\cdot}3 \times 10^2$
215	$0{\cdot}83 \times 10^2$	$0{\cdot}66 \times 10^2$	$0{\cdot}59 \times 10^2$
220	$0{\cdot}42 \times 10^2$	$0{\cdot}31 \times 10^2$	$0{\cdot}31 \times 10^2$

copolyester-esters can be injection-moulded and extruded. However, extrusion of a TDI-coupled product requires the introduction of a trifunctional component in order to increase the melt viscosity under processing conditions.

5.3. Processing Practice

In practice the following techniques are used for processing polyether-esters and polyester-esters:

1. injection moulding;
2. extrusion;
3. rotational moulding;
4. melt casting;
5. blow moulding.

5.3.1. Injection Moulding

Both polyether-esters and polyester-esters can be processed by injection moulding. In principle, both types of polymer can be processed on all types of injection moulding machines. Preference is given, however, to screw-plasticising injection moulding machines because their use gives rise to satisfactory homogeneous plasticised material.

Three-zone screws having a length of 15–24 D and a compression ratio of 1 to 3–3·5 are suitable for the injection moulding of both types of polymers.

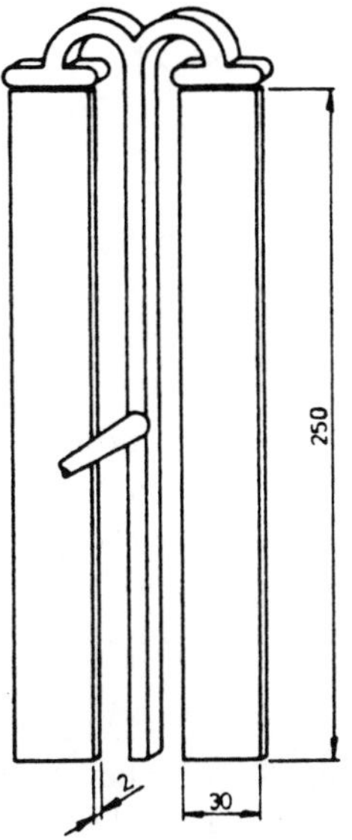

FIG. 19. Injection-moulded double strip.

TABLE 13
MOULDING CONDITIONS FOR COPOLYETHER-ESTERS[62]

Moulding	strip, double (see Fig. 19)				
Dimensions	250 × 30 mm				
Weight	20 g per strip; total weight including sprue 60 g				
Wall thickness	2 mm				
Gating system	flash gate				
Gate dimensions	length 1 mm, width 30 mm, height 0·8 mm				
Machine data:					
maximum shot capacity	190 cm³				
maximum injection pressure	165 MPa				
screw diameter	45 mm				
mould clamping force	1500 kN				
Hardness (Shore D)	40	46	55	63	74
Cylinder temperature (°C):					
zone 1 (hopper)	200[a] (170)[b]	210	220	220	225
zone 2	200 (180)	210	220	225	230
zone 3	210 (190)	215	225	230	235
Nozzle temperature (°C)	215 (200)	220	230	235	240
Melt temperature (°C)	220 (205)	220	240	245	250
Mould temperature (°C)	50	50	50	50	50
Injection pressure (MPa)	80	80	85	100	120
Dwell pressure (MPa)	50	50	60	80	100
Back pressure (MPa)	6	6	6	6	6
Injection + holding time (s)	15	15	9	7·5	7·5
Cooling time (s)	15	13	15	13	13
Plasticising time (s)	12	12	10	10	10
Total cycle time (s)	38	38	28	25	25
Screw speed (min^{-1})	75	75	75	75	75

[a] Figures for Arnitel from Akzo Plastics.
[b] Figures for Hytrel–4055 from du Pont.

To ensure an even granule feed and optimum plasticisation of the melt, wide fluctuations in cylinder temperatures should be avoided.

The moisture content of the granules must not exceed 0·1 %. When exposed to the air, the granules absorb some water; therefore to avoid processing problems and any adverse effect on the quality of the mouldings, drying may be necessary.

Typical injection moulding conditions for copolyether-esters (POTM–PETP, POTM–PBTP, POE–PBTP and POP–PBTP) and copolyester-esters are described in Tables 13 and 14.

TABLE 14

COMPARISON OF MOULDING CONDITIONS OF COPOLYETHER-ESTERS AND COPOLYESTER-ESTERS

Moulding	strip; 150 × 40 × 2 mm
Shot volume	14 cm^3
Machine data:	
maximum shot capacity	67 cm^3
maximum injection pressure	172 MPa
screw diameter	30 mm
mould clamping force	500 kN

	POTM–PBTP	*POTM–PETP*	*POE–PBTP*	*POP–PBTP*	*PBA–MDI–PBTP*
Hardness (Shore D)	55	53	56	55	50
Cylinder temperature (°C)					
zone 1 (hopper)	210	245	225	225	190
zone 2	215	250	225	225	190
zone 3	220	255	230	230	200
Melt temperature (°C)	240	260	240	235	210
Mould temperature (°C)	20	140[a]	25	40	50
Screw speed (min^{-1})	100	100	100	100	75
Injection + holding time (s)	10	10	15	15	10
Cooling time (s)	15	30	15	15	20
Total cycle time (s)	30	45	33	33	33

[a] Mould temperature can be lowered from 140 °C to 90 °C without detracting from the mechanical properties when use is made of a nucleating agent (e.g. sodium benzoate).

TABLE 15

TYPICAL PROCESSING CONDITIONS FOR 55 D POTM–PBTP COPOLYETHER-ESTERS[63]

Extrusion product	*Tube*	*Flat film*	*Blown film*
Process	Vacuum calibration	Chill roll method	Vertical upwards
Product dimensions (mm)	Outer diameter = 11 Inner diameter = 8·4	Width = 800 Thickness = 0·050	Width = 500 Thickness = 0·05
Screw type	Three-zones	Three-zones	Three-zones
diameter (mm)	45	60	45
L/D ratio	24	21	25
compression ratio h_1/h_2 (mm)	3	3	2·5
length of feed zone	9D	6D	10D
length of compression zone	5D	6D	5D
length of metering zone	9D	9D	10D
Die	Tubular die: diameter outer die = 21·0 mm diameter mandrel = 15·9 mm	Adjustable lip: die lip width = 800 mm die lip opening = 0·8 mm	Centre-feed head diameter mandrel = 80 mm die opening = 0·7 mm
Screens (mesh, in^{-1})	60 + 120 + 60	60 + 120 + 60	60 + 120 + 60
Calibration	Tubular vacuum sizing	—	—
Take-off system	Caterpillar	Rolls	Nip rolls
Temperature hopper dryer (°C)	95	95	95
Barrel temperature, hopper to die (°C)	200/205/210/220	220/230/240/240	200/200/210/210
Die temperature (°C)	220/210/210	230/230/230/230	220/220/220/220
Polymer temperature (°C)	225	230	220
Polymer pressure (MPa)	13	3	20
Screw speed (min^{-1})	32	28	32
Blow-up air pressure (mm H_2O)	—	—	110
Frost line height (cm)	—	—	20
Take-off height (cm)	—	—	190
Take-off speed (m min^{-1})	4	8	20
Chill roll temperature (°C)	—	15	—
Second roll temperature (°C)	—	15	—
Chill roll tangent (°)	—	45	—

5.3.2. Extrusion

Most of the copolyether-esters and copolyester-esters can be processed by several extrusion techniques, provided that the melt viscosity is sufficiently high. The POTM–PETP copolyether-ester is less suitable for extrusion because of its unduly low rate of crystallisation.

Suitable extruders are those which are used in the extrusion of polyamides and polyolefines. Best results are obtained with the use of three-zone screws, having a length to diameter ratio (L/D) of at least 20.

Typical extrusion conditions for the production of tube, flat film and blown film from POTM–PBTP copolyether-esters (55 D) are described in Table 15.

5.3.3. Rotational Moulding

As stated in Section 5.2, the low melt viscosities of copolyether-esters and copolyester-esters at low shear rates permits them to be processed by rotational moulding. To that end it is preferred that the materials should be milled.

Data for rotational moulding of 55 D POTM–PBTP copolyether-ester are exemplified in Table 16. The rotomoulding machine used is of the McNeil type.

5.3.4. Melt Casting

The copolyether-esters and copolyester-esters can be very satisfactorily processed by melt casting, especially when small series of items (prototypes), or items having a complex shape are to be obtained. Aluminium

TABLE 16
ROTATIONAL MOULDING WITH POTM–PBTP COPOLYETHER-ESTER

Polymer	POTM–PBTP—55 D[a]
Particle size	150–500 μm
Rotomoulded object	Fuel tank
Dimensions	38 × 30 × 26 cm
Weight	2·3 kg
Temperature of air oven	375 °C
Residence time in oven	12 min
Cooling time in air	1 min
Cooling time water spray	5 min
Revolutions, A	7·5 min^{-1}
(A ⊥ B), B	10 min^{-1}

[a] Arnitel-ER 550 from Akzo Plastics, Arnhem, Holland.

TABLE 17
CHARACTERISTICS OF BLOW MOULDING POTM–PBTP COPOLYETHER-ESTERS

Property	*Conventional copolyether-ester*	*Blow moulding copolyether-ester*
Melt index (°C min^{-1})[a]	10	1·1
Melt viscosity (Pa . s)[b]	660	2790
Melt tension (g)[c]	0·2	3·1
Parison sag (cm)[d]	380	25
Die swell (%)[e]	1	43

[a] 2160 g load and 230 °C.
[b] At 59 s^{-1} and 225 °C.
[c] At 225 °C.
[d] At 230 °C.
[e] At 19 kPa and 230 °C.

or steel moulds of the type commonly used for the open casting of liquid polyurethane polymers can be employed, except that the mould cavity must be enclosed. Vent holes opposite the feed opening prevent air from being entrapped within the mould cavity. To melt cast the polymers, a conventional plastic extruder is used which is provided with a nozzle fitting in the feed opening of the mould.

Demoulding can be accomplished as soon as the polymer has sufficiently hardened (depending on the thickness of the moulded article).[44,45]

5.3.5. Blow Moulding

The standard extrusion grades of copolyether-esters and copolyester-esters cannot be blow moulded because the melt viscosity, melt strength and die swell are too low. However, a new type of copolyether-ester (POTM–PBTP) is commercially available—the blow moulding elastomer HTG-4275 of du Pont.[46] Some properties of the new product, compared with those of the conventional 55 D material, are given in Table 17 and typical conditions for its processing are summarised in Table 18.

5.4. Compounding

Although all of the above types of copolyether-esters and copolyester-esters should be suitable for compounding, examples are known only for the POTM–PBTP types.

5.4.1. Blends

A segmented elastomer can be blended with a polymer that is compatible

TABLE 18
PROCESSING CONDITIONS FOR EXTRUSION BLOW MOULDING[46]

Barrel temperature (from the rear, °C)	230/220/210
Die bushing temperature (°C)	215
Melt temperature (°C)	215
Mould temperature (°C)	45
Screw speed (min^{-1})	100
Extruder head pressure (MPa)	45–53
Blow air pressure (MPa)	0·56
Cycle (s)	
extrusion	1–3
blow	10
exhaust	2–5

with one of its segments. As long as the (physical) crosslinking remains, the blends will continue to exhibit elastomeric behaviour.[57]

Blends are known of POTM–PBTP copolyether-ester with PVC[38,47] (giving improvement of resistance to abrasion and good acoustic damping), with ethylene copolymers[49] (for reducing hardness), with acrylonitrile-butadiene copolymer[50] (for reducing hardness and improving blow mouldability), and with radial teleblock SBS-type copolymers[51] (giving blends with unexpectedly high elastic recovery).

Blends of POTM–PBTP with PBTP are used to improve the impact resistance of the pure PBTP.[52–5] Future trends in blending may be directed to the improvement of certain properties of various plastics either containing copolyether-esters or copolyester-esters.

Further, the rather high price of copolyether-esters and copolyester-esters may be brought down by blending with relatively inexpensive polymers like polyolefines, without detracting from the properties desired.

5.4.2. Plasticisers

In order to reduce the hardness and modulus of the copolyether-esters and copolyester-esters, the copolymers may be compounded with a great variety of plasticisers. Examples of POTM–PBTP copolyether-ester compounded with up to 50 phr of plasticiser, have been described.[56] As polymer substrate, a 40 D material was chosen; the harder grades of POTM–PBTP copolyether-ester have a much higher fluid resistance, as a result of which the plasticisers are generally less compatible with those polymers.

TABLE 19

GLASS FIBRE-FILLED POTM–PBTP COPOLYETHER-ESTER

Material	Flexural modulus (MPa), ISO 178	Tensile strength (MPa), DIN 53504	Elongation at break (%), DIN 53504	Charpy impact strength ($kJ m^{-2}$), ISO R 179			Heat distortion temperature (°C), ISO 75	
				23°C	−20°C	−40°C	Method A	Method B
40 D[a]	53	21	700	n.b.[d]	—	n.b.	—	—
46 D	85	22	715	n.b.	—	n.b.	—	—
55 D	185	27	460	n.b.	—	n.b.	49	110
63 D	330	32	440	n.b.	—	n.b.	57	147
74 D	830	45[c]	360[c]	n.b.	—	n.b.	48	171
40 D-G[b]	870	35[c]	17[c]	n.b.	n.b.	90	123	165
46 D-G	1 460	39[c]	10[c]	n.b.	n.b.	80	133	164
55 D-G	1 800	50[c]	8[c]	61	60	—	176	197
63 D-G	2 750	64[c]	6[c]	54	50	—	191	205
74 D-G	4 700	93[c]	4[c]	44	40	—	205	218

[a] Figures for Arnitel copolyether-ester, Akzo Plastics; hardness in Shore D in accordance with ISO R 868.
[b] G indicates 30 % glass fibre-reinforcement.
[c] In accordance with ISO R 527.
[d] No break.

5.4.3. *Glass Fibres*

Glass fibre reinforcement can be used to raise the flexural moduli of the elastomeric copolymers to the level of those of non-reinforced thermoplastic polyesters and polyamides.[57,58] The heat distortion temperature is also raised dramatically. However, the glass fibre-reinforced elastomers show considerable decrease in tensile impact strength and an extremely severe loss of elongation. Some relevant data on glass fibre-filled POTM–PBTP copolyether-esters are shown in Table 19.

5.4.4. *Carbon Black*

Pigmentation of all kinds of polymers with carbon black is common practice. However, addition of electrically conductive types of carbon

TABLE 20
ELECTRICAL RESISTANCE OF POTM–PBTP WITH THE ADDITION OF 8% EC-BLACK; INFLUENCE OF DEGREE OF DISPERSION

Type of screws	*Screw speed (min^{-1})*	*Surface resistance, Ω*[a]
Low shear	50	40
Low shear	150	70
High shear	80	2×10^3
High shear	200	2×10^{10}

[a] DIN 53482.

black (e.g. Ketjen black EC[48]) is a special feature for rendering plastics antistatic or, depending on the percentage added and the degree of dispersion, semi-conductive. The degree of dispersion of the EC-type carbon black dramatically influences the electrical resistance of the polymer, as can be seen in Table 20 from the results obtained after adding 8% carbon black to 55 D POTM–PBTP, using a twin screw extruder.

6. APPLICATIONS

As only POTM–PBTP copolyether-ester has yet been commercialised, we will confine ourselves to describing some main uses realised for this polymer. Possible uses of other copolyether-esters and copolyester-esters will be briefly discussed.

6.1. POTM–PBTP Copolyether-esters

The field of application of copolyether-ester is divided into the following market sectors:

1. hose and tubing;
2. sports goods;
3. mechanical items;
4. automotive components;
5. miscellaneous.

The application of POTM–PBTP for these various purposes has become interesting in view of the following typical polymer properties:

(i) rather high load-bearing capacity;
(ii) high flexural fatigue endurance;
(iii) low and high temperature performance;
(iv) good chemical and weathering resistance;
(v) easy and efficient processing.

6.1.1. Hose and Tubing

In these products use is made of the economies of processing techniques typical of thermoplastics. Compared with traditional rubber hose and tubing, no post-curing is required and extrusion in continuous lengths can be practiced.

Furthermore, it is possible to obtain high strengths in products of relatively low wall thicknesses, and flexibility over a large temperature range (especially at low temperatures). As compared with competing nylon compounds, no plasticiser is required for flexibility, so the material remains inherently flexible for a very long time. The POTM–PBTP copolyether-ester finds use, for example, in hydraulic hose. The hose is mostly reinforced by, say, a polyester fibre carcass sandwiched between the inner and the outer polymer ply.

Another use is in tubing for motor vehicles. The harder grades (63 and 74 Shore D) of this polymer are particularly suitable for employment as material for petrol tubing of motor vehicles. These POTM–PBTP copolyether-ester grades fulfil all requirements in accordance with DIN 73378 and show even better performance than the more expensive nylons 11 and 12; moreover, the latter are less resistant to the action of a methanol/petrol mixture, i.e. the fuel which will most probably be used in motor vehicles in the near future.

6.1.2. Sports Goods

The excellent flex properties of POTM–PBTP copolyether-esters over a wide temperature range, combined with a high modulus which affords high strength per unit weight, have contributed to their effective utilisation in high performance sports goods. Examples are soles for cross-country ski boots and football boots, and carcasses for footballs and rugby balls (these last-mentioned articles are manufactured by rotational moulding).

The slight modulus change of copolyether-esters over a broad temperature range provides the same comfort in warm and in cold weather.

6.1.3. Mechanical Items

Due to the high vibration damping behaviour, the great flexibility over a wide temperature range, the good wear resistance and the very low heat build-up, the POTM–PBTP copolyether-esters find application in transmission components such as flexible couplings, V-belts, etc. These elastomers have also proved very suitable for use on steel conveyor drive cables where they prolong the service life by minimising wear and reducing the internal friction. To reduce noise, as compared with that of the hard engineering plastics, POTM–PBTP copolyester is used for gears in clocks and mechanical toys.

6.1.4. Automotive Components

The good resistance to petrol and grease, excellent flexibility over a large temperature range (especially the low-temperature flexibility down to −40 °C), high flex fatigue, and good tear- and puncture-resistance allow these elastomers to be used in bellows, fuel lines (see Section 6.1.1), back-up rings in shock absorbers, etc. The use of POTM–PBTP elastomer film membranes for fuel pumps is very interesting, this polymer being resistant to the petrol/alcohol mixture of the near future.

Further automotive uses of the elastomer are in off-the-road tyres and fuel tanks (both made by rotomoulding), in bumpers, and in seat-belt locking devices.[58] For heavy duty trucks, use is made of coiled electric cables and coiled pneumatic tubing (air brakes) made from copolyether-ester.

6.1.5. Miscellaneous

This category of applications includes telephone retractable cord (good elastic memory), butterfly valve liners (excellent chemical and abrasion resistance),[58] pipe clamp liner dampers (to eliminate metal-to-metal

FIG. 20. Applications of polyether-ester elastomer. Drive shaft bellows (blow moulding); castor wheels (injection moulding); reinforced hose (extrusion coating).

FIG. 21. Applications of polyether-ester elastomer. Cross-country ski boot sole (injection moulding); football (rotational moulding); compressed air line (extrusion).

contact and vibration problems), housings for electrical domestic appliances (excellent impact resistance), bottle caps, beer stoppers, snowmobile tracks, mandrels (used as a core for the production of rubber hose; the advantages are: high flexibility, no plasticiser migration and good recovery from bending).

Various applications of POTM–PBTP copolyether-esters are illustrated in Figs 20 and 21.

6.2. Other Copolyether-esters and Copolyester-esters

Except for the POTM–PETP copolyether-ester, which is less suitable for extrusion due to its low crystallisation rate, other copolyether-esters and copolyester-esters can, in principle, be employed in the same fields in which use is made of thermoplastic polyurethanes and of POTM–PBTP copolyether-esters. However, no other members of these families have found commercial applications as yet.

REFERENCES

1. Asahi Chem. Ind., Japanese Patent 69/20472, 1969.
2. du Pont de Nemours & Co., US Patent 3 775 375, 1973.
3. SANGEN, O. *et al.*, *Sen'i Gakkaishi*, 1971, **27**, 153.
4. CELLA, R. J., *Encyclopedia of Polymer Science and Technology*, supplementary vol. II, Wiley, New York, 1977.
5. COLEMAN, D., *J. Polym. Sci.*, 1954, **14**, 15.
6. ICI Ltd, British Patent 682 866, 1950.
7. duPont de Nemours & Co., British Patent 779 054, 1953.
8. NOSHAY, A. and MCGRATH, J. E., *Block Copolymers*, Academic Press, New York, 1977.
9. IWAKURA, Y. *et al.*, *J. Appl. Polym. Sci.*, 1961, **5**, 108.
10. Goodyear Tire and Rubber Co., US Patent 2 691 006, 1954.
11. CHARCH, W. H. and SHIVERS, J. C., *Text. Res. J.*, 1959, **29**, 536.
12. Toyobo Co. Ltd, Japanese Patent 73/00 991, 1973.
13. Teijin, Belgian Patent 834 004, 1975.
14. LENZ, R. W. and GO, S. J., *J. Polym. Sci., Polym. Chem. Ed.*, 1973, **11**, 2927.
15. LENZ, R. W. and GO, S. J., *J. Polym. Sci., Polym. Chem. Ed.*, 1974, **12**, 1.
16. Monsanto Co., US Patent 4 102 868 (1978).
17. Bayer AG, German Patent 2 706 297, 1978.
18. Akzo NV, European Patent 13 461, 1980.
19. Dainippon Ink & Chem. Inc., Japanese Patent, pre-publication, 50-156386, 1975.
20. du Pont de Nemours & Co., US 3 023 192, 1962.
21. ICI Ltd, British Patent 1 066 162, 1966.

22. Bayer AG, Belgian Patent 590 810, 1968.
23. Kurashiki Rayon, British Patent 1 060 401, 1964.
24. HOESCHELE, G. K. and WITSIEPE, W. K., *Angew. Makromol. Chem.*, 1973, **29**/**30**, 267.
25. NISHIMURA, A. A. and KOMOGATA, H., *J. Macromol. Sci.-Chem.*, 1967, **A1**(4), 617.
26. KNOX, J. B., *Gummi, Asbest und Kunststoffe*, 1974, **27**, 438.
27. GRIFFIOEN, D., *Plastica*, 1978, **31**(3), 66.
28. WELLS, S. C., *Handbook of Thermoplastic Elastomers*, chapt. 4, ed. B. Walker, van Nostrand Reinhold, New York, 1979.
29. HOESCHELE, G. K., *Chimia*, 1974, **28**(9), 544.
30. WOLFE, J. R., JR., *Rubb. Chem. Technol.*, 1977, **50**, 688.
31. GOODMAN, I. *et al.*, *Br. Polym. J.*, 1975, **7**, 329.
32. COFFEY, D. H., *Rubb. Chem. Technol.*, 1957, **30**, 238.
33. GRIEVESON, B. M., *Polymer*, 1960, **1**, 499.
34. CUSANO, C. M. *et al.*, *J. Polym. Sci.*, *C*, 1963, **4**, 743.
35. ONDER, K. *et al.*, *Polymer*, 1972, **13**, 133.
36. WOLFE, J. R., *ACS*, *Polym. Prepr.*, 1978, **19**(1), 5.
37. LILAONITKUL, A. and COOPER, S. L., *Rubb. Chem. Technol.*, 1977, **50**, 1.
38. HOURSTON, D. J. and HUGHES, I. D., *J. Appl. Polym. Sci.*, 1977, **21**, 2093.
39. MODY, P. C. *et al.*, *ACS Polym. Prepr.*, 1979, **20**(2), 539.
40. BONART, R., *Polymer*, 1979, **20**, 1389.
41. van KREVELEN, D. W., *Properties of Polymers*, Elsevier, Amsterdam, 1972.
42. du Pont de Nemours & Co., Hytrel Bulletin, I-16, 1976.
43. NEUMANN, W. *et al.*, *Proc. 4th Rubber Conf.*, *London*, 1962, p. 738.
44. du Pont de Nemours & Co., Hytrel Technical Bulletin, Hyt-409.
45. ZAISER, G. L., *Plast. Eng.*, 1980, May, 61.
46. GOODMAN, A. L. *et al.*, *Plast. Techn.*, 1976, August, 43.
47. du Pont de Nemours and Co., Belgian Patent 783 652, 1972.
48. Akzo NV, Ketjen black EC, Amsterdam, Holland.
49. du Pont de Nemours & Co., US Patent 3 963 802, 1976.
50. du Pont de Nemours & Co., US Patent 4 124 653, 1978.
51. Eastman Kodak Co., US Patent 4 011 286, 1977.
52. du Pont de Nemours & Co., Belgian Patent 823 490, 1974.
53. Eastman Kodak Co., German Patent 2 338 615, 1973.
54. Toyobo Co., Japanese Patent 0048-059, 1973.
55. Bayer AG, German Patent 2 363 512, 1975.
56. du Pont de Nemours & Co., Hytrel Technical Bulletin, Hyt-302.
57. TEBERGHE, J. E., *et al.*, *31st Ann. Techn. Conf.* Reinforced Plastics/Composites Institute, SPI Inc., Section 13-E, 1, 1976.
58. KANE, R. P., *J. Elast. Plast.*, 1977, **9**, 416.
59. GOLDBERG, E. P., *J. Polym. Sci.*, *C*, **4**, 1964, 707.
60. BERGER, J. and HUNTJENS, F., *Angew. Makromol. Chem.*, 1979, **76**/**77**(1116), 109.
61. Goodyear Tire and Rubber Co., Dutch Patent 6 510 146, 1966.
62. Akzo NV, Injection moulding of Arnitel, Technical Bulletin, Arnhem, Holland.
63. Akzo NV, Extrusion of Arnitel, Technical Bulletin, Arnhem, Holland.

Chapter 8

DEVELOPMENTS IN POLYURETHANE BLOCK COPOLYMER SYSTEMS FOR REACTION INJECTION MOULDING

D. C. ALLPORT, C. BARKER and J. F. CHAPMAN

ICI Ltd, Organics Division, Manchester, UK

SUMMARY

The Reaction Injection Moulding (RIM) of large shaped objects is a recent technological adaptation of the fast chemistry of block copolyurethane formation from liquid precursors. RIM products already have an established importance as components for automotive construction, and are likely to find extensive uses in many other fields.

The development of RIM, and of its newer variant incorporating glass fibre reinforcement (RRIM), has required the parallel and integrated evolution of reactant formulation, process and machinery engineering, and an understanding of the relevant product structure–property relationships.

The chapter provides a critical survey of the numerous factors involved and describes the properties, applications, economics and energy conservation aspects of RIM and RRIM processes and products.

1. INTRODUCTION

Reaction Injection Moulding (RIM) is an advanced technological process usually used for producing large polymeric mouldings very rapidly directly from liquid intermediates. The reactants are mixed intimately by a high pressure feed impingement technique in a static mixer and injected immediately at moderate downstream pressure into a mould where the polymer-forming reaction becomes substantially completed in a very short period of time (5–120 s) to produce a 'dry' moulding which can be removed from the mould within a few minutes, and sometimes in less than 1 min.

Although RIM can be applied to a variety of polymer types, including epoxy and vinyl systems, it has in large measure developed around polyurethanes, with which we are solely concerned here. Amongst other reasons, this is because of the convenient physical nature of the precursors—mainly low viscosity liquids—and because of the wide and convenient low-to-medium temperature ranges at which fast isocyanate reactions proceed. The reactions progress vigorously to ultimate completion in a manner which is sufficiently tolerant of minor changes in operating conditions and reactant ratios to give a robust manufacturing process.

Nomenclature presents a minor problem in that RIM is also referred to by various synonyms, notably LIM (Liquid Injection Moulding) and LRIM (Liquid Reaction Injection Moulding) although some authors do not agree with these definitions.[8] RIM will be used exclusively throughout this chapter to signify the process whereby polyurethane elastomers are made by Reaction Injection Moulding.

Reinforcement of polymers and commercial plastics is a well established technique for further modifying properties and is used, for example, in thermosetting phenolic resins and in thermoplastic nylons. Glass fibre, amongst a range of organic and inorganic fibres and fillers, can now be incorporated into RIM polyurethanes, to give reinforced materials by Reinforced Reaction Injection Moulding which will be denoted here as RRIM.

The RIM polyurethanes of current commercial interest are high density (specific gravity = 0·8–1·1) microcellular elastomers. The background work on polyurethane elastomers was carried out in the period 1937–1945 in Germany—notably through the work of Professor O. Bayer, in Britain and the USA. This resulted in the development of flexible polyurethane coatings and of elastomeric casting compounds; the latter were soon found to have certain unique mechanical properties as compared with other rubbers then available. Wright and Cummings[1] have summarised the broad knowledge on solid polyurethane elastomers up to about 1968. Allport and Mohajer[2] have reviewed polyurethane block copolymers with special emphasis on the property–structure relationships in various elastomers, and Redman[3] has updated the previous review and set out recent ideas on the morphology of polyurethane elastomers (see also Chapter 6 of this book).

Many of the investigations reported in the above texts concerned linear or slightly branched polyurethanes. However, as is discussed below, RIM polyurethanes are, in general, not linear polymers. Nevertheless, they exhibit mechanical behaviour similar to that of the thermoplastic polyurethane elastomers (TPUs), which are an important group of non-vinyl block copolymers. RIM polyurethanes are also phase-separated, microcrystalline polyurethane block copolymers, even though they are branched.

A recent publication by Becker[4] provides a useful review of various facets of RIM and RRIM technology. The reader is referred to the above four monographs for an historical background to the subject.

The more limited aim of this chapter is to provide an entrée to the principles, current ideas and literature on RIM polyurethanes and, specifically, to report recent developments in the glass reinforcement of RIM. RIM and RRIM polyurethanes are the logical extension of over 40 years of progressive development of polyurethane chemistry, machinery and applications and they represent one of the important current growth sectors in commercial polymer exploitation.

2. CHEMISTRY AND COMPONENTS

The key chemical reactions in polyurethane technology are those of isocyanates with various reactive hydrogen compounds, as set out below. The simple urethane-forming reaction (I) naturally predominates.

$R'NCO + ROH \rightarrow R'NH.COOR$ urethane (I)

$R'NCO + R''NH_2 \rightarrow R'NH.CO.NHR''$ substituted urea from amine extender (II)

$2R'NCO + H_2O \rightarrow R'NH.CO.NHR' + CO_2$ substituted urea from water: the blowing reaction (III)

$R'NH.CO.NHR' + R'NCO \rightarrow R'NH.CO.NR'(CO.NHR')$ biuret-forming branching reaction (IV)

$R'NH.COOR$ (urethane) $+ R'NCO \rightarrow R'NCOOR(CO.NHR')$ allophanate-forming branching reaction (V)

In addition to reactions ((I)–(V)) which occur to varying extents during the RIM process, a catalysed polymerisation of isocyanate groups to carbodiimide and uretonimine structures is also of importance in the prior conversion of pure MDI (VII) into derivative isocyanates having a functionality above 2·0. This is done externally to the RIM process as a step in the manufacture of specific MDI variants.

$$2\,R'NCO \rightleftharpoons R'N{=}C{=}NR' + CO_2 \quad \text{Carbodiimide}$$

$$\Big\updownarrow R'NCO \qquad \text{(VI)}$$

$$\begin{array}{l} R'N{-}C{=}NR' \\ \;\;|\quad\;\; | \\ \;\;C{-}NR' \\ \;\;/\!\!/ \\ O \end{array} \quad \text{Uretonimine}$$

$$OCN{-}C_6H_4{-}CH_2{-}C_6H_4{-}NCO$$ Diphenylmethane-4,4′-diisocyanate (MDI) (VII)

R' represents the residue of a di- or higher-functional isocyanate; chain propagation therefore occurs through similar reactions occurring with other NCO groups in the polyisocyanates employed as reactants.

Each of the above reactions differs in sensitivity to temperature and to specific catalysts and the concentration thereof. However, unlike the case of conventional low density polyurethane foams, the optimum catalyst choice and usage in a RIM system is dictated less by the need to balance the blowing reaction with the polymer-forming reactions ((III) and (I)), than by the controlling of the relative rates of chain growth with hard block formation ((I) and (II)), coupled with the desire to obtain a minimum demould time and a maximum demould toughness in the products. The catalyst system commonly used is a mixture of a tertiary amine (e.g. triethylene diamine) and a metal-based catalyst of the dialkyltin dicarboxylate type, for example, dibutyltin dilaurate.

Whatever the system or catalysts in use, the known kinetics of the above reactions predict that a variable reaction temperature will lead to a variable product. Accordingly, standardisation of all the conditions in the RIM process is essential if consistent end products are to be obtained. This has major implications for the control of temperature in the feed chemicals and in the mould, including the method of dissipating the exotherm from one repeat moulding operation to the next, and of the temperature gradient within the moulding—from skin to core—which generally cannot be predesigned. The keynote, therefore, is consistency even if the standard operating conditions have to be set empirically.

The three main coreactants in a RIM polyurethane system are:

1. a polyhydric alcohol (functionality ≥ 2) of high M.W.;
2. a low M.W. chain extender—a diol or diamine;
3. an MDI-based isocyanate.

Components 1 and 2 may each be mixtures of more than one individual compound of that class, but 3 is most commonly a single commercial isocyanate product, though this may be a prepared complex mixture.

The specific chemical building blocks of RIM polyurethane are not unique to this technology. Most are well-established components of other mainstream polyurethane technologies, particularly those of high performance solid and microcellular elastomers (non-RIM type) and of MDI-based polyurethane flexible foams.

The most important substances in practical use are listed below followed by comments on some implications for the macrochemical structure of the RIM polyurethane polymer matrix produced therefrom:

1. High molecular weight polyhydric alcohols:
   (a) Polyether triols made by polymerising propylene oxide with a triol initiator followed by reaction with ethylene oxide. These are the so-called EO-tipped triols having high primary hydroxyl end group contents. Molecular weights in the range 3000–6500 are usually selected.
   (b) Analogous ethylene oxide-tipped polyether diols, M.W. 2000–4000 (sometimes used mixed with polyether triols).
   (c) Polyethers containing dispersions of other polymers, such as vinyl copolymers, which are usually produced by *in situ* polymerisation in the polyether.
   (d) Linear and lightly branched saturated polyesters. These are capable of giving good RIM formulations, but they are not currently the focus of interest through impingement mixing machines because of constraints of viscosity, compatibility and price.
2. Chain extenders:
   (a) Short-chain aliphatic diols, for example, ethylene glycol or 1,4-butanediol.

(b) Diols centred on an aromatic nucleus for greater chain stiffness. Examples:

$$HOCH_2CH_2O-C_6H_4-OCH_2CH_2OH$$

1,4-bis(2-hydroxyethoxy)benzene

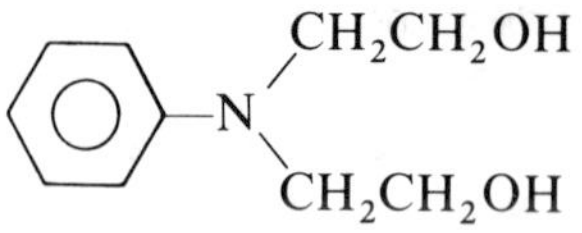

$$C_6H_5-N(CH_2CH_2OH)_2$$

N,*N*-bis(2-hydroxyethyl)aniline

(c) Sterically hindered aromatic diamines.

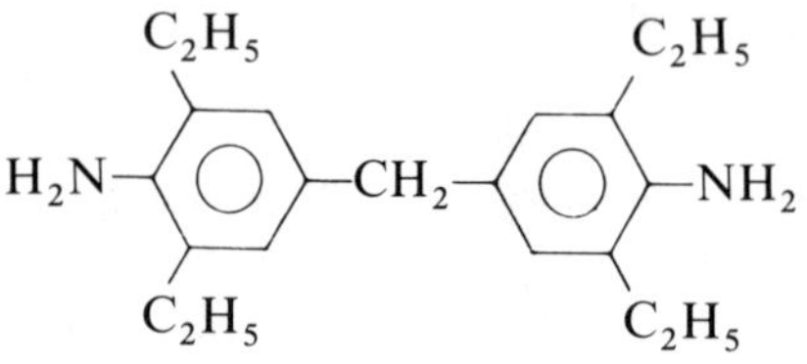

$$H_2N-C_6H_2(C_2H_5)_2-CH_2-C_6H_2(C_2H_5)_2-NH_2$$

4,4′-diamino-3,3′; 5,5′-tetraethyldiphenylmethane[5]

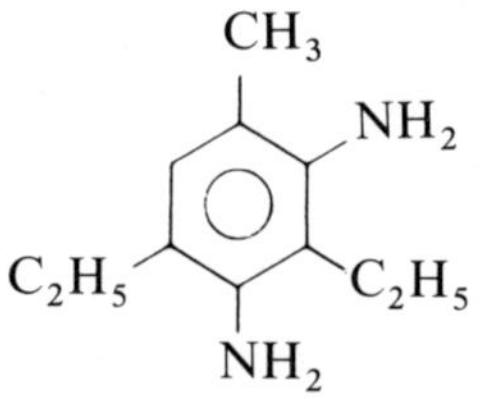

$$CH_3C_6H(NH_2)_2(C_2H_5)_2$$

2,4-diamino-3,5-diethyltoluene[5]

3. Isocyanates:

The MDI types in Table 1 are those currently of importance in practical RIM technology. The MDI variants selected have been developed to give optimum mechanical and physical properties in the RIM elastomer, whilst being stable liquids at ambient temperatures. Liquidity minimises the formation of the insoluble MDI uretidione dimer and related species which form spontaneously in solid pure MDI on standing. (Pure MDI melts at 42 °C.) Additional advantages are the avoidance of melting-out facilities at the production site or the use of sophisticated temperature-controlled storage vessels, which would be required if pure MDI alone were employed. The chemical modifications chosen in the manufacture of MDI variants for RIM usually maximise the 4,4′-MDI content for the best ultimate elastomer properties, whilst ensuring adequate liquid stability in the variant.

In addition to the three main coreactants, a RIM polyurethane formulation contains minor weight proportions of catalyst(s), water and other blowing agents to produce the modest volume expansion required and various additives which control the physical form and stability of the microcellular elastomer.

In practical terms, components 1 and 2 and all the catalysts and additives are combined into a mixture called the 'resin blend'. The isocyanate, selected from class 3, forms the other reactant stream. Accurate and consistent ratio control of the two parts of the system is crucial.

2.1. Chain Branching

The following centres for chain branching in RIM polyurethane polymers can now be identified:

1. Triol functionality in the polyether polyol.
2. A low uretonimine (triisocyanate) content in the MDI-based isocyanate component: this has typically a functionality of about 2·1, but isocyanate with a value of up to 2·3 may be used in lower performance systems.
3. Biuret formation from isocyanate and the water added to cause blowing, or possibly from the urea derived from an amine chain extender when used.
4. Allophanate-forming reactions, which proceed to a low level in all urethane systems reacting at above 90–100 °C. The peak temperature in a RIM moulding depends on thickness and formulation, and may exceed even 140 °C. Allophanates are also formed when post-curing of RIM elastomers is employed after demoulding to ensure the attainment of ultimate properties.

Branching functions 1–3 distinguish RIM polyurethane elastomers from thermoplastic polyurethane elastomers, which are formed from exclusively difunctional components. Despite this significant difference in network structure there is now convincing evidence that RIM polyurethanes, like thermoplastics, are microphase-separated block copolymers in which molecular alignment and hard aggregation contribute in a major way to the physical properties.

2.2. Polyether Choice

The M.W. of the preferred polyethers for RIM, whether diols or triols, is such that the separation between the urethane links in the soft block is a minimum 2000 units of M.W. This chain length, and the associated low T_g of the soft block, is an important factor influencing the low-temperature flexibility of the resulting polymer. Other factors governing the structure of polyethers suitable for RIM relate to the relative proportions and positions of oxypropylene and oxyethylene units in the chains. Some oxyethylene content is needed to help the miscibility of the major components of the mix; too much introduces moisture sensitivity. Primary hydroxyl ends help systems of low reactivity whilst secondary hydroxyl ends are sometimes introduced to moderate excessively reactive mixtures.

3. PROPERTIES AND PROPERTY–STRUCTURE RELATIONSHIPS

Polyurethane block copolymer elastomers have been available in a variety of forms for many years as spandex fibres, castable elastomers, elastomers for injection moulding, extrusion and film formation and as solutions for spraying or coating. The bulk elastomers may be expanded, if required, to form foams or to give microcellular compositions. All of these elastomeric materials have two important molecular features in common:

(a) a soft block or segment, usually a polyether or a polyester oligomer, having a T_g well below room temperature;
(b) a hard block or segment, formed by the reaction of a diisocyanate with a short-chain diol or diamine.

The hard blocks link together the soft blocks in a repeating $[AB]_n$ structure.

Because of these structural similarities it is not surprising that many of the physical properties of this wide variety of elastomers fall within bands of values which are typical of these materials. The low-temperature properties of the polymers are largely determined by the T_g of the soft block. The strength, modulus and tear properties are dependent upon the hard block structures, the polar interactions between hard and soft blocks and the ability or otherwise of the soft blocks to stress-crystallise reversibly upon extension. The hard blocks are phase-separated from the soft blocks as small, partially ordered and irregularly shaped microdomains having diffuse boundaries and they act both as crosslinking sites of large volume in the polymer network and simultaneously as hard fillers. Their presence increases the elastic modulus and raises the hardness of the bulk polymer. Since polyurethane block copolymer elastomers are polar molecules, they show excellent swelling resistance to non-polar solvents but swell, and if linear they dissolve, in highly polar solvents such as dimethylformamide. The strength properties of these elastomers depend critically upon the polar interactive forces between hard blocks and on hard block–soft block interactions, not primarily upon covalent crosslinking sites in the polymer network. Hence the thermal dissociation of the partially-ordered hard block segments largely controls their high temperature behaviour. Under preparative conditions which allow optimum polymer morphology to develop, polyurethane block copolymer elastomers can be formed which exhibit complicated microstructures in which some crystalline or para-crystalline behaviour can be detected[20–24] by X-ray diffraction, usually accentuated at high degrees of extension of the elastomer. Large scale spherulitic organisations composed of hard block arrays[23,25] are also often observed. Thermal analysis studies show complicated transitions which depend partly upon the polymer structure and partly upon the thermal history of the specimen under examination. Three broad groups of transitions are found as sample temperatures are increased, (a) those at low temperature, typically below −20 °C, which are associated with the softening of the flexible block, (b) transitions above 200 °C which reflect the melting and dissociation of hard block crystallites if present, and (c) a series of transitions above 140 °C which are associated with the dissociation and re-organisation of hard block aggregates of different sizes. If the soft block crystallises, additional thermal transitions associated with this process may also be found.

There have been many reports of the structure–property relationships[1–3,18,19] in polyurethane block copolymer elastomers and there is now a good understanding of many of the key interrelationships.

RIM and RRIM formulations differ somewhat from many of the compositions on which much of the fundamental structure–property work has been based. These differences are due largely to the exacting demands of the technological processing conditions demanded of RIM polyurethanes.

Firstly, liquid ingredients of moderate viscosities at the processing temperatures are highly desirable in order to facilitate rapid pumping and mixing of the reactant streams. Reactant blends which are solid at room temperature are not favoured because they would necessitate melting-out or other specialised heating facilities at the production site and would give rise to many problems due to solidification in feed lines and in the processing machines during shut-down periods. Polyethers based on propylene oxide meet the preferred requirements for the soft blocks, whilst the hard blocks derived from MDI are based on liquid MDI compositions in which some pre-reaction with diols, or low degrees of branching brought about by partial reaction to uretonimine is necessary to maintain liquidity.

Secondly, the economic requirements for fast injection/reaction/demould cycles have led to two devices to obtain fast production cycles: (a) the employment of some chain branching in both soft and hard blocks since this contributes to an earlier generation of adequate green strengths for demould, (b) the use of very high levels of urethane catalysis. The high catalyst levels promote very rapid reactions between the isocyanate and hydroxyl reactant streams. These developments have led to extremely fast cycle times which have been accelerated further by the use of amine chain extenders. However, some of the modifications introduced for technological reasons lead to elastomers which are less well-defined structurally than many of the materials on which the published body of structure–property work is based. Not only do the RIM elastomers often contain branching both in the soft and hard blocks, but the rapid reaction rates, with the consequent high exotherms, will lead to a proportion of secondary branching reactions. Furthermore, despite the fact that most RIM mouldings are made in thin sections using metal moulds having temperature-controlled surfaces, there remain significant temperature gradients from the surface to the centre of mouldings. This may result in different morphological characteristics throughout the thickness of a given specimen[25] and it is highly probable that the extremely fast reactions used in RIM technology do not allow the formation of those polymer organisations which have the ultimate physical properties theoretically possible under more ideal preparative conditions. Nevertheless, RIM

TABLE 1

PURE MDI AND VARIANTS FOR RIM POLYURETHANES

Product (Suprasec[a] range)	*Chemical type*	*Isocyanate value: % NCO*	*Functionality*
MPR	Pure 4,4′ MDI ($T_m = 42\,°C$)	33·6	2·0
VM02	Prepolymer from pure MDI	23	2·0
VM10	Hybrid type	26	2·07
VM20	Uretonimine-modified MDI	29	2·13
VM30	Hybrid prepolymer type	30	2·3

[a] Suprasec is a trade mark of ICI Ltd.

polyurethane elastomers have highly attractive mechanical properties which are already making them extremely competitive in some applications. The processing/property compromises which have been sought have concentrated on good processability (adequate flow with quick reaction profiles from liquid ingredients) giving stiff, elastic polymers showing adequate-to-good absorption of impact forces, with low modulus/temperature profiles over the temperature ranges normally associated with the in-service conditions of typical automobiles, viz. −30–+70 °C. The incorporation of some blowing in the reaction moulding process aids flow, contributes to the virtual elimination of surface defects in a smooth mould, and gives added benefit in the weight reduction of the moulded parts.

Typical properties found in RIM polyurethane elastomers are summarised in Table 2 which are based on an MDI variant of the type shown in Table 1.

From the one isocyanate, it can be seen that RIM elastomers of varied stiffnesses can readily be obtained by suitable choices of the polyol blends. At a given hardness, wide variations in elongations at break, flexural moduli and heat sag properties are available (compare formulations 1 and 2; 3 and 4). The heat sag properties are especially relevant to the use of RIM

TABLE 2
SOME RIM MOULDINGS OF INCREASING STIFFNESS (SOURCE, ICI Ltd)

Property	*Polyol blend number (isocyanate in parentheses—see Table 1)*					
	1(VM10)	*2(VM10)*	*3(VM10)*	*4(VM10)*	*5(VM10)*	*6(VM10)*
Density ($kg\,m^{-3}$)	←		1000–1100			→
Hardness (Shore D)	48	52	58	60	65	80
Tensile strength (MPa)	17	18	22	27	27	29
Elongation at break (%)	300	150	160	220	170	50
Angle tear ($kN\,m^{-1}$)	64	59	106	94	144	120
Flexural modulus (MPa):						
room temperature	120	160	260	330	750	940
−30 °C	330	320	540	610	1540	1700
+70 °C	81	80	190	250	360	525
Ratio, −30 °C/+70 °C	4·1	4·0	2·8	2·4	4·3	3·2
Heat sag (mm):						
30 min at 120 °C	4	5	5	0	2	3
60 min at 120 °C	7	8	7	1	3	5
Linear coefficient of expansion ($K \times 10^{-6}$)	136	140	130	143	96	94

Typical values on 3·5 mm thick sheets.

elastomers in the automotive industry where resistance to elevated temperatures during painting processes is of critical importance. The changes in flexural modulus with temperature can now be restricted within the narrow limits desirable for automotive exterior part applications. As harder elastomers are produced (increasing hard block concentrations) lower elongations at break are often obtained, although this parameter can depend critically on the particular formulation chosen. Samples 2 and 4 made use of an amine chain extender, the other materials all having glycol chain extension. The main benefits of amine chain extension are not shown in Table 2 although the elongations at break are improved. The advantages are found particularly in very fast processing cycles coupled with good strengths at the demould time, i.e. early attainment of strength properties. It will be apparent from Table 2 that when the use of alternative isocyanate variants is also taken into account the range of elastomer properties available to a user is very broad indeed. The balance of the particular processing and physical properties which is desirable can only be determined when a full analysis of the end use is made; the inevitable compromises between conflicting priorities can then be selected.

Some useful comparisons with similar polyurethane elastomers having little or no intentional crosslinking in either the hard or the soft blocks can be seen in Table 3. The elastomers in this series are made[26] from pure MDI, a soft block polyether diol of M.W. 2000 having 30·4 wt % poly(oxyethylene) as an end block on poly(oxypropylene) chains, with 1,4-butanediol being used exclusively as the chain extender. None of the

TABLE 3
PROPERTIES OF SOME COMPARABLE POLYETHER/MDI/BUTANEDIOL THERMOPLASTIC ELASTOMERS[26]

Hard block content (%)	50	60	70	80
Hardness, Shore D	50	64	74	77
Tensile strength (MPa)	22·7	26	33	53·5
Elongation at break (%)	478	407	287	50
Die C tear (kN m^{-1})	106	162	217	—
Flexural modulus (MPa):				
+23 °C	140	344	755	1690
−29 °C	273	960	2100	3100
+75 °C	107	239	410	705
Ratio, −29 °C/+75 °C	2·55	4·0	5·1	4·4

Elastomers were made with 0·02 parts of dibutyltin dilaurate as catalyst, and were cured at 100 °C for 16 h. Test specimens were 3·2 mm thick.

elastomers are blown. The compositions thus differ in several respects from the others discussed in this section: (a) in the method of preparation, (b) because of the absence of crosslinking in hard and soft blocks, (c) in the chain extender used. Nevertheless, the technological materials made under fast RIM conditions show properties quite similar to those of the elastomers in Table 3. The tensile strengths and elongations of the butanediol-extended thermoplastic polyurethane elastomers are usually better than those of the RIM elastomers. The same broad trends are observable, however, in the thermoplastic elastomers: as the hard block content increases, the hardness, tensile strength, tear strength and flexural moduli all increase. The elastomers in Table 3 were shown by DSC and dynamic mechanical studies to exhibit the two-phase structures expected of this type of block copolymer. At high hard block contents (above 60 wt %) evidence was found which suggested that increased phase mixing or even phase inversion was taking place, consistent with a change from a tough elastomeric material to a more brittle, high modulus plastic.

An analysis has recently been published[27] of the effects of changes in MDI variants on RIM elastomer properties (Tables 4 and 5). The formulations reported do not represent commercial products, but serve to illustrate the effects of some MDI variant structural changes at comparable hard block contents. Increasing the hard block content results in a stiffer,

TABLE 4
RIM FORMULATIONS EVALUATED USING DIFFERENT MDI VARIANTS

Property	*Formulation number*				
	1	*2*	*3*	*4*	*5*
Hard block content (%)	60	50	60	60	60
Composition:[a]					
Polyol	40·0	50·0	40·0	40·0	40·0
Ethylene glycol	10·0	8·2	8·2	8·6	9·2
Uretonimine variant(A)	50·0	41·8	—	—	—
Prepolymer variant(B)	—	—	51·8	41·7	28·9
Polymeric MDI(C)	—	—	—	9·7	21·9

[a] Parts by weight.
Polyol: ethylene oxide-tipped poly(oxypropylene) glycol, equivalent weight, 1935.
Catalysts (parts per 100 parts polyol): 0·075 dibutyltin dilaurate, 0·30 triethylene diamine.
Isocyanate index: 100.
Post-cure: 60 min at 120 °C.

TABLE 5
PROPERTIES OF RIM POLYURETHANE PANELS

Property	*Formulation (isocyanate in parentheses—see Table 4*				
	1(A)	*2(A)*	*3(B)*	*4(B + C)*	*5(B + C)*
Functionality (approximate)	2·15	2·15	2·10	2·15	2·3
Specific gravity	1·06	1·02	1·10	1·00	1·09
Hardness (Shore D)	64	50	62	62	66
Tensile strength (MPa)	24	14	21	19	23
Elongation at break (%)	200	220	240	165	110
Flexural modulus (MPa):					
at +22 °C	370	110	350	370	510
at −30 °C	810	260	890	840	1140
at +70 °C	170	60	150	150	210
Ratio, −30 °C/+70 °C	4·7	4·2	5·9	5·5	5·5
Heat sag (mm), 60 min at 120 °C	7·5	10	19	13	10
Die C tear ($kN\,m^{-1}$)	94	61	110	79	86

harder polymer, more resistant to the effects of sag at elevated temperatures. In these examples (formulations 1 and 2, Table 4) the increase in hard block content also results in an increase in crosslink density since the MDI variant has a functionality above 2. When a difunctional prepolymer MDI variant replaces the uretonimine variant (formulations 1 and 3), at the same hard block content, a polymer more resistant to tear and less resistant to deformation at 120 °C is produced. In the view of the authors[27] this reflects the lower crosslink density in the prepolymer variant (B) and the absence of the stiff uretonimine ring structure. This interpretation is supported by the ratios of flexural moduli at different temperatures, the stiffer and branched uretonimine being more resistant to thermal deformation than the hard blocks which are more dependent on polar attractive forces. In composition 4, no apparent overall functionality or hard block content changes occur as compared with the MDI variant used in composition 1. However, the MDI variant used in composition 4 contains polymeric MDI, a commercial MDI blend of many species having functionalities up to 7. The presence of the high functionality species reduces the tear strength and the elongation at break. When a blend of polymeric MDI and the prepolymer variant is compounded at the even higher functionality of 2·3 (formulation 5), yet still at the same hard block content as formulations 1 and 3, the flexural modulus is increased and the tensile elongation at break reduced because of the increased crosslink density. The tear strengths are apparently in-

TABLE 6
EFFECT OF ISOCYANATE INDEX ON RIM ELASTOMER PROPERTIES (SOURCE, ICI Ltd.)

Property	*Isocyanate index*				
	90	*95*	*100*	*105*	*110*
Density (kg m^{-3})	1180	1210	1180	1130	1170
Hardness (Shore D)	61	65	66	69	70
Tensile strength (MPa)	21	22	23·5	22·5	14·1
Elongation at break (%)	120	110	140	130	130
Tear strength (MPa)	75·5	91	99	92	98
Flexural modulus (MPa):					
room temperature	210	240	340	360	460
−30 °C	510	680	820	1050	1330
+70 °C	130	150	180	190	230
Ratio, −30 °C/+70 °C	3·9	4·5	4·6	5·5	5·8

All the samples were tested on 3·8–3·9 mm thick specimens. The elastomers used were all based on ICI Suprasec VM10 and the same glycol-extended polyol blend.

fluenced by a number of interacting factors which cannot be distinguished clearly in this analysis.

An additional formulation variable available in RIM polyurethane compositions is that of isocyanate index, the percentage ratio of isocyanate to hydroxyl + amine groups. Significant deficiencies in the numbers of isocyanate groups over active hydrogen atoms do not necessarily produce catastrophic falls in mechanical properties as can be seen in Table 6 for a series of RIM elastomers in which the isocyanate index varies from 90 to 110. As the index rises, increases in hardness and flexural modulus are produced. The tear strengths rise to what is probably a plateau region. Tensile strengths are sensitive to the isocyanate index only at the extremes of the bands studied, consistent with the production of a limited polymer network at low index and an excess of secondary branching reactions disrupting the hard blocks at high index.

4. FIBRE-REINFORCED RIM (RRIM)

Reinforced RIM (RRIM) has been a natural progression from unreinforced, all-organic RIM. It is the most recent example of a polymer–fibre composite designed to extend the property range achievable with the polymer alone. RRIM is distinguishable from all other composites in that the reinforcement—most commonly glass fibre—is predispersed in one or both of the two reactants *before* the polymer formation step is initiated by mixing. By contrast in sheet moulding compound (SMC), the mixture of organic components of the matrix is complete before it meets the glass and in glass-reinforced nylon, for example, a true thermoplastic high polymer is fully formed before the reinforcement process starts. Surface effects between fibre reinforcement and the RIM polymer are very important in RRIM polyurethane and data illustrating the effect of coatings on glass fibre are cited below.

The practical development of RRIM polyurethane has been application-led, principally by the demands of the automotive industry, which saw a means of extending the current use of microcellular RIM elastomers into body panel applications as a replacement for steel and aluminium. Not only would such materials be lighter and more corrosion-proof than their metal analogues, but they could be deformed to a significant degree in minor impacts and return undamaged to their original shape and they could retain a first class surface appearance. The key reasons for needing to reinforce the RIM polyurethanes are to reduce the linear thermal coefficient of expansion of the polyurethane closer to that of steel, and to improve the resistance to deformation at elevated temperatures. At the same time, and rather coincidentally, greater stiffness is imparted to the product; this can be a more economical route to higher modulus than increasing the hard block content of the unreinforced RIM matrix. Unlike glass reinforced polyester (GRP) for example, fibre reinforcement of RIM polyurethanes is not undertaken to improve impact strength properties. The basic polyurethane matrix, being elastomeric, is already exceptional in this respect; reinforcement actually reduces its impact resistance.

In the following account of some recent developments in RRIM polyurethane, matters of block copolymer structure are not central; property–structure data available to date are largely empirical, not surprisingly in view of the complex macrophase interactions and the difficulty of investigating the separate contributions of glass and polymer matrix to composite properties. Nevertheless, we believe that the inclusion of RRIM in this chapter as a development in polyurethane block copolymer technology is fully justified by events. RRIM is already a major industrial development within the field of polyurethanes and can be expected to attract fundamental study designed to provide a basis for predictive system development.

4.1. Reinforcement: Mechanism and Agents

The monograph by Wake[9] provides an account of the types and functions of reinforcing agents and fillers in industrial polymers and discusses the principles of reinforcement. One function of the polymer in all cases is to transfer stress from the outside surface (the applied load) to the individual fibres or particles of reinforcing agent. However, whereas in GRP this is the primary function of the polymer matrix (which in this case has very poor intrinsic elongation at break), in the case of RRIM the polyurethane matrix contributes directly to overall strength as evidenced by the excellent mechanical properties of the matrix alone, albeit having lower modulus.

Specifically reviewing reinforcement by glass, Parvin[10] has listed the main types of glass fibre available and illustrated how surface coating of the glass—most importantly by silanes—is effective in promoting both the chemical bonding of the polymer matrix to glass fibre and/or improved physical adhesion, either of which increases the pull-out stress of fibre from the matrix and improves the efficiency of reinforcement. Milewski[11,12] has surveyed the potential of a wide range of organic and inorganic fibres, including glass, as polymer reinforcing agents and has considered the optimum length/diameter (aspect) ratio and surface/volume ratio for reinforcing fibres.

The introduction of a particulate filler, for example, silica, in addition to glass fibre, can be advantageous in helping the processing of the longer fibres by a kind of 'ball-bearing' effect whilst at the same time introducing cheap materials which can significantly lower the overall materials cost.

The above three papers and those listed in the bibliography provide a theoretical background to the role of the specific glass fibre types in current use in RRIM polyurethane, as described below.

4.2. Glass Fibres for RRIM

Most of the established techniques for RRIM have been developed around hammer-milled glass fibre[6,7,13–17] which is specified according to the screen size through which it passes during comminution (typically 1/16 in (1·5 mm) screen size). However, milled fibres have a wide fibre length distribution, from a few millimetres (longer than the screen aperture) down to dust, with an average in the 0·1–0·3 mm region. They have the following disadvantages:

1. The fibres suffer damage during manufacture and the strength of individual filaments may be lower than that of chopped strand fibres of comparable thickness.
2. Because a large proportion of hammer-milled fibre is shorter than the critical length necessary for reinforcement, only the longer

fibres are actually reinforcing the composite efficiently. The critical fibre length (CFL) is that below which fibres with good adhesion to the matrix tend to pull out on failure. Above the CFL fibres tend to fail in tension, contributing their full available strength.

3. Because of 2, the efficiency of reinforcement of the RRIM composite with hammer-milled fibre is low; as a result a large proportion of hammer-milled fibre is needed to produce a given increase in modulus, so risking impairment of the surface quality of the RRIM composite.[14]

Nevertheless, RRIM containing hammer-milled glass fibre (1/16 in screen) has made headway in the USA in prototypes in the automotive industry, and short production runs of automotive wings (General Motors, Omega sports) and wings and panels (Porsche Carrera) have already taken place. In contrast, conventional chopped-strand glass fibre, as used in GRP composites (non-sheet techniques), consist of stiff bundles of glass filaments (usually several hundred in one bundle) held together by a size coating. Strands have typical dimensions of 3–100 mm length and 0·5–2 mm diameter. These are unsuitable for the RRIM process because they will not pass through the small orifices of the static mixing head. Whilst the individual filaments in this conventional chopped-strand glass are only 10–20 μm thick and would, if separated, pass through the head, these bundles are not specifically designed to break up (filamentise); thus they stay largely intact in a RRIM polyol resin base, and so cause machine blockages.

Recognising the nature and importance of the above obstacles to achieving an efficient utilisation of chopped-strand glass fibre in RRIM, Fibreglass Ltd, UK (a division of Pilkington Glass), working in cooperation with ICI Polyurethanes Group, have developed a novel short-fibre chopped strand glass specifically for RRIM polyurethane. The material, designated WX 6450 (formerly WX 6012), comprises uniform length (1·5 mm) bundles of very fine glass fibres (up to 2000 per bundle) held together by a polymer size which is so compatible with the polyurethane base resin that filamentisation occurs rapidly on dispersing WX 6450 in polyether using a simple low-speed propeller mixer. The development of this unique glass fibre for RRIM has been described fully by Chisnall and Thorpe[28] and associates.[29,30]

It is interesting that the fibres are all strongly aligned in the direction of flow, both in the ingredient liquid stream(s) during processing and in the subsequent moulding. Because this results in anisotropy in the physical properties of the product, the selection of the best direction for flow and optimum material entry point into the mould are important factors to be considered when an item is being designed for manufacture in RRIM polyurethane. Benefit can be gained in some cases by arranging to align the reinforcement in a direction in which the greatest stiffness or lowest coefficient of linear thermal expansion is most beneficial.

No difference is reported in the degree of alignment found in mouldings produced using hammer-milled glass (average length 0·1–0·3 mm) from those made using 1·5 mm chopped glass strand (WX 6450).[31]

4.3. Properties of Typical RRIM Polyurethane Composites: Comparison of Chopped-Strand Glass (WX 6450) and Hammer-Milled Glass

There is no optimum formulation for all RRIM composites. Each application requires a specific balance between modulus and tensile and impact strength. For a selected polymer matrix the elongations at break of derived glass-reinforced composites decrease as the proportion of fibre is increased to raise the modulus, which being the main factor controlling apparent stiffness (once the design of the component is finalised) is frequently a principal formulation criterion. Practical formulations and properties illustrating the range of RRIM polyurethane composites which can presently be formulated from ICI Polyurethane RIM matrices reinforced with the preferred chopped-strand glass fibre (WX 6450, Fibreglass, Ltd) are shown in Table 7. Comparative data for similar composites containing hammer-milled (1/16 in) fibre are cited to illustrate the relative efficiencies of reinforcement.

4.4. Fibre Attrition

In the RRIM process physical work is done on the fibre both during mixing/dispersion into the polyol and, more particularly, during passage through the fine-clearance mixing head. Attention has recently been focused, therefore, on effects leading to fibre attrition and hence on the real distribution of fibre size in the final mouldings.[30,39]

Fibre breakdown rate was measured by circulating a 10 wt % slurry of WX 6450 1·5 mm chopped-strand glass in polyol through a Krauss-Maffei 40/80 RRIM machine and head. This slurry is a typical RRIM polyol, which, when combined with an appropriate isocyanate, yields a representative low-to-medium modulus RRIM moulding containing overall approximately 5 % of chopped glass fibre.

The number of passes of slurry through the mixing head was monitored and the weight-average fibre lengths measured at different stages (Table 8).

TABLE 7

TYPICAL PHYSICAL PROPERTIES FOR RRIM POLYURETHANE SYSTEMS

System	Filler wt/wt% on composite	Density ($kg\ m^{-3}$)	Hardness Shore D (15 s)	Tensile strength (MPa)		Elongation at break (%)	Angle tear ($kN\ m^{-1}$)	Flexural modulus (MPa)						Ratio, $-30°C/+70°C$	Sag, 120°C (mm)		Coefficient of linear expansion ($K \times 10^{-6}$)
								+20°C		−30°C		+70°C			30 min	60 min	
				↑↑	↑→	↑↑	↑↑	↑↑	↑→	↑↑	↑→	↑↑	↑→	↑↑	↑↑	↑↑	↑↑
1.	*Unfilled:*	1 200	65	23	—	130	106	230	—	540	—	190	—	2·8	5	7	130
	8% Chopped strand glass[a]	1 200	64	23	20	40	87	817	480	1 690	940	660	280	2·6	1	1	42
	25% Hammer-milled glass[b]	1 200	69	22	19	40	78	800	540	2 120	870	610	210	3·5	1	1	45
2.	*Unfilled:*	1 100	65	27	—	170	145	750	—	1 540	—	360	—	4·3	2	3	93
	6% Chopped-strand glass[a]	1 180	81	37	31	20	149	1 700	1 000	2 400	1 800	700	400	3·5	1	1	56
	25% Hammer-milled glass[b]	1 140	79	39	28	15	108	2 050	1 080	3 600	2 400	960	410	3·8	1	2	68
3.	*Unfilled:*	1 150	66	26	—	180	125	430	—	910	—	390	—	2·3	4	4	135
	10% Chopped-strand glass[a]	1 150	70	32	23	15	140	1 600	600	2 200	930	1 410	450	1·6	1	2	50
	35% Hammer-milled glass[b]	1 250	65	27	17	25	105	1 000	620	2 020	1 240	900	450	2·2	2	3	31

↑↑, Parallel to flow.

↑→, Perpendicular to flow.

[a] 1·5 mm chopped-strand glass WX 6450 (Fibreglass Ltd).

[b] P117B hammer-milled glass (Owens Corning Inc.).

Test sheets of 1 m × 0·3 m × 3·5 mm produced using Krauss Maffei 40/80 dispensing machine.

System 1 uses hybrid type isocyanate (VM10, Table 1), glycol chain extender and triol polyether; system 2 uses hybrid type isocyanate (VM10, Table 1), glycol chain extender and diol polyether; system 3 uses hybrid type isocyanate (VM10, Table 1), diamine chain extender and triol polyether.

Source of data: ICI Ltd.

TABLE 8
FIBRE ATTRITION IN THE KRAUSS-MAFFEI MIXING HEAD

Sample	*Average number of passes through mix head*		*Weight-average fibre length (mm)*
	In recycle mode	*In mix mode*	
Initial slurry	0	0	1·5
Slurry after calibration	1	0	1·4
Moulding after calibration	1	1	1·4
Moulding after further recycle	2	1	1·2
Slurry after third recycle	3	0	0·9
Moulding after third recycle	3	1	1·0

This measurement is comparatively simple in the case of chopped-strand glass because almost all fibres start at close to 1·5 mm length.

Clearly some fibre attrition does occur. Passing the complete contents of the tank (polyol/glass fibre slurry) at high pressure three times through the mixing head reduced the average fibre length to two-thirds of the original value. However, this amount of recirculation is far in excess of that which would occur in practice. Developments in low pressure/by-pass circulation and in multi-station RRIM installations should reduce the average head contact to less than one recycle plus dispense.

Physical property measurements on the moulded products confirmed that the effectiveness of the glass reinforcement had been retained even throughout the excessively severe treatment applied. This is recorded in Table 9. The interpretation of the results is interesting and confirms the important feature involved to be the maintenance of the weight-average fibre length above the critical length of the composition. In the example cited the critical fibre length (CFL) was calculated as 0·4 mm and double this is commonly adopted as a security measure. The optimum length of fibre to seek for this RIM system would thus be 0·8 mm, when the greatest ease of filler incorporation would coincide with good reinforcing efficiency. Chopping glass strand to such short lengths is not currently a commercial reality.

4.5. Criteria For Automotive Applications

RRIM polyurethane as a material of construction for motor vehicle exterior parts must compete with steel and, selectively, with aluminium. Eventually this is a cost/effectiveness situation and some aspects will be

TABLE 9

PHYSICAL PROPERTIES OF RRIM MOULDINGS AFTER VARYING RECIRCULATION OF POLYOL SLURRY

Property	*Unfilled*	*Normal M/C preparation*	*Plus 1 contents recirculations*	*Plus 3 contents recirculations*
Average fibre length (mm):				
polyol tank	—	1·4	—	0·9
moulding	—	1·4	1·2	1·0
Density ($kg\,m^{-3}$)	1100	1100	1050	1050
Hardness (Shore D)	58	65	67	64
Direction of test		↑↑	↑↑	↑↑
Tensile strength (MPa)	22	22	23	21
Elongation (%)	160	30	35	45
Angle tear ($kN\,m^{-1}$)	105	97	100	87
Flexural modulus (MPa):				
room temperature	260	800	790	750
−30 °C	540	1300	1100	1100
+70 °C	190	610	540	490
ratio +70 °C/−30 °C	1:2·8	1:2·1	1:2·0	1:2·2
Sag test at 120 °C (mm):				
30 min	5	0	0	0
60 min	7	0	1	0

discussed in a later section. A real difficulty which exists in translating RRIM polyurethanes into live usage is the large gap which exists between laboratory physical test data and the performance of real parts in service. Attention has been drawn to this in the literature,[34] and a number of authors have then attempted to use their own experience to define the characteristics they feel will provide RRIM polyurethane composites of good enough quality to meet the performance requirements of body panelling. Papers by Mikulec (Ford, USA),[6,7] Simpkins (General Motors, USA),[14] Nelson and Dabrowski (Davidson Rubber, USA)[15] and Charlesworth (British Leyland, UK),[32] have all sought to clarify what is needed and which technical areas need attention. Some of the main points raised are summarised in Table 10 in general terms. It must be remembered that optimum properties for any single application may differ markedly from those for a different part, e.g. a wheel arch surround is likely to be preferred if it is of low modulus and shows rubbery recovery, even from substantial deflections: a door panel is likely to be preferred in a higher modulus composition, requiring more modest deflection recovery characteristics.

RRIM polyurethane technology is still new and is currently developing

TABLE 10

SOME DESIRABLE PERFORMANCE DATA FOR RRIM POLYURETHANES IN AUTOMOTIVE APPLICATIONS

Property	*Level required*	*Reason*
1. Flexural modulus	Varied for the outlet; at room temperature 250–800 MPa unreinforced: 600–2 500 with glass fibre incorporation	Parts must have adequate stiffness and be self-supporting with about the same number of fixing points as the corresponding steel part; the type of fixing can be specified
2. Impact resistance	Good over the range −30 °C–+70 °C; equal to or better than SMC based on tests carried out on the actual moulded part	Part must not shatter under specified light collision conditions at −30 °C and must return to original shape
3. Thermal stability: heat sag	As high as possible: no degradation or permanent set either (a) during service or (b) painting	Self evident; part must be as durable as the rest of the vehicle
4. Thermal expansion coefficient (CLTE)	Low: as close to that of steel as possible (CLTE steel = 11×10^{-6} cm/cm/°C)	RRIM panels must be compatible with the steel mainframe and fixings
5. Surface quality	Automotive class 'A'	Final painted RRIM part must match steel surfaces in quality and colour; this requires good substrates; this factor (5) interacts with 6 below
6. Painting characteristics	Accept same body coats as suitable for rest of vehicle; aim for painting on vehicle; developments now leading to lower bake temperatures	Must not need painting off the vehicle as a separate operation (a special keycoat primer may be permissible)

dynamically, with progress being made towards achieving good performance in all the areas listed. It is, nevertheless, sufficiently well-defined to have made possible ambitious consumer trials on limited scales of production, some of which will be mentioned in Section 6. A compromise choice of composition for a given application will always need to be made eventually, encompassing cost, processing features and property selection, and the only true criterion of successful design and composite structure will be favourable performance in service. This is why the numerous on-vehicle testing programmes are being mounted and keenly followed. Limiting factors at present appear to be high-temperature stability and possibly impact resistance at low temperatures (−30 °C).

5. PROCESS ENGINEERING

Process engineering is a vital aspect of the industrial application of RRIM polyurethanes. This section discusses some of the basic principles.

RIM polyurethanes differ from most other industrial polymers in that liquid chemicals are converted in one step *in situ* and very rapidly (often in seconds) into a shaped solid or expanded polymer which has ceased to have flow properties. The whole chemical development of industrial polyurethanes has depended upon the parallel evolution of machinery for accurately dispensing and mixing the chemical components and for shaping and moulding the product, whether it be foam or elastomer. The machinery and ancillary equipment has had to match the unique nature of the isocyanate-based reactions, namely the highly exothermic addition process, in most cases accompanied by volume expansion which may vary from 10 to 3000 % of the original component volume.

RIM polyurethane machinery represents a logical development of the well-proven low- and high-pressure dispensing machines used for polyurethane foams and microcellular elastomers. Major recent reviews of such polyurethane machines have been published.[35,36] All major manufacturers of machinery for polyurethane foam produce RIM units and most have, or are developing, add-on dosing modules to handle polyol–glass fibre slurry for RRIM.

Machinery packages vary from basic laboratory evaluation equipment to sophisticated multi-station self-contained units complete with safety screens to protect nearby workers and water curtains to take release-agent spray away. Microprocessor control systems eliminate metering stages and permit close and continual monitoring of the operations in progress and the

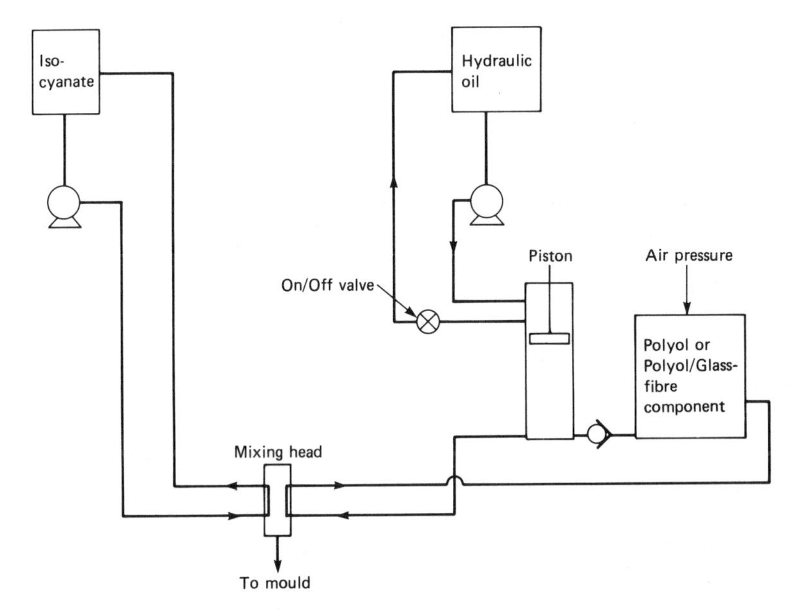

FIG. 1. Schematic representation of the RIM/RRIM machine having a piston metering unit on the polyol side to allow accurate, reliable dispensing of abrasive glass-containing slurries.

self-correction of any process variables which begin to change. However, all machines have at their heart (i) a reliable means of maintaining ingredient streams in a uniform consistency at a desired predetermined temperature, (ii) a means of propelling controlled proportions of the two streams to dynamic equilibrium round the equipment and (iii) a system of valving to switch the circulating chemicals into a self-cleaning mixing head and thence into the mould. The manner of achieving these actions varies from machine to machine; the following description relates to a Krauss-Maffei 40/80 machine in a non-automatic laboratory installation suitable for processing RIM and RRIM polyurethane mouldings up to the size of a family saloon car wing. Diagrammatically the machine may be represented as shown in Fig. 1.

It comprises the following main plant items:

1. a stirred and jacketed pressure vessel for the polyol blend or polyol blend/glass slurry;
2. a similar vessel for the isocyanate;
3. high pressure metering pumps; these are commonly rotary axial

piston pumps having variable displacement and hence variable output;

4. a dosing unit for the polyol/glass fibre slurry comprising a vertical single throw hydraulic piston pump energised by a separate hydraulic oil unit; a duplicate unit is frequently used on the isocyanate side to also enable fillers to be incorporated in that stream;
5. an impingement mixing head;
6. the control console.

5.1. Dosing Unit for the Polyol/Glass Slurry

Slurries of glass fibre in polyol cannot be pumped and metered directly by rotary pumps because of rapid wear of the pump components by the glass and damage to the glass fibres by the pump. This applies equally to chopped-strand and to hammer-milled glass. Metering must be effected by indirect displacement and such units have recently been developed based upon the principle of a single-throw piston pump backed by a hydraulic metering unit. The accuracy of delivering material to the mixing head from a piston displacement, with the simultaneous longer life afforded to the pumps by their handling hydraulic oil rather than reactive chemicals, has led some machinery manufacturers to favour this new approach even when there is no desire to introduce fillers to the component(s).

In practical operation, polyol–glass fibre slurry, which has been prepared and held as a homogeneous dispersion in one of the pressure vessels, is transferred pneumatically to the vertical cylinder. From here it is metered to the RIM head by hydraulic oil flow at constant rate according to the feed rate and component ratio selected.

5.2. Mixing Head

The impingement mixing head is a critical process element and much has been published on its design. Figure 2 shows the cross-section of such a unit which operates on the counterflowing injection principle. It has no agitator and is self-cleaning, requiring no solvent flush after the shot.

In the recirculation mode the chemical streams can be brought to dynamic and thermal equilibrium, and a single movement of the piston then diverts the head to the dispensing mode. In this active or dispensing mode the piston is retracted and the two components streams, fed at high pressure, impinge in the small end chamber at high velocity; mixing occurs turbulently and the mixture passes at reduced velocity and lower pressure out of the head and virtually directly into the mould on which the head is

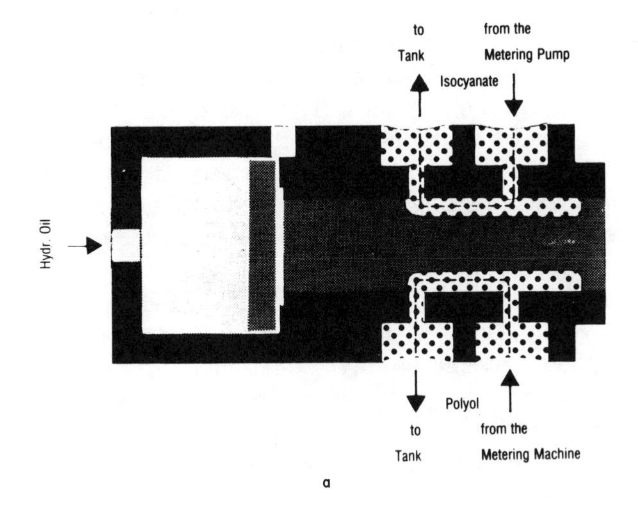

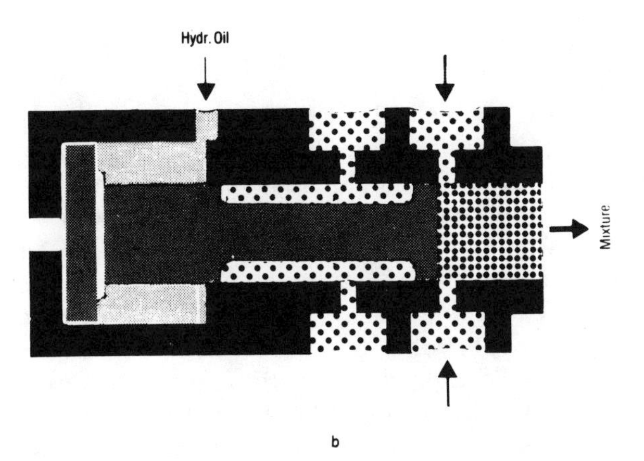

FIG. 2. Cross section of a high-pressure RIM mixing head (Krauss-Maffei, AG). Sections in (a) recycle and (b) dispense modes.

mounted. Separating the head from the mould cavity and gate there may be a second static mixer, of labyrinth construction, and a sprue feed, both of which form an integral part of each mould.

The portions of the mixing head where the glass slurry passes at high velocity through restricted regions must be made of specially hardened steel.

Not illustrated are the two remaining essential modules of any RIM installation, namely the mould and mould carrier clamp unit.

5.3. Mould Design and Construction

This is a highly specialised technology within polymer engineering as a whole. For information relevant to RIM, the reader should refer to the papers of Misitano[37] and Knipp and Becker.[36] The basic construction usually employs a horizontal split line: one version has the mixing head mounted on the lower part while the top section lifts on the clamp unit to release the formed part after moulding. Built into the mould is a static aftermixer, sprue runner and gate.[38]

RIM moulds may be fabricated from various materials depending upon the number of mouldings required from them. Mineral-filled, cast epoxy resin gives good service in prototypes and moulds for short production runs. Metallisation of such a resin mould may prolong its life perhaps fivefold. For full-scale production, all-metal moulds are preferred, but they can be of generally lighter construction than an equivalent unit for high pressure injection moulding using a thermoplastic polymer such as modified polypropylene.

5.4. Clamp Unit

The operating pressure within a RIM/RRIM mould is comparatively low—typically 0·35 MPa (50 lb in^{-2}). Nevertheless, because the mouldings being made are often of large surface area (for example, 0·5–1·0 m^2) the clamping force needed between the press platens is around 20–40 tonnes. Standard clamp units offered for RRIM processing cater for forces up to 200 tonnes. Typical of these are the well-proven Kannegiesser presses having clamping forces of 30–80 tonnes and daylight area of up to 3 m^2. The platen unit carrying the mould may be either static (lower portion) or rotatable in one or two dimensions to allow optimisation both of the orientation at filling, to obtain best flow within the mould, and of the best position for demould. Recent installations have called for a book opening mechanism which often facilitates the demould operation where finished products are lifted from the female half of the mould.

5.5. Part Design

In all injection moulding processes part design is important for its influence on (a) the ease with which the liquid polymer can flow from the gate to the outermost extremities of the mould and (b) the strength of the moulding at all times, from demould through to in-service conditions. RIM systems present a particular problem because the viscosity is increasing rapidly during injection and flow, yet the mix must ultimately fill all the corners and minor cavities before setting. RIM processing of polyurethanes is assisted by the volume expansion which is designed into the system. Thus the shot volume (calculated as liquid mix) is normally only 90% of the mould volume, and final filling results from the foaming reaction. This greatly facilitates the production of a smooth surface finish which is essential for many of the emerging applications for RRIM urethanes.

Part design for RIM has been evaluated by US workers[4,38] and all have stressed the importance of the following points:

1. correct design of sprue and gate to give entry at the lowest point and to encourage non-turbulent flow of the reacting liquid mixture over all main surfaces, especially those required to be of class A surface quality; with RRIM polyurethanes the geometry of flow must also be considered in order to optimise the anisotropic physical properties for the application;
2. absence of sharp edges and angles; generous radii at the corners will minimise stress concentration and improve mouldability;
3. absence wherever possible of undercuts and sections requiring loose or moving parts in the mould, which increase both the complexity and the cost of the mould and increase the cycle time;
4. careful design of holes, slots and ribs to avoid knit lines or potential sink areas (behind ribs) in important surfaces.

Part design for RIM and RRIM urethanes is an evolving art which in the future will be an important element governing acceptability of the mouldings as replacements for pressed metal.

Various workers[13,28] have drawn attention to the very limited experience which yet exists of the durability, and hence effective life of all-metal moulds when subjected to the highly abrasive conditions in processing glass-filled RIM urethane. Only production experience will resolve these uncertainties, but early indications are that problems are encountered only in the after mixer/sprue gate region in areas of high velocity flow of material.

6. PRESENT AND FUTURE APPLICATIONS OF RIM AND RRIM POLYURETHANES

RIM polyurethanes have found their main outlet in the automotive and mass transport industries as resilient interior and exterior mouldings of non-structural function. Glass reinforced RIM (as yet in its infancy) has its greatest volume potential in the same industries, as the basis for exterior

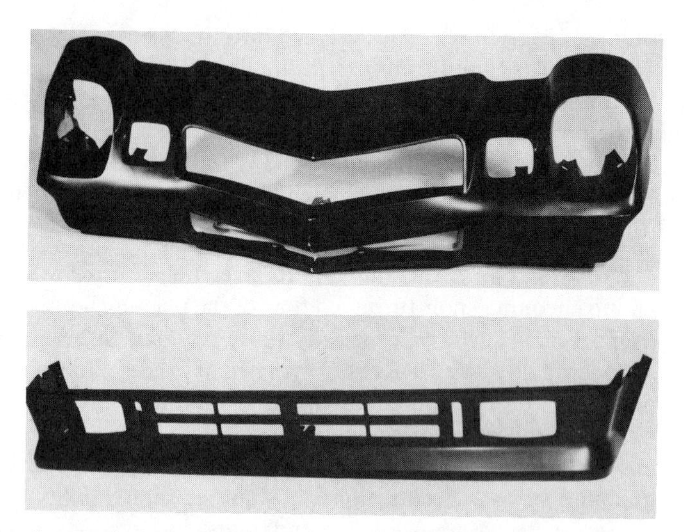

FIG. 3. Typical automobile front-end mouldings in RIM polyurethane.

body panels and other rigid components which form an integral part of the vehicle structure, but not for the inner high-stressed frame.

Low modulus, low performance RIM mouldings, exemplified by integral skin flexible polyurethane foam of overall density typically 400 kg m^{-3}, form the established basis for almost all interior trim mouldings in passenger cars, for example, crash pads, arm rests, steering wheel cladding and instrument panel fascias. These mouldings are microcellular elastomers with low-to-medium levels of physical properties and are of a type not extensively studied for their fundamental properties. They represent an extrapolation of the high-resilience flexible foam technology of the 1960s and are now based on an MDI variant of the hybrid polymeric/pure MDI type, such as ICI's Suprasec VM30. Arm rests and other multiple repeat mouldings in mass transport vehicles (buses, trains, etc.) are also frequently made from integral skin flexible polyurethane foam by the RIM process.

Unreinforced microcellular RIM polyurethane elastomers of higher density (600–1000 kg m^{-3}) based upon a pure MDI variant (e.g. ICI's Suprasec VM10) are used for applications demanding higher levels of physical properties. Such elastomers form the basis for high-performance resilient exterior mouldings for cars and trucks. Typical of these are the large soft bumpers on some sports and premium saloon cars and in the USA the so-called soft front ends which are microcellular elastomer RIM mouldings made as a single piece, weighing up to 9 or 10 kg. Much has been

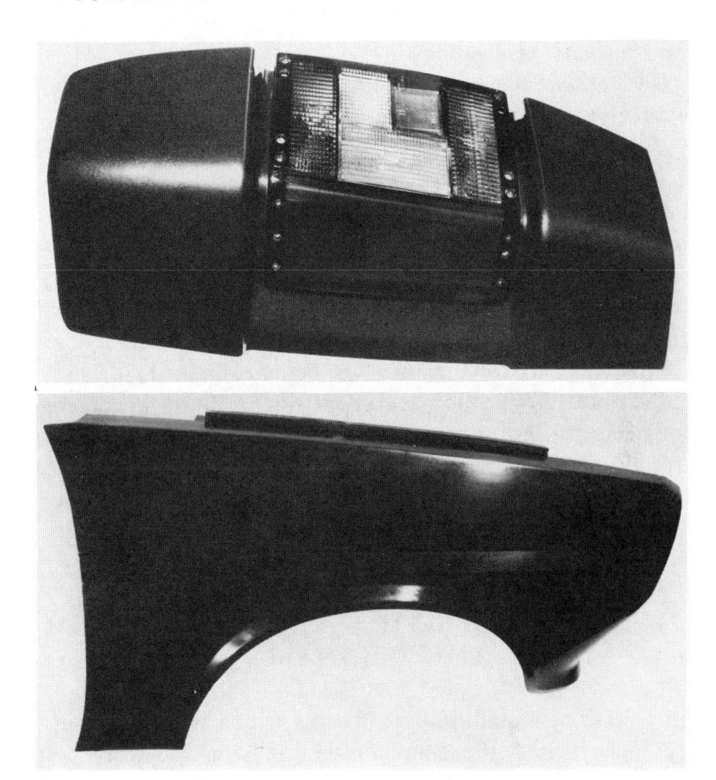

FIG. 4. Two experimental reinforced RIM (RRIM) mouldings.

written about the design and function of such energy management parts. Figure 3 shows typical flexible RIM parts for the applications described above.

RRIM is currently at an advanced stage of development, although certain vehicle parts, e.g. the spoiler of the Firebird, are already being manufactured this way. Very few mass production vehicles as yet incorporate any major RRIM parts and accordingly all the forecast areas of application are unproven. Nevertheless, the years 1979–80 have seen a high level of application test activity in glass-reinforced RIM. All applications utilise the high stiffness-to-weight ratio of RRIM combined with its properties of good resistance to corrosion, low thermal expansion and high resistance to mechanical impact. Typical examples of experimental RRIM parts are wing panels, wheel arch 'eyebrow' surrounds, bumpers (either as total mouldings or as cladding for steel), air dams and spoilers. Two mouldings are illustrated in Fig. 4.

6.1. Non-transport Applications

RRIM polyurethane has potential application wherever an easily mouldable, relatively rigid plastic is required. It can be expected to compete with unsaturated polyesters (GRP/SMC), various high-modulus, high-performance thermoplastics, for example, ABS and polypropylene (including blends), as well as with other reactive liquid systems such as reinforced epoxy resins. Any item requiring large area, stiff yet not unyielding mouldings with a thickness of about 3 mm, probably with stiffening ribs at strategic points, could be considered for RRIM polyurethane processing. Computer housings, machinery covers, shipping containers and crates, as well as items and parts of furniture are being examined.

In all the above applications excellent surface finish and comparative ease of painting are desirable features and RRIM polyurethane commends itself strongly by comparison with competing systems such as SMC and polypropylene thermoplastic. Furthermore, RRIM presents fewer design limitations than alternatives, including, in many cases, steel.

7. THE COST–BENEFIT ANALYSIS OF RIM/RRIM POLYURETHANES

Reaction Injection Moulding is a process which is low in energy consumption.[32] In Fig. 5 the energy used up from the world supply is considered and compared for a variety of plastics and metals. The comparison takes into account two important factors:

1. that plastics, as replacements for metals, are usually needed at three times the thickness of the metal;
2. that naphtha feedstock converted into plastics could have been used to generate power; this is denied to the world if it is used up in plastics production.

Taking both these factors into the computation still puts RIM polyurethane among the lowest overall energy users; it is far and away the lowest energy user during the actual moulding operation. It has been calculated[4] that the process energy for converting a polyol stream + isocyanate feed into a shaped RIM or RRIM part uses only 4 Btu in^{-3} (258 J cm^{-3}). This is exceedingly small and far less than is needed for processing other plastics or metals. In the interests of conserving world energy, therefore, RIM and RRIM polyurethanes deserve promotion.

The economy in energy for converting monomers by RIM is understand-

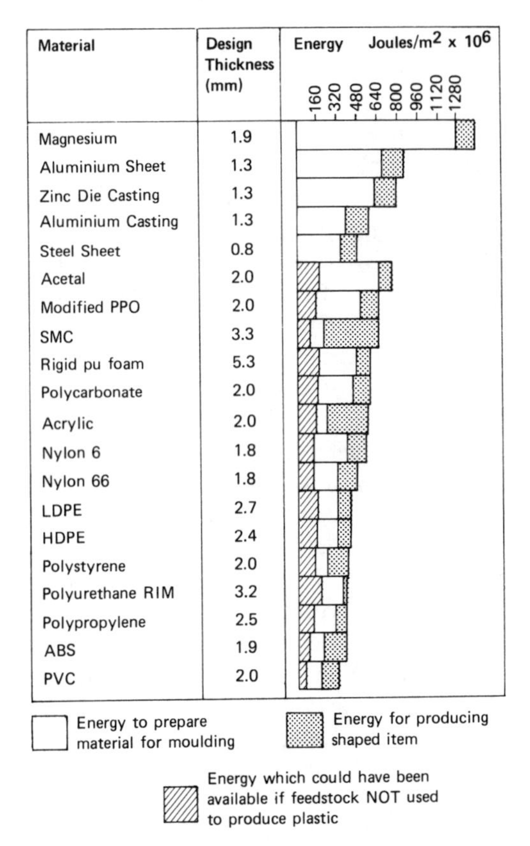

FIG. 5. Energy requirements of various materials at their normal thickness.

able in view of the nature of the process—liquid ingredients at room temperature react exothermically to give a polymer of any shape dictated by the mould, and flow into those shapes at low pressure.

This ease of processing is reflected in the modest capital features which are characteristic of RIM/RRIM polyurethane manufacture. In the UK in 1981 a basic manufacturing facility was estimated to cost between £100 000 and £150 000. This would comprise a RIM/RRIM dispensing machine (0·5–3·0 kg s^{-1} output), coupled to a mould carrier (press) having a closing pressure of 50 tonnes and versatile ability, together with ancillary equipment such as mixers for preparing the components, extraction units and model test moulds. Furthermore, the buildings needed to house this equipment are normal 'industrial estate' type premises without specially strengthened floors or huge power supplies being needed. Yet this facility would be capable of moulding items as large as bumpers, wings and door panels for a family saloon car.

Such modest, relatively lightweight equipment contrasts with the requirements when metals or many plastics are processed and shaped. Murchie (Ford, UK), for example,[33] has compared RRIM polyurethane with SMC, modified polypropylene and polycarbonate for manufacturing mass produced automotive front- and rear-end mouldings. Relative press equipments (in tonnes) were 50 (polyurethane); 800 (SMC); 1750 (polypropylene) and 3000 (polycarbonate). The study of the potential weights and costs of mouldings from the respective plastics concluded that RRIM polyurethane was worthy of consideration on both counts—the only plastic to be so.

In another study,[30] of the cost of moulding an automotive wing at the rate of 225 000 per annum in RRIM polyurethane, it was suggested that such items might be produced at 1·5–2·0 times the cost of the polyurethane ingredients. This calculation did not include working capital or profit.

As the number of items being produced per annum decreases, so the value of the RIM/RRIM polyurethane process becomes greater compared to alternative processes. Indeed, in comparison to other popular methods of moulding, for example, high pressure thermoplastic injection moulding, RIM polyurethanes tend to be relatively high in chemical costs but low in conversion costs (moulds + equipment + labour). The most favourable economics for RIM/RRIM polyurethanes lie in cases where 2000–10 000 units per annum are to be moulded. Many items required in this range, for which mould costs would be prohibitively high if manufactured by thermoplastic injection moulding, become viable in RIM/RRIM polyurethane. The cost of moulding at such rates depends strongly on the particular conditions of individual cases.

Charlesworth[32] has quantified the substantial weight savings to be made if plastics parts are used in place of metals in a number of automotive applications. These are illustrated in Table 11. The desire to reduce the weight of automobiles makes good sense from a world energy conservation aspect and also in meeting Federal targets: lighter parts mean that less fuel is used throughout the life of the vehicle.

The benefits of the RIM of polyurethanes are not restricted to economic factors.[29] It is convenient and economical to take delivery of raw material components from the manufacturers, in liquid form in tankers, to store them in tanks fitted with contents-recording devices, with automatic feed facilities to pump the materials to day tanks and thence to the dispensing

TABLE 11
WEIGHT SAVING USING PLASTICS IN PLACE OF METAL

Part	*Weight (kg)*		
	Metal	*Plastics*	*Saving by use of plastics*
Bonnet	16	8·2	7·8 or 48%
Door	16	7·3	8·7 or 54%
Boot lid	11·4	5·9	5·5 or 48%
Bumpers (2)	36·4	18·2	18·2 or 50%

machines. Indeed the whole system is versatile and can be tailored to allow a moulder to produce polymers at different stiffnesses from the same basic raw materials. Such a facility can reduce both the raw material storage requirements in a factory and the raw material inventory.

Probably the greatest value of RIM/RRIM polyurethanes as a process compared to alternatives is the possibility of trying out prototype mouldings quickly and inexpensively. Short runs of mouldings can be made in tools produced in epoxy resin (or, better, spray metal cavity backed by epoxy) and these have progressed in working practice from an artist's drawing to representative moulded items in less than a four-week period. The moulds to test the validity of the part design, and to allow adjustments to be made where necessary to the volume production tools, cost only a fraction (e.g. 10%) of the cost of the final production tools. The latter in turn normally cost less than half that of tools of a standard suitable for injection moulding comparable parts. RIM/RRIM polyurethane mouldings from good quality moulds have excellent sealed surface finishes and can be readily painted, especially using polyurethane paints, to give automotive class finishes.

Finally, nearly all the RIM/RRIM process can be automated if desired and the total procedures become even lower in labour content.

RIM/RRIM of polyurethanes as a process has many attractive features and compares favourably with thermoplastic injection moulding on the one hand and SMC/DMC/hand lay-up GRP processes on the other.

9. CONCLUSIONS

RIM polyurethanes are elastomeric addition block copolymers composed of polyether (or less usually polyester) soft segments, and polyurethane or polyurea segments derived from diols or diamines reacted with MDI. They exhibit phase separation phenomena which are reflected in their property–structure relationships.

Like the historically earlier developed cast and thermoplastic solid polyurethane elastomers, RIM polyurethanes represent the successful application of the chemical principles of elastomer design to major industrial product applications.

Though the physics of RIM polyurethane elastomers are as yet only partially understood, current work on the morphology of microcellular elastomers and on the understanding of the effects of formulation variables offers the prospect of systematising our knowledge of RIM polyurethanes in a similar manner to that of other polyurethane block copolymer elastomers.

RRIM polyurethanes containing glass fibre are a technical reality, now beginning to be applied to industrial practice. Through reinforcement, the already wide range of mechanical properties achievable in RIM polyurethanes can be extended greatly in the direction of higher modulus, lower thermal expansion and the mechanical stability of the composite. Glass fibre length and fibre/polymer matrix interactions are key factors which are being actively investigated.

RIM and RRIM polyurethanes are cost-effective, energy-efficient and economical materials formed by one-stage processes from liquid chemicals. They are already viable alternatives to some thermoplastic articles produced by injection moulding. As a process, RIM presents fewer design limitations and offers considerable scope at low capital cost for the manufacture of large plastics mouldings.

For the future, developments must proceed in the process engineering of RRIM polyurethanes to allow the reliable handling of longer fibres and/or of higher overall fibre loadings through machinery proven to be capable of withstanding the wear implicit in the continuous production of RRIM mouldings. Automation of RIM and RRIM processes, already foreshadowed, will enable shot time and overall cycle time to be reduced to levels comparable with the high-pressure injection moulding of thermoplastic polymers and incidentally allow for better control of the recycle functions in current processing techniques, so reducing the risk of attrition of longer fibres when present.

Physico-chemical investigation of glass fibre/urethane polymer interactions will continue and hopefully point to ways of improving the efficiency of reinforcement by glass and of optimising the type and amount of glass employed. Concurrently, developments in the organic matrices for

RIM and RRIM polyurethanes can be expected to yield polymers having better mechanical and processing properties and improved high temperature mechanical and chemical stability.

Improved test methods for RRIM composites must relate the behaviour in laboratory tests to real world failure conditions to provide a secure basis for the selection of new products for high-performance industrial applications. The availability of relatively cheap moulds and a wide spectrum of RIM and RRIM polyurethane formulations will also encourage the parallel development of a wide spectrum of alternative applications outside the automotive industry, especially those requiring complex shapes and short total production runs.

REFERENCES

1. WRIGHT, P. and CUMMING, A. P. C., *Solid Polyurethane Elastomers*, McClaren, London, 1969.
2. ALLPORT, D. C. and MOHAJER, A. A., in: *Block Copolymers*, chapter 8, ed. D. C. Allport and W. H. Janes, Applied Science Publishers, London, 1973.
3. REDMAN, R. P., in: *Developments in Polyurethane—1*, chapt. 3, ed. J. M. Buist, Applied Science Publishers, London, 1978.
4. BECKER, W. E. (Ed.), *Reaction Injection Moulding*, Van Nostrand Reinhold, New York, 1979.
5. BAYER, A. G., British Patent 1 534 258, 1967.
6. MIKULEC, M. J., RRIM: A new process for the automotive industry. *Proc. 34th Ann. Techn. Conf., Reinforced Plastics/Composites Inst. of the SPI*, USA, 1979, section 11–B.
7. MIKULEC, M. J., Refining of the RRIM process. *Proc. SPI/FSK Conf. on Cellular and Non-Cellular Polyurethanes*, Strasbourg, France, June 1980, section IVa, pp. 137–48.
8. EMMERLICH, W., *Plastics Technology*, 1980, **26**(4), 91.
9. WAKE, W. C., *Fillers for Plastics*, chapt. 1, IPC, London, 1971.
10. PARVIN, K., in: *Fillers for Plastics*, chapt. 7, ed. W. C. Wake, IPC, London, 1971.
11. MILEWSKI, J. V., *Plastics Compounding*, 1979, November/December, 17–37.
12. KATZ, H. S. and MILEWSKI, J. V., *Handbook of Fillers and Reinforcements*, Van Nostrand Reinhold, New York, 1978.
13. SCHULTE, K. W., BODEN, H., SEEL, K. and WEBER, C., *Eur. J. Cell. Plast.*, 1979, April, 61.
14. SIMPKINS, D. L. (General Motors), *SAE Congress*, Detroit, Technical Paper No. 790165, 1979.
15. NELSON, G. V. and DABROWSKI, A. J., *SAE Congress*, Dearborn, Technical Paper No. 800513, 1980.
16. LEIS, D. G., *J. Elast. and Plast.*, 1979, **11**, 301.
17. ISHAM, A. B., *SAE Congress*, Detroit, Technical Paper No. 760333, 1976.

18. DINBERGS, K. and SCHOLLENBERGER, G. S., *Adv. Urethane Sci. Technol.*, 1974, **3**, 36.
19. SEEFRIED, C. G., KOLESKE, J. V. and CRITCHFIELD, F. E., *J. Appl. Polym. Sci.*, 1975, **19**, 2493, 2503, 3185.
20. BONART, R., *Polymer*, 1979, **20**, 1389.
21. BLACKWELL, J. and GARDNER, K. H., *Polymer*, 1979, **20**, 13.
22. BLACKWELL, J. and NAGARAJAN, M. R., *Polymer*, 1981, **22**, 202.
23. FRIDMAN, I. D. and THOMAS, E. L., *Polymer*, 1980, **21**, 388.
24. HESKETH, T. R., VAN BOGART, J. W. C. and COOPER, S. L., *Polym. Eng. Sci.*, 1980, **20/3**, 190.
25. FRIDMAN, I. D., THOMAS, E. L., LEE, L. J. and MACOSKO, C. W., *Polymer*, 1980, **21**, 393.
26. ZDRAHALA, R. J., CRITCHFIELD, F. E., GERKIN, R. M. and HAGER, S. L., *Proc. SPI/FSK Conf. on Cellular and Non-Cellular Polyurethanes*, Strasbourg, France, June 1980, p. 243.
27. FERRARINI, J., FOWLER, R. and SPATAFORE, N., *Proc. 36th Ann. Techn. Conf. Reinforced Plastics Composites Inst. of the SPI*, USA, February 1981, section 6-F.
28. CHISNALL, B. C. and THORPE, D., RRIM: A novel approach using chopped fibre glass. *Proc. 35th Ann. Techn. Conf. Reinforced Plastics/Composites Inst. of the SPI*, USA, 1980, section 22-A.
29. CHAPMAN, J. F. and FORSTER, J. M. W., *Reinforced Plastics*, 1980, **24**(1), 14.
30. CHAPMAN, J. F., RRIM: Processing, properties and applications. *Plastics and Rubber Inst. Symp.*, Solihull, UK, February 1981, paper 12.
31. JOHNSON, A. E. and JACKSON, J. R., Effect of milled and chopped glass fibres on the anisotropy of RRIM composites. *Plastics and Rubber Inst. Symp.*, Solihull, UK, 1981.
32. CHARLESWORTH, D., *Plastics and Rubber Intern.*, 1980, **5**, 189.
33. MURCHIE, R. J., Polymers for bumpers: an automotive evaluation. *Proc. Plastics and Rubber Inst. Conf. on Polymers on the Road*, Kenilworth, UK, July 1980.
34. CHASTAIN, C. E., *Machine Design*, 1975, 23 January.
35. BLACKWELL, J. B. and RUBATTO, R., in: *Developments in Polyurethane—1*, chapt. 9, ed. J. M. Buist, Applied Science Publishers, London, 1978.
36. KNIPP, U. and BECKER, W. E., in: *Reaction Injection Moulding*, chapt. 8, ed. W. E. Becker, Van Nostrand Reinhold, New York, 1979.
37. MISITANO, G., *Plastics Engineering*, 1979, February, 27.
38. MCBRAYER, R. L., (BASF Wyandotte Corporation), Elastomeric RIM urethanes. Paper presented to *Plastics Design Forum*, October 1977.
39. DENO, L., THORPE, D. and WALKER, M. G., Some recent advances in polyurethane technology. Paper presented at *NVFK Technical Days*, Utrecht, October 1980.

BIBLIOGRAPHY

LEE, L. J., Polyurethane reaction injection moulding: a review. *Rubb. Chem. and Technol.*, 1980, **53**, 542.

NIERZWICKI, W. and WYSOCKA, E., Microphase separation and properties of urethane elastomers. *J. Appl. Polym. Sci.*, 1980, **25**, 739–46.

KONG, W. S. W. and WILKES, G. L., Observations regarding the effect of deformation on the domain structure in segmented polyurethanes. *J. Appl. Polym. Sci., Polym. Lett. Ed.*, 1980, **18**, 369.

COTGREAVE, T. C. and SHORTHALL, J. B., The mechanism of reinforcement of polyurethane foam by high-modulus chopped fibres. *J. Mat. Sci.*, 1977, **12**, 708.

METHVEN, J. M. and SHORTHALL, J. B., Developments in reaction injection moulding. *Eur. J. Cell. Plast.*, 1978, January, 27.

GAMACHE, A. R. and CARLETON, P. S., Rheological study: Isocyanate slurry behaviour. *Proc. 35th Ann. Techn. Conf., SPI*, St. Louis, USA, 1980, section 22-D.

CASTRO, J. M., MACOSKO, C. W., STEINLE, E. M. and CRITCHFIELD, F. E., Kinetics and conversion monitoring in a RIM thermoplastic polyurethane system. *J. Appl. Polym. Sci.*, 1980, **25**, 2317.

MACGREGOR, C. J. and PARKER, R. A., Controlling the physical properties of RIM urethanes with non-organic reinforcement. *SAE Congress*, Detroit, Technical Paper No. 790166, 1979.

RICE, D. M. and DOMINGUEZ, R. J. G., RIM elastomers having superior high-temperature performance. *Polym. Eng. Sci.*, 1980, **20**(18), 1192–6.

MCBRAYER, R. L., Variables in reinforced RIM. *Elastomerics*, 1980, July, 33.

ANON., Reinforced RIM: processing know-how progresses. *Process Engineering News*, 1978, November, 15.

BIESENBERGER, J. A. and GOGOS, C. G., Reactive polymer processing. *Polym. Eng. Sci.*, 1980, **20**(13), 838.

LEE, L. J., OTTINO, J. M., RANZ, W. E. and MACOSKO, C. W., Impingement mixing in RIM. *Polym. Eng. Sci.*, 1980, **20**(13), 868.

CASTRO, J. M., MACOSKO, C. W., CRITCHFIELD, F. E., STEINLE, E. C. and TACKETT, L. P., RIM: filling of a rectangular mould. *J. Elast. Plast.*, 1980, **12**, 3.

FERRARI, R. J., Automotive RIM products: trends and new technology. *J. Cell. Plast.*, 1980, November/December, 338.

SNELLER, J., Where RIM gets its new momentum. *Modern Plastics Intern.*, 1980, May, 28.

ANON., Improving RIM technology multiplies the options. *Modern Plastics Intern.*, 1980, February, 37.

JACOBS, K. F., Economics of a RIM production. *J. Cell. Plast.*, 1977, March/April, 133.

METZGER, S. H., PROPELKA, D. J. and SEEL, K., New RIM elastomers for automotive exterior body panels. *Proc. SPI/FSK Conf. on Cellular and Non-Cellular Polyurethanes*, Strasbourg, France, June, 1980, section IVB.

GLENN, W. B., Plastics for weight reduction: glass/mineral reinforced resins for body panels. *SAE Congress*, Dearborn, Technical Paper No. 800814, June 1980.

INDEX